Charles Henry Fielding

Memories of Malling and Its Valley

With a Fauna and Flora of Kent

Charles Henry Fielding

Memories of Malling and Its Valley
With a Fauna and Flora of Kent

ISBN/EAN: 9783337234348

Printed in Europe, USA, Canada, Australia, Japan

Cover: Foto ©berggeist007 / pixelio.de

More available books at **www.hansebooks.com**

MEMORIES

OF

MALLING AND ITS VALLEY;

WITH

𝔄 𝔉𝔞𝔲𝔫𝔞 𝔞𝔫𝔡 𝔉𝔩𝔬𝔯𝔞 𝔬𝔣 𝔎𝔢𝔫𝔱.

BY

REV. C. H. FIELDING, M.A.,

SECOND-CLASS LAW AND HISTORY, OXFORD; EXHIBITIONER OF TONBRIDGE SCHOOL
AND LINCOLN COLLEGE, OXFORD; CHAPLAIN OF MALLING UNION, KENT.
AUTHOR OF "SYMBOLISM," "HANDBOOK OF HIGHAM," ETC.

DEDICATED BY PERMISSION

TO THE

RIGHT HONOURABLE EARL STANHOPE.

HENRY C. H. OLIVER,

PUBLISHER, WEST MALLING, KENT.

LONDON AGENTS:

E. MARLBOROUGH & Co., 51, OLD BAILEY, E.C.

1893.

TO THE RIGHT HONOURABLE EARL OF STANHOPE.

My Lord,—

I have much pleasure in dedicating this book to your lordship, because of the lively interest you take in all matters historical and archæological which concern England in general, and, more especially, this county of Kent. It would take a lifetime to describe accurately all that can be learned in these matters in our whole county; more particularly if, as I have done, one attempted anything like its flora and fauna : I have, therefore, selected a district of Kent known to me from childhood, which affords us many monuments and records of individuals, who have lived in the varied scenes of English History, that teach us by the means of the small valley of Malling how one of the many similar districts into which our country can be divided gives existing proofs of what has happened in it from the very earliest times. Scattered about us, though frequently unnoticed among the rough flint stones that bestrew our paths, are the uncouth though sharp tools of the earliest races that lived in our island ; but hereabouts the prehistoric period is still more plainly marked by monuments like Kit's Coty House, the Countless Stones, the Coffin Stone, the fallen stones at Addington, and the Dolman at Coldrum.

Leaving this period behind, we find ourselves amidst written annals and archæological facts, that record in this valley the different scenes of English history from the earliest times to the present day.

The British period is marked by weapons, coins, and ornaments; perhaps by the paved causeway under the Medway at Aylesford, and the grass road popularly known as the Pilgrims' Path.

The Roman period can be traced by relics found at Holboro', Snodland, and in most of our parishes, which prove beyond a doubt that the world-subduing race once trod this valley.

The Saxon period is distinguished by weapons, by ornaments, by coins, by roads, by ancient battle-fields, and by documents that record the grants of many of the Saxon princes of this country, such as those of Egbert, king of Kent, Offa, king of Mercia, Egbert, Ethelwulf, Ethelstan and Edmund, and perhaps by some of the oldest parts of the church of Trottescliffe.

The Norman period has left us most of our churches, Malling abbey, St. Leonard's tower, and the castles of Allington and Aylesford.

The Barons' wars, during the reigns of John and Henry III., are exemplified here by the history of Roger de Leybourne and his compeers.

The Crusades are marked by the heart shrine in Leybourne church of Sir Roger de Leybourne and by the friars at Aylesford, built for the Carmelites by Lord Grey of Codnor on his return from the Holy Land.

The Scotch wars of Edward I. are called to our memory by Sir William de Leybourne and his contemporaries.

Edward II.'s reign is connected with this district by the disgrace of one of the abbesses of Malling.

We had some of our landowners holding castles in France ; while the Black Death clearly stopped the church-building about here, as well as carried off some of our clergy; and it would appear that Jack Straw, who joined Wat Tyler with a ruffianly mob, was a native of Offham.

Though not one of the battles of the Wars of the Roses was fought in Kent, still, the Nevills, who at this period obtained a settlement in our valley, as all the world knows, were well to the front, and no doubt drew away large numbers of the inhabitants of this part to fight for York. The rebellion of Cade was aided specially from the neighbourhood of Malling.

Tudor Times are marked by the history of the Wyatts. At Allington lived Sir Henry Wyatt, the poet, and Sir Thomas, the rebel leader.

The conspiracy of Babington found partisans in this neighbourhood, and in Pole's " Register " we learn that a vicar of East Malling was presented by the serene princes *Philip* and *Mary*, *King* and *Queen* of *England*. This period is farther marked by the dissolution of Malling abbey and the Carmelites at Aylesford, and by changes in property and by the commencement of some of our registers.

In the times of the Stuarts the Kentish people from Bluebell hill watched Fairfax and his army proceeding from Gravesend to Maidstone ; and at East Malling, Judge Twisden, who tried the regicides, and Colonel Tomlinson, who took Charles I. to his trial, lie side by side. We have many matters of interest belonging to this eventful time brought to our notice by the parish registers, and the records of the Twisden family. At this period we have the first mention of hopping and paper-making.

Though the Georgian Era has less points of interest connected with this valley than the preceding ones, still, many entries in our parish registers speak of our foreign wars.

The present period is marked by the railways and the great impulse given to population and manufactures by the rise of the cement trade in the parishes of Birling, Halling and Snodland, and in Burham and Wouldham, across the river, all of which form one large town that may be fitly named Cementopolis.

From the time of Edward III. down to the present day we have monuments and inscriptions telling of various persons.

I have refrained from crossing the river except for Aylesford, or recording anything about places that cannot be seen from Holly or Punish (Povenashe) hill. I have added to this book a list of gentry, and of the older names of other persons, and their occupations. Besides, I have given the entries in the registers of the chief families who have occupied the valley, and have traced the clergy, from the earliest times, of the parishes of

Addington,	Halling,	Ryarsh,
Allington,	Leybourne,	Snodland,
Aylesford,	Offham,	Trottescliffe,
Birling,	Paddlesworth with	West or Town Malling.
Ditton,	Dode, or Dodecirce,	
East Malling,		

I have also given Kentish proverbs and the names of animals and flowers in the Kentish dialect. To these I have added the birds, beasts, flowers, and fish to be found in Kent, principally obtained from fourteen years' research into the Natural History of the county.

In conclusion, I will only add that on page ix. I have stated whence I obtained assistance from friends and from the writings of others; and I trust this book will be found useful to my readers in studying the story of Kent, and the history of England.

CECIL HENRY FIELDING.

BOOKS CONSULTED.

"Burns on Parish Registers."
"Archæologia Cantiana."
Domesday (Lambert Larking).
Dugdale's "Monasticon."
Tanner's "Monasticon."
Thorpe's "Ancient Monuments of the Rochester Diocese."
Thorpe's "Custumale Roffensis."
Manuscripts of the Antiquarian Society.
Stalschmidt on Bells.
Records of Bradbourne.
Greenwood.

Harris.
Hasted.
Hussey.
Lambarde.
Philpott.
Burke's "Peerage."
Debrett's "Peerage."
Morris's "British Birds."
Yarrell's "British Birds."
Yarrell's "British Fishes."
Babington's "British Botany."
Hooker's "British Botany."

The Records of Lambeth and Rochester.
The Registers of—

Addington	East Malling	Ryarsh
Allington	Halling	Snodland
Aylesford	Leybourne	Trottescliffe
Birling	Offham	West Malling
Ditton		

I have received much valuable assistance from the kindness and courtesy of the clergy of the various parishes, from the Dean of Rochester, from G. Knight, Esq., and from S. W. Kershaw, Esq., librarian of Lambeth Palace, and from the Archbishop of Canterbury, who have allowed me to consult the ancient records under their charge, and I must also add that I have obtained useful aid from the following ladies and gentlemen :—
Mr. Allchin, librarian of Maidstone museum; Rev. J. G. Bingley, rector of Snodland; Rev. A. L. Brine, curate of Town Malling; Mrs. Cator, late of Malling Abbey ; * Lord Clifton ; Miss Dudlow, West Malling ; Rev W. F. S. Fraser, rector of Offham ; * Mr. Green, Rainham, Kent ; Rev. J. Guise, rector of Addington ; Rev. C. C. Hawley, rector of Leybourne ; Rev. E. B. Heawood, rector of Allington ; † Mr. Hepworth, Rochester ; † Professor Holmes, Pharmaceutical Society ; H. St. John Hope, Esq., Sec. Antiquarian Society ; Rev. G. P. Howes, vicar of Halling ; Rev. Canon Knollys, rector of Wrotham ; * Mr. Lamb of Maidstone ; Rev. G. M. Livett, minor canon of Rochester ; G. Payne, Esq., Sec. Kent Archæological Society ; Rev. Canon Scott Robertson, vicar of Throwley ; Rev. C. W. Shepherd, rector of Trottescliffe ; Rev. J. H. Timins, vicar of West Malling ; Rev. J. Twisden, Bradbourne ; H. D. Wildes, Esq., West Malling.

* Information on Kentish Birds.
† Information on Kentish Flowers.

x

The following places are mentioned in the book :—

Addington (near Maid-
    stone).
Allhallows (Hoo).
Allington (Kent).
Altesdon.
Ash (next Wrotham).
Ashburnham (Bucks).
Aylesford, Battles of, etc.
Barming.
Barnstone.
Bearsted
Beatrich's den, i.e., Be-
    thersden.
Beddington.
Bermondsey.
Bexley.
Betchworth (Surrey).
Betshanger.
Bickley.
Birling.
Birmacham (Birming-
    ham).
Bookham, Great (Surrey).
Bluebell, Lower.
Bluebell, Upper.
Boreham.
Boroughbridge, Battle of.
Boston.
Boughton.
Boughton Monchelsea.
Boughton under Blean.
Brampton.
Brandbridges.
Brenchley.
Brixton.
Burham.
Burwash.
Caerlaverock, Siege of.
Caermarthen.
Canterbury.
Canterbury (St. Mary
    Magdalen).
Canterbury (St. Peter).
Chatham.
Cheswick.
Chilham.
Champigny (near Melun).
Chislett.
Clifton (in Bristol).
Cobham.
Coldrum.
Colebrooke.
Comp.
Cossington.
Cowden.

Cumbwell, Priory of.
Dartford.
Deptling.
Ditton.
Docle or Dodeciree.
East Farleigh.
East Malling.
East Peckham.
Eatonbridge   or   Eden-
    bridge.
Eccles.
Elmsted.
Elmley.
Estburgate.
Euston.
Evesham, Battle of.
Excete (Chichester).
Exeter.
Eyhorne.
Farningham.
Faversham.
Fetcham.
Field of the Cloth of Gold.
Frant.
Frindsbury.
Fringe.
Frittenden.
Fullmere.
Genes or Genoa.
Gillingham.
Goodnestone.
Goudhurst.
Grayne, Isle of
Greenwich, West.
Hackington.
Hadstocke.
Hardres, Lower.
Harrietsham.
Halling.
Hastings.
Higham (near Rochester).
Hoathe.
Holboro.
Hollingbourne.
Holly Hill.
Horstead.
Horsmonden.
Hunton.
Ightham.
Itchen (Hants).
Kemsing.
Keston.
Kingsdown with Mapis-
    comb.
King's Plain.

Kit's Coty House.
Lampeter-upon-Severn.
Langley.
Larkfield.
Laver Parva.
Lea (Gloucester).
Lee (Kent).
Leeds, Castle of (Kent).
Lessness Abbey.
Lewes.
Leybourne.
Long Melford.
Longsole Chapel.
Luddesdown.
Lullingstone.
Lynn (Norfolk).
Lynstead.
Maidstone.
Malling Deanery.
Maltham.
Marden.
Margate (St. John's).
Mayfield.
Medway.
Mendlesham.
Meopham.
Mereworth.
Merton.
Milton.
Nettlestead.
New Hythe.
New Romney.
Northampton.
Northfleet.
Norton.
Offham.
Orpington.
Otham.
Paddlesworth.
Penenden Heath.
Penshurst.
Perth (Western Australia).
Pilgrims' Path.
Preston Hall.
Preston (next Wingham).
Rainham.
Reigate.
Ridley.
Rochester(St. Margaret's).
Rochester (St. Nicholas).
Rochester Castle.
Rochester Deanery.
Ruckinge.
Ryarsh.
Rye.

Saint Blasius.
St. Lawrence in Halling.
St. Leonard's Street.
St. Mary Graces.
St. Martin's Outwich.
St. Michael's, Lewes.
St. Paul's, London.
Saltwood.
Sandwich (St. Clements).
Seale.
Sheldwich.
Sheppey
Shorne.
Snargate.
Snodland.
Snoreham (Essex).
Southborough.
South Malling.

South Mimms (Middlesex).
Speldhurst.
Stepney.
Stopham.
Sutton Deanery.
Swindon (Stafford).
Tenterden.
Teston.
Tettenhall (Stafford).
Tewkesbury, Battle of.
Thanet.
Tonbridge.
Tonbridge School.
Tottington.
Trottescliffe.
Union of Malling.
Veles.
Vigo.

Wadhurst.
Warlesworth.
Welcomestow.
Welcombe.
Wenlakesbarn.
Westgate.
West Farleigh.
West Malling.
West Peckham.
Westminster.
Windsor (Berkshire.)
Winwick.
Woking (Surrey).
Wouldham.
Worth.
Wrotham.
Yalding.

CONTENTS.

MEMORIES OF MALLING AND ITS VALLEY.

CHAPTER I.

INTRODUCTORY REMARKS.

A MONGST the many lovely valleys of which Kent can boast, none perhaps excels in beauty the Vale of Malling, which can be so well viewed from the neighbouring chalk hills. Not only its beauty, but its varied flora and fauna, and above all its history, make it repay the lover of nature or antiquity, or the student who prefers the chronicles of his own native land to those of the Continent. In the county of Kent (as will be seen by the appended lists of natural history specimens at the end of this book) are to be found a great number of the most beautiful of the various members that form the fauna and flora of the British Isles, most of which may be found in this fertile vale.

From Holly Hill (642 ft.), or the Vigo (680 ft.), the eye may wander over this lovely valley, from the ground beneath one's feet to where the greensand hills, clothed with the Malling, Great Comp, and Mereworth woods, shut out the scene; and from the banks of the muddy, meandering Medway, to where the valley is almost closed near Wrotham. The parishes inclosed in this tract are West or Town Malling, East Malling, Ditton, Aylesford, Allington, Snodland, Halling, Paddlesworth, Birling, Leybourne, Ryarsh, Trottescliffe, Addington and Offham. These parishes were at one time all of them in the bishopric of Rochester, saving East Malling, which was a peculiar of the archbishop; but during the recent changes of that see, all except Aylesford, Halling and Snodland (whose fortunes have been attended by Paddlesworth, now no longer considered, though really a separate parish) have been transferred to the archbishopric of Canterbury. Addington, Birling, Ditton, East and West Malling, Leybourne, Offham, Ryarsh and Trottescliffe continue in the deanery of Malling, to which deanery East Malling was added. Allington was during these last twelve months transferred

1

to the Sutton deanery, without any regard to geography or history. While Aylesford, Snodland and Halling, which belonged to the old Rochester deanery, now make part of the division of Rochester diocese which is known as the Cobham deanery. These parishes are all in the jurisdiction of the Malling division of magistrates, with the exception of Halling, which is in the Rochester division; and with the same exception they all form part of the Malling Union. They are all in the Lathe * of Aylesford. By the different Reform Bills these parishes have been transferred to the Parliamentary Divisions of West Kent, Mid Kent and Medway, according to the different ways into which the county has been divided.

Though these parishes have never been very populous, they have maintained their own through every period of English history, owing to the bountiful supply of water and the fertility of the district. We subjoin a table of their population in the order of the last census :—

PARISH.	1841.	1851.	1861.	1871.	1881.	1891.
Snodland with Paddlesworth	500	900	1078	1844	2826	3187
Aylesford	1344	1487	2057	2100	2800	2937
East Malling	1578	1741	1974	2326	2383	2380
Halling	448	550	760	838	1273	2091
West, or Town Malling . .	1784	2081	2086	2077	2242	2028
Birling	511	620	662	718	884	1384
Ditton	244	235	255	287	336	878
Ryarsh	431	449	447	475	552	514
Offham	358	372	411	448	358	397
Trottescliffe	305	253	293	288	296	286
Addington	208	220	211	261	264	273
Leybourne	255	268	289	271	277	270
Allington	37	51	66	93	147	130

This district has been always well supplied with water, for not only does the Medway bound the valley, but also several other streams that spring from the foot of the chalk hills in the Gault on one side, and from the overflow of the reservoirs that lie in the greensand range on the other flow through it. The most important of these streams takes its rise at Nepicar, in Wrotham, and is fed by one of those peculiar underground springs known in Kent as Nailbournes (a word of uncertain origin), which overflow at certain periods in a number of years, and finding its way though Addington Park and past Ryarsh Church, is joined by a considerable rivulet that rises near St. Leonard's Tower, in Malling, where there is an archway and a paved bottom (perhaps this once formed a baptistery to St. Leonard's Church), passes under the Malling and Tonbridge road, forms an ornamental

* Kent is divided into five divisions, called Lathes: these Lathes, proceeding from west to east, are Sutton at Hone, Aylesford, Scray, Shepway, and St. Augustine.

pond, and proceeds onwards to the abbey grounds, where it once fed the fish ponds of the nuns; leaving them by the cascade built by Mr. Foote in the year 1810, so well known to all visitors to Malling for its picturesque appearance, it runs through the bottom of what are called Banky Meadows, where it formerly was utilised for the tanyards which once gave Malling the importance it possessed. The two streams together unite to form the Leybourne mill pond, just below which the rivulet—no doubt in ages past—was used for the mote to Leybourne Castle, traces of which may still be seen. Keeping not far from the road it runs to Snodland, before reaching which place a considerable stream from Birling joins it, one of the heads of which is an iron spring, said to have been one of the constituents of the Birling drink,—a bottle of this is reported to have been an infallible cure for the bite of a mad dog. At Snodland this rivulet supplies the water to the paper-mills, and falls into the Medway after a winding course of about seven or eight miles. This stream was once, no doubt, the river that formed the valley.

Another brook that rises in the uplands of East Malling, and turns the corn- and paper-mills there, once formed a mote round Bradbourne House, where there is still a pond ; and then, passing across the London and Maidstone roads, it falls into the Medway a little below Aylesford. On the other side of the river the Boxley stream and Cosington spring are both well known.

This great supply of water has, from time immemorial, been most serviceable for corn- and paper-mills, for which East Malling and Snodland have been famous for two hundred years. At the former place, we are told, the cardboard called millboard was first manufactured by Mr. Barling some forty or fifty years ago. The plentiful moisture has made this land valuable for orchard purposes, and at Leybourne and Snodland forms fine pasture land ; at the former place were bred the famous horses that won the Derby for the late Sir Joseph Hawley, viz.: Musjid, Teddington, Bedesman and Blue Gown.

CHAPTER II.

THE valley which we have already described, which forms the subject of our history, was filled in the earliest days by the great forest of Andredswolde, which appears to have covered the whole country between the North and South Downs, and was one hundred and thirty miles in length. We may here remark, in passing, the statement that the beech not being a native of Britain, resting as it does on the sole authority of Julius Cæsar, who says that timber of every kind which is found in Gaul also grows in Britain except the beech and the silver fir,* can hardly be believed, since the short time that that commander stayed in this country was not sufficient to explore this mighty forest.

At the foot of the chalk hills this great forest was bounded by what is most probably the oldest road in this country, which is popularly known as the Pilgrim's Path, and which, entering this county from Surrey, runs on to Canterbury. This road passes into our district from Wrotham at Trottescliffe, and traverses that parish and Birling to Snodland, then on to Halling, where a branch road went on to Rochester; but the main road crosses the Medway, and proceeds under the Downs by Debtling and Hollingbourne. As might be expected, this road forms the basis of all discoveries of an early period. At Wrotham have been found a number of British bronze celts. At Trottescliffe, not far from this road, still stands the ancient cromlech called Coldrum : this not only has three upright stones remaining, but one, though broken, still on the top—the front one has gone; besides this, part of the circle round it continues, consisting of stones to the number of nineteen, which, though fallen prostrate, are in their original places. When the Coldrum monument was re-discovered some years ago, two young gentlemen found under it a skeleton, which was removed and buried in the churchyard of Meopham by the vicar of that parish. Upon this the Rector of Trottescliffe, in which parish Coldrum stands, wrote to ask the Vicar of Meopham what he meant by stealing his oldest

* Materia cujusque generis ut in Gallia est practer Fagum et Abietem.

4

parishioner. At a mile or so distant, farther off from the old road, are a number of stones of the same kind, in Addington Park. Wright speaks of the elliptical group as a circle : the heap I do not hesitate to consider a fallen dolmen. In the parish of Ryarsh many of these stones are scattered about, and some have been built into the walls of different buildings, as the one so especially noticeable in the foundation of Trottescliffe Church. At Tottington, in the parish of Aylesford, many of these stones are found ; one of them has been named the " coffin stone " by Dr. Stukely, and this name was sealed to this stone by the learned Thorpe. This monument is not far from the well-known Kit's Coty House, which consists of three upright stones and one above. There is another group of stones in a field under some trees, called the " countless stones " or numbers, locally, which are evidently remains of another dolmen. At Tottington may be seen the ruins of a fourth, and Mr. Shaw, of Eccles, assures me that he has had a large number of sarson stones, of which these erections are formed, ploughed up nearer the Medway on his farm at Rowe's Place.

Now what do these stones mean ? It would seem, by comparing Coldrum and Kit's Coty House, we can come to a fair conclusion, when we think over the different customs of races and nations at the same time. It requires no great stretch of imagination to make the centre-piece of Coldrum into a perfect Kit's Coty House : all we have to do is to mend the stone that stood at the back of the two, and the one on the top. But if Kit's Coty House was thus like Coldrum, why should it not have a circle round it of stones like Coldrum ? If we look at the last monument, the bank and stones have been gradually got rid of in days gone by,* and this advises us that the same has been done most probably, or done more completely, at Kit's Coty House.

The stone in the foundation of Trottescliffe Church, those in the walls of buildings in Ryarsh village, and pieces in the walls of Birling Place,—once known as Comfort, the ancient seat of the Earls of Abergavenny,—and those scattered about at Aylesford, all advise us that this work of destruction went on to no small extent during the Middle Ages. Superstition alone, perhaps, saved the relics that are left. Thus the field of the " countless stones " near Aylesford, the writer can remember, when a boy, was said to be impossible to clear ; and that several who had taken the farm were ruined because they wickedly attempted to move the stones. Again, whoever attempted to measure Kit's Coty House was frustrated in his task, as he always forgot the size of the three stones when he measured the fourth.

But to proceed. If these monuments have been tampered with, surely we may go a little further and compare them with others, which we find were made with four upright stones and one on the

---

* I am pleased to state that my friend G. Payne, Esq., Secretary of the Kent Archæological Society, has prevented Coldrum being destroyed any more, by having obtained it to be placed under the Ancient Monuments Act.

top, and around them stood a circle of stones; but whether with others laid upon them, as Stonehenge, it were difficult to say.

What was the purpose of these erections? Persons have disputed about them as temples, altars, and tombs, and each has rejected the other's theory; still those who deny them to be tombs must explain how skeletons were found beneath them as at Coldrum. The fact is, these circles were probably temples, and the dolmen was very likely the altar. When some great chief died they buried him there; and thus perhaps to our Celtic forefathers we and other Christians owe the custom so long prevalent of burying in our churches—a practice that was certainly not borrowed from the Jews, who would have been disgusted with the idea, as all unclean objects (of which dead bodies, both animal and human, were considered the worst) were never allowed within the sanctuary.

As regards Kit's Coty, which I believe means Catigern's Stone House—quoits being still used for the game originally played with stones—or (as some like to think that the word means wood) Catigern's Wood House, the latter word simply tells us where the monument stood, or what it was made of, which is immaterial: it is to the word Kit, which is the name that has always clung to this monument, that I look for an explanation that Catigern was buried in it or near it. We learn that the Saxons and Britons had a severe contest at Aylesford, A.D. 465, and tradition says that Catigern and Horsa fell in hand-to-hand conflict. If this was so, where did they bury him? Where more likely than in the sepulchre of the kings, as we so often find expressed in the Bible,—"they buried him in the sepulchre of the kings" or, "with his fathers," or not, according as a prince had deserved well of the nation. Thus, the fact of Catigern's burial at Kit's Coty House does not preclude any former burials at the place, or cause it to have been erected at the time he fell, or suggest to us that it was then used as a temple and an altar. I believe it points out to us that he was placed here in an honourable burying place, which the Britons knew of, handy to the battle; and though they had become serfs, still the conquered race pointed to the tomb of the man who died for his country. And when the Celt mingled his blood with his Saxon conqueror, we cannot tell whether out of mockery, or from the inconvenient length of the word, Catigern got shortened into Kit; but the word has survived to point out the grave of the hero, while not far off the name of Horsted is given to a place traditionally declared to be the grave of his equally brave Saxon antagonist Horsa. These two places appear to me to contradict the sceptics who would deny these men a position in the history of our island, and show us that Celt and Saxon were as worthy ancestors of the English race as Dane and Norman.

These groups of stones lead us to consider that this wooded valley must have been much frequented by the earlier races in these islands, and the finds which have been made of objects of archæological interest point us to the same fact. At Ryarsh, British gold coins of a period before the Christian era have been found, and the same at Chequer Lane in Oftham, and St. Leonard's Street in Malling, as well as coins elsewhere in the latter place. At Aylesford drift implements have been found, as well as late Celtic instruments; and in the Medway itself, British torques and armillæ of gold. Moreover, a paved ford was discovered at Aylesford, some 30 ft. wide, formed of boulders. These matters all point out to us that there were some important early settlements here. At Halling the cremated remains found in the quarries point us to the Roman custom of burial near their roads. The recent discovery of a Roman villa at Snodland, the treasures of Roman times found at Holborough, and the sepulchral deposits there, declare to us the fact that the Rochester and Malling road follows one of great antiquity. The roads from Wrotham to Maidstone, and Malling to Mereworth, and Wrotham to Teston, were probably other passes through this wooded valley of an early date : it would seem that the latter was the *via militaris*—an ancient Roman road.

In September and October 1892, under the road near St. Leonard's Tower, were found two Roman urns of Upchurch ware (one standing in a Samian saucer), and broken pieces of others; and a little farther off bones of cattle, principally the parietal bone and the horns, the jaw of a sheep, and the tusk of a pig. The urns contained cremated human bones, which Dr. Tannahill of Borstal identified thus : (1) The smaller urn contained portions of the skull, ribs, vertebral column, arms and legs; the whole weighing 3·75 oz. (2) The larger urn contained portions of the skull, vertebral column, sacrum, ribs, sternum, arms, and legs, and was part of the body of a man; the whole weighing 8·5 oz. The bones of the animals so close to water, and a cemetery, seem to point to sacrifice. Not far off was discovered some time ago a paved way, and this, with the fact that the old road ran past St. Blaise's Church, points out that St. Leonard's Street and Kent Street once joined the straight road through Malling, and form part of the old Roman *via* from Rochester.

Besides the relics we have already spoken of, a British burial at Allington; foundations of Roman buildings, instruments, and coins at Eccles; a Roman urn found in the Hermitage Wood at Malling, probably pointing out a burial; together with Roman foundations and a tile tomb at Allington, all declare to us the early history of this part, and show us that when Saxon and Briton joined battle at Aylesford the Saxon was fighting to gain an important British ford and settlement, which, on his victory, gave him the key to open his way into the centre of this country.

The walls of East Malling chancel are constructed almost entirely of materials taken from the Roman buildings which seem to have largely existed in these parts; and a great number of the other churches about here bear proofs that those who built them used very lavishly the materials thus placed ready at hand, without the slightest regard to the damage they were doing posterity by destroying the relics of the past. Since the author commenced this work a number of Roman urns have been found this spring (1892) near his house at Larkfield, by Mr. Wigan, which appear to point out that there was a Roman cemetery there also.

CHAPTER III.

THE SAXON ERA (465—1066).

PROCEEDING onwards from the first battle of Aylesford, we come to that time when Christianity was introduced amongst the Saxons by St. Augustine, in the year 596; and in the year 604 the see of Rochester was founded. When any of our churches in this neighbourhood were first built it were difficult to say, but the parish churches of Meallinges, Meallingetes, Leleburne, Dictuna, Eddituna, Riesci, Offham, Birling, Allington, Trottesclive, Aylesford, Halling, Esnoiland, Paddlesworth, Dodecirce and Tottington appear all to have been in existence when—

"Conquering William brought the Normans o'er."

Besides divers relics in the shape of Saxon coins and weapons, and also perhaps parts of the churches being Saxon, and the names of most of these places, but not all, being wholly or partly of Saxon origin (for I am amongst those who consider that Aylesford is Eglwysford, or the ford of the church which I think Eccles points us to, and thus is a name partly British). We have certain grants of lands for ecclesiastical purposes from the Saxon kings. The earliest of these records is that of the parish of Halling, from which Egbert, king of Kent (770-785), gave ten ploughlands, according to the *Textux Roffensis*, to Dioran, bishop of Rochester, and the Church of St. Andrew there; this grant is signed by Egbert, king of Kent, and the Archbishop Jaenberht. The next oldest is that of Offa, king of the Mercians, who gave six ploughlands in the parish of Trottescliva, or Trottescliffe, for the relief and safety of his soul, in the year of Our Lord's Incarnation, 788; the boundaries of this land are, on the east and south Birling, on the west Wrotaham, and on the north Meapaham. His grant is signed by the king himself, his queen Cynedrith, Jaenberht the archbishop, who put on it the sign of Christ's Cross, Hygeberht (archbishop,) and Ceolwulf (bishop), amongst others. This grant is very interesting, inasmuch as it shows the power a Mercian king possessed in Kent, which, in consequence, was not likely at that time to have been a separate kingdom; and also because it mentions the adjoining parishes of Wrotham,

Meopham, and Birling. After this we learn of Egbert, king of the
Angles, "Hic dedit ecclesiæ Christi villam quæ Meallinges vocatur."
He gave to the Church of Christ a villa, which is called Malling—the
date is 827, at the earliest, but probably a year or two later. In 836
Egbert and Ethelwulf jointly gave to Bishop Beornmode four ploughlands in Snoddingland, and Holanbeorge, or Holboro', in the royal
bica called Frericburn, and a mill on the torrent called Holborough
burn, and on the king's mountain (*mons regis*) fifty carrabas of wood,
four being added at Denberri, Hwetonstede, Heahthen, Helffe,
and Helmanhurst; this is signed by Egbert, Ethelwulf, Archbishop
Ceolnod, Bishop Beornmode, of Rochester, and Bishops Eahlstan,
Eadhun, Cynred, and Ceolberht, and others. An interesting fact
connected with this grant is the mention of the great antiquity
of the mill on the stream at Holborough, this brook having been
utilised, we thus learn, for its water power, over one thousand
years ago. After this, Ethelwulf, in 841, styling himself king of
the West Saxons, gives the same Bishop Beornmode two ploughs in
that part which is called Holanbeorges tuun ; this gift is signed by
Ethelwulf, King Ethelstan, and the same archbishop and bishop of
Rochester as the last, and Bishop Helmstan. The next event is
Alfred's decisive triumph over the Danes, which took place almost
on the very ground at Aylesford where his ancestors defeated the
Britons. No doubt the Danes, as the Saxons before them, having
poured in from Sheppey, found the Medway hard to cross below, and
so came over the hills to seek a ford. This victory appears to have
put an end to the depredations of Hastings about the year 895.

But of all the charters of this period, the one that had the most
influence upon the future of this district was the charter of King
Edmund, which ran as follows :—

"In the name of the Supreme God, and our Saviour Jesus Christ. Who
Himself reigns for ever, and disposes all things pleasantly : Wherefore I,
Edmund, king moreover of the Angles and Mercians, grant to my beloved
bishop, by name Buhric, so much of my land which is called Meallingas as
three ploughs. for the relief of my soul, for an everlasting heritage, to increase
his monastery, which is dedicated in honour of Saint Andrew the Apostle,
brother of St. Peter, and his companion in suffering ; with all things pertaining
to it, with fields, woods, pastures. Moreover fowling also ; and this also with the
advice of my chiefs and princes, whose names are written beneath, from hence I
swear in the name of Jesus Christ, who is the just judge of all, that this land
be freed from all royal service for ever. If any one shall presume to diminish
or defraud this donation, may he be separated from the assembly of the saints,
so that while living he be deprived of the blessing of God, and be damned in
the lowest hell, unless he shall have by full satisfaction atoned before his death
for what he hath unfairly done, although this still by no means should remain
in force. And he who would at any time increase our gift, may the Lord
increase to him His heavenly gifts, and give him eternal life."

The above is in Latin : then follows, in Anglo-Saxon, a description
of the boundaries :—

"From the south bounds to the King's Plaine, and from thence to the bounds of the ville of Offham, and thence to the military way and along the said way over Lillieburn to the bounds of the parish of East Mallinges, and so southward from the east of the cross or gallows to the broadway towards the south in a direct line along the said way to the King's Plaine."

To which the king added certain denberies for the pannage of hogs. Then follows :—

" I, EDMUND, king of the Angles, have confirmed this by the sign of the Cross.
" I, EDRED, the king's brother, have corroborated this.
" I, EADGISE, the king's mother, was present.
" I, ARCHBISHOP ODO, have subscribed.
" I, ÆLFEH, bishop, have agreed.
" I, DEODRED, bishop.   I, ÆLFRED, bishop.
" I, WULFSTAN, archbishop.   I, ÆLFRIC, bishop.
" I, CENWALD, bishop.   I, ÆTHELGAR, bishop.
" I, ÆGILFU, the king's wife, was present.
" Duke Wulgar, Duke Athelstan, Duke Edmund, Duke Senla, Sigferd, minister, Wulfric, minister, Ealdred, minister, Ælfstan, minister, Ordcah, minister, Edward, minister, Wulfric, minister, Odda, minister, Ælla, minister, Duke Offa, Whitgar, minister, Wulfsige, minister, Birthwald, minister."

Though this charter is undated, since Edmund reigned only from 941 to 946 we know within a very few years the date of the charter. The charter is very interesting in many ways. In the first place it informs us that the parishes of Offham, Leybourne, and East Malling, as well as West Malling, were then in existence. The King's Plaine seems to have been, from the other places mentioned, somewhere towards the south, and near the Union-house. We find the name of King's Mill still survives; probably a relic connected with the old name.

Thorpe is of opinion that the military way here mentioned was the road from Town Malling through St. Leonards to Teston, etc., and so on to Cranbrook. The word street added to St. Leonards may confirm this statement; but "along the said way, over Lillie-burn" from Offham to East Malling, appears to me to point rather to the present London and Maidstone road. From these facts, however, we learn the antiquity of the roads in this district.

The property which Edmund granted to the Church did not continue in the hands of the see of Rochester, but was lost in the wars with the Danes. Upon the succession, however, of William I. to the throne, it became the property of the rapacious Odo, bishop of Baieux, who was afterwards compelled to disgorge it; and we shall have occasion to mention it as being the origin of Malling Abbey.

Ælfstan, Buhric's successor in the see of Rochester, witnessed the will of a Saxon gentleman and his wife, giving a firma of two days from Birling and their property in that parish to Wulfege, to pay 1000 denarii to the church of St. Andrew, in Rochester, besides giving their property in Snodland to the same church. This gentleman and lady, Brihtric and Ælfwitha, have in this way descended to posterity as the oldest private landowners known by name in this district. This bequest adds to the parishes we have already heard of no new ones, since Birling and Snodland are both mentioned previously.

The continual landing of the Danes at the mouth of the Medway, and their incursions into this part of Kent from Sheppey and Sitting-bourne, must have caused many changes in the parishes of which we are writing, as is shown by their having wrested from the monks of Rochester the lands granted by Edmund in Town Malling. In the early part of the eleventh century they attempted once more to occupy this part of Kent, and appear to have again tried to ford the Medway at Aylesford, but were defeated by Edmund Ironsides, 1016. The Saxon period has not left behind it too many annals, but it is strange that the village of Aylesford should be the scene of three decisive battles, if there were not some great strategic point to be gained by coming this way when Rochester was too well defended to allow of their gaining a passage there. With this last victory at Aylesford our annals so far as Saxon times are concerned close; as from this time up to the days of William the Conqueror, we have no account of the doings of our predecessors in the Vale of Malling. We do, however, know that Allington was the possession of the fourth son of Godwin, Gurth, at the time of the Conquest.*

* A beautiful relic of Saxon times was shown me by Col. Luck, J.P., of the Hermitage, West Malling : it was discovered in his grounds. The ring is of silver, on it are three figures placed transversely, not upright, as in ours. The upper figure is a bishop, the centre a king, and the lower a prince, as would appear from what looks like a cap of maintenance. It was not the property of an ecclesiastic of Malling Abbey, as Mr. Surtees surmised, since it was not opened at this date, but may refer to the grant of the land (near which it was found), by Edmund to Buhric, which was witnessed by the king's brother Edred.

CHAPTER IV.

NORMAN TIMES (1066—1216).

THE acquirement of the throne by William of Normandy, through conquest, naturally led to a redistribution of lands; as the conqueror would of course remove those who were in power who were likely to oppose his authority, as the Saxon landlords, and replace them by Normans, who, with their retainers, would keep the country in awe. The consequence is we find at this time the landowners have Norman and French names; and this is the great period of building monasteries and castles throughout the country, as well as the time of improving and enlarging our churches.

This was the time, as we shall find, during which the nunnery of Malling was built, as well as the castle of Leybourne and the keep of St. Leonards; and the churches round here show distinct signs of early and late Norman.*

All these parishes that we are considering are mentioned in Domesday, and all, with the following exceptions, appear to have become the prey of the unscrupulous, ambitious, and greedy Odo of Baieux (who had been created Lord Warden of the Cinque Ports, and must have been a well-known man in Kent, since he fortified the castle of Tonbridge against Rufus), viz.: East Malling, which continued the property of the archbishop, and Aylesford, which was a royal demesne. Lanfranc, the Archbishop of Canterbury, however, succeeded, at a meeting in 1076 on Penenden Heath, in obtaining from the king an order to make Odo give up West Malling and Halling, and it seems uncertain whether he ever really obtained Snodland and Trottescliffe. At any rate, soon after, when William II. disgraced the bishop and marched on Tonbridge (perhaps through this very valley), all these places changed hands once more.

It was in the year 1077 that the famous architect Gundulf was appointed Bishop of Rochester, and it is to him we owe the building of the nunnery at Malling, and St. Leonard's Tower, which Parker declares to be the oldest Norman keep in existence; and where the

---

* Behind Mr. Jarvis's and Mr. Carman's shops in Malling High Street, are the remains of a Norman house outside the abbey wall, but probably ecclesiastical. It may have been the home of "prebendarius magnæ missæ in monasterio de West Malling."

curious may examine the strange way in which the Normans laid their courses of masonry, and how carefully they mixed their mortar and lined the corners of their buildings with tufa. It is a plain, massive, square building, with clasping buttresses, showing all the characteristic features of early Norman work in general, and of Gundulf's in particular. The walls are between 7 and 8 ft. in thickness; the material is Kentish rag and tufa, the latter used for all quoins and faced work, and running in bands round the building at the set-offs. The building is divided into two stages; a covered staircase, running up the north-east corner of the building (entered from the bottom on the inside), communicating with the upper stage by a lobby leading out on the roof—the latter, with the upper part of the tower, was dismantled. Each stage is lighted by a narrow round-headed opening, splayed internally in the lower stage, but not so in the upper. The dividing floor formerly existed at a height of 24 ft. from the foundation level. The total height of the tower in Thorpe's day was 71 ft., taken by a theodolite; last year it was considered to have been somewhere about 60 ft. There appears to have been no entrance or doorway to the original building, entrance being probably gained through one of the lights by means of ladders or ropes. The east side of the tower is adorned with an arcade, and a greater number of lights than the other sides. The now blocked opening in the basement, on the north side of the east wall, appears to be not contemporaneous with the original building, but slightly later. Probably steps led down to St. Leonard's Chapel from it.

When this chapel was built is uncertain. In Thorpe's day the following information could be ascertained about the chapel:—it was 70 ft. long and 33 ft. broad; and we are further told that the return of the east wall of the church was visible for about 3 feet. The fields beyond and part of the wood above being known as Abbey Fields and Abbey Wood, it is evident that Gundulf built this tower to defend his newly-acquired domains towards the south, and perhaps to over-awe the Saxons in that direction; as on the north the rising towers of Leybourne promised to protect him, and his own keep which he was building at Rochester offered further assistance in that direction.*

When the chapel of St. Leonard's was built we cannot say, but it is mentioned in the *Textus Roffensis* as important enough to pay six denarii for chrism to the see of Rochester.

Malling Abbey was built about the same time as St. Leonard's Tower, though its tower was not completed till much later, as it exhibits early Norman at the lower, succeeded by later Norman in the second story, and again by early English in the upper part; this latter part, however, may have been built because the upper part was destroyed in the great fire. The cloisters of Malling Abbey, now built

---

* In the books of the Manor of Malling, dating from Queen Elizabeth, St. Leonards is known as the precinct of. Ewell.

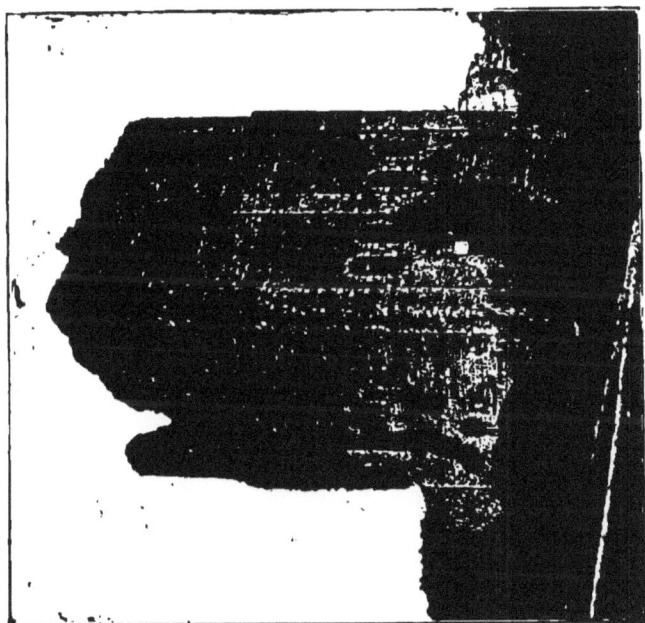

ST. LEONARD'S TOWER.

into the modern dwelling house, are unique. The windows show that there were five bays consisting of three separate lights—in one case four. There are none to be found like them elsewhere in England.

Philpott and Hasted tell us the abbey was founded in 1090, while Leland gives in his *Collectanea* the date 1106 for the opening of the abbey—this date is very unlikely, as Gundulf died March 7th in that year. At the time of the opening of the abbey Gundulf bestowed the government of it on the nun Avicia, or Avice, from whom he extorted an oath of fidelity and subjection. We are told " On that day on which Gundulf, Bishop of Rochester, gave the Abbey of Malling to the nun Avicia, the same nun swore fidelity and subjection to the same bishop and his successors, and to the Church of Rochester, and that she would not try by her own means, or by any other person, to break this subjection. When this oath was made, the aforesaid Avicia promised the bishop by a firm and sure compact that, without his consent and licence, she would neither place nor displace a prioress in the abbey given to her, nor receive any nun, nor grant nor take away thence any land." The witnesses of this deed hearing and seeing her were :—

Radulf, abbot of Belli (Battle); Orduvinus, prior; Paulinus, secretary; Alured; Andrew, doctor; Arnulf, chaplain of the bishop; John; Goisford; Alberic; Odo; Beringar, sub-prior; William; Hunfrid; Erngrin; Hugh, nephew of the bishop; William, a young man; Radulf, a clergyman; Ansfrid, a clergyman; Godard, a clergyman; Robert, keeper of the closet; Hugh, keeper of the closet; Ansfrid, steward; Hunsfrid, porter, and many others.

The nuns of Malling were to be held in high esteem, for it is ordered that, " amongst what we ought to do for our dead brethren, that the same be done for the nuns of Malling as for a monk of Christchurch, Canterbury, viz. : seven full offices, and for thirty days *mea verba*, and one priest says seven masses each, and others the psalms." Tanner recites the charters of William II., Henry I., and Henry II., concerning the subjection of Malling to the bishopric of Rochester; and Hasted informs us that the donation was confirmed by William II., Henry I., Stephen, and Henry II., as well as by Archbishops Anselm, Ralph, William, and Thomas A'Becket, and by Bishops Ralph, Arnulf, John, and Ascelin, who confirmed to the monastery Mallinga's Parva, a market, St. Leonard's Chapel, and the Church of St. Mary. Anselm, in the time of William II., gave the nunnery the church of East Malling. In the year 1190 both the abbey and town were much damaged by fire. Soon after this, we learn from Dugdale, " John conceded by this charter, and confirmed to God and to the church of Malling and the nuns worshipping God there, the church of East Malling with all pertaining to it, which Lord Hubert, archbishop, restored

to them and converted into their proper use, as the charter of
the same archbishop devises. Wherefore I wish. . . . Given by
hand Hugh de Welles, Archdeacon of Wells, at Rummenall, April 5th,
seventh year." This would be in the year 1206.

At a little distance away from St. Leonards, in the woods, about
three-quarters of a mile to the left of the Wrotham and Teston
road, stood a chapel, which lasted into this century as a ruin; but
you can now only perceive that there has been a building on the
spot. Thorpe describes this as " Blair's Chapel, or rather, perhaps,
cell, for the accommodation of a father confessor of the abbey." It
was probably dedicated to St. Blasius, whose feast is on February 3rd,
as a black-letter day in our Prayer Book ; for the wood is still known
as Blase's wood, and as it appears to have been under the hospital
at Strood, most probably had nothing to do with the abbey.

Besides the tower of the abbey there are considerable remains
of the church behind, showing its high pitched roof and the fine
arch leading into the refectory and the cloisters, which, with
a more modern dwelling house, form a quadrangle. In the square
are three stone coffin lids, erroneously thought to be tombstones.
They were probably the covers of the depositories of the bodies of
some of the nuns, who were buried, as was so customary, in the
cloisters; this would not, of course, necessitate that there should have
been a burying ground at the spot.* In the square court there are
two angels with scrolls, on one of which is " Benedictus Deus in
domo ejus " ; on the other, " et in omnibus operibus suis." Over a
passage in the western wall, in the same character, is " R. Morton."
Strange to say, Rose Morton was one of the nuns who received, at
the dissolution, an annuity. In another place there is the inscription " Orate pro dominâ benedicta." At the western entrance, on
shields, are the instruments of the crucifixion, a heart distilling blood
on a shield ermine, and a crozier on a bend sinister ; and on a chef,
three annulets. Just by the entrance is the gateway chapel of the
abbey : this has been a carpenter's shop, and in 1773 it was used as
a meeting house. It was renovated by the owners, and has been
used for daily service since All Saints' Day, 1858. It is about
22 ft. long and 13 ft. broad, and there is an ante-chapel behind the
screen about 15 ft. long. Above the doorway is a recess for the
images, and at the side of it a place for the holy water. We
mention these matters as connected with the existing building of
the abbey, but of course many of them are of far later date than
the eleventh and twelfth centuries. Soon after the beginning of the
thirteenth century we find the chapelries of Longsole, near Barming
Heath, of St. Laurence in Halling, and of New Hythe on the Medway
mentioned ; but as we have no record of their building, and very little
of their remnants, we can hardly do anything more than suppose

* Lately another has been found (and buried there), together with a pillar
support of a crowned angel and a ring, and certain pieces of carved stones.

them to have been built about this time. In the year 1115, Ernulf, Bishop of Rochester, composed the *Textus Roffensis*, in which we learn that all the parishes whose history we are writing contributed nine denarii for chrism to the see of Rochester, and in addition to them Paddlesworth and Dode as parish churches ; and St. Leonards also, as already mentioned, gave six denarii. Henry I. granted to Gundulf, and the church in Rochester of St. Andrew, the churches of Dartford and Aylesford, and all the whales found in the bishopric of Rochester.* When the chapel of St. Laurence in Halling was founded, as already stated, we cannot find out ; but certain we may be that it was originally built for the convenience of pilgrims to good St. Thomas' shrine about this time, as they came along the old British path, from them called the Pilgrims' Way, since it conveniently stood near that track after they had passed Paddlesworth church, and was possibly placed near one of the turnings where they left the road to cross the river Medway. The Normans built their churches from material found close at hand, till they imported Caen stone later on. Thus the sarson stones, greensand rock, the latter some-times used when coated with iron, perhaps for ornament, and tufa, which has been found even lately at East Malling, formed their materials in Kent. . Though many people trace their family to the Norman Conquest, it is a fact that in this district we have but few names amongst those who held the manors to record in this chapter, which reaches down to the time of John.

The manor of Addington at the time of the Conquest was in the possession of Odo, Bishop of Baieux, and was held by William de Gurnay from him.

The manor of Allington was the property, as has been already stated, of Harold's brother Gurth at the time of the Conquest ; from him it passed to Odo, from whom Anschitil held it ; it then came in succession the property of William de Warenne and of Lord Fitzhugh, whose daughter conveyed it to Sir Gyles de Allington. Warenne is said to have fortified it. The place is already stated to have possessed a castle in Saxon times, which was razed to the ground by the Danes. Philpott declares the castle to have been the erection of William de Columbarius in Stephen's time, but we are inclined, from circumstances which we shall speak of in the next chapter, to hold this castle of later date.

Aylesford manor was the property of the king. It is said to have been held by Estrangea in 1157, a name wonderfully like *a stranger*, and not ill-fitted to the men of Norman and French times. In the ninth year of King John it was held by Osbert de Giffard, whose father appears to have held it before him from 1175—1185.

---

* Whales are still caught in the bishopric, one being taken near Graves-end in 1883, and one at Gillingham on August 30th, 1888, as shown by the published records of the Rochester Naturalist Club, and as mentioned in our appendix.

Birling also became the property of the Bishop of Baieux, from whom it was held by Ralph de Curbespina (Crookthorn).* This family continued till the reign of Henry II., and were succeeded in the estate by Magminot de Wakelin, who died without children in 1191, when Alice, his sister, carried the property to Geoffry, second son of William de Say; but the living was granted to the monks of Bermondsey, with whom it continued till the Reformation.

Ditton, which consisted of the manor of Sifletone as well as that of Ditton, with the appendant manor of Brampton, was all held by the Bishop of Baieux, of whom Hamo, the sheriff, held Ditton and Brampton, and Vitulus, Sifleton. These manors reverting to the crown soon afterwards, appear to have become the property of the Earls of Gloucester.

The ancient manor of Tottington, in Aylesford, we are told, in these days was held by Robert Malgerius de Rokesley and Robert de Rokesley, who paid tithe to the monks of St. Andrew; their landlord was Odo. Richard Fitz Turold held Eccles, which soon after came into the same hands as Tottington. Odo also possessed the manors of Paddlesworth and Ryarsh. As the latter was held by Hugh de Port, and the former by Hugo, it may be that the same person had them both. Early in John's reign we find William de Crescie held the manor of Ryarsh. Paddlesworth became the possession of the Chetwynds, who changed it with Hamo de Gatton, of Throwley, who sold it to Sir William de Huntingfield, whom we find Lord Warden of the Cinque Ports in the time of King John. Offham was another manor that belonged to the king's half-brother, Odo, who let this also to Hugh, but afterwards to Anschitil; it reverted to the crown, together with other manors, after Odo's disgrace.

At Preston, near Aylesford, stands an old barn dated 1102, with " T. C. " on it—the initials of Thomas Colepepyr; but the arms quartered, as well as the writing, are later than Edward III. We, however, find Sir Thomas Colepepper was a judge in the time of King John. East Malling, as has been already stated, was in the hands of the archbishop. West Malling, after being held a short time by Odo, was given up to the see of Rochester, and its fortunes we have followed. That prelate also possessed the manors of Halling, Trottescliffe, and Snodland or Esnoiland.

The palace of Halling was another construction of Gundulf's, and was built in 1077; it was rebuilt by Gilbert de Glanville in 1185. It is possible that when Richard, Archbishop of Canterbury, died there in 1184, from fright it is said at what he had seen in a dream, if the place were out of repair, the howling of the north-east wind in the gorge of the Medway may have had something to do with it.

Gundulf had the manor of Trottescliffe to support his table, and as part of the bishop's share of the spoils when Odo was compelled to restore the lands of the monks of St. Andrew's in 1293. The bishop

    * Crookhorn wood still keeps up the name.

being called upon to prove the right he had to the manor, claimed
it from time beyond memory; the palace was in ruins when Gilbert
Glanvil came to the see, but it was then rebuilt, and became a
favourite dwelling-place of the Bishops of Rochester. The manor of
Leybourne was possessed by Odo, Bishop of Baieux, and soon after it
reverted to the king, who had given in marriage Amy, daughter and
heiress of Robert Fitzgerald. Charles Seymour, who wrote a survey
of the cities, towns, and villages in the county of Kent under
Leybourne, tells us that "Sir William Arsick was the owner of
Leybourne in the time of William the Conqueror, and Lord Leybourne,
of an ancient and illustrious family there, was at the same time
possessed of the castle built by his ancestors: it was a place of
strength." It would, however, appear that Leybourne Castle is not
older than Norman times. In the year 1194, we find Sir Philip
de Leybourne died in possession of the manor, which then passed
to his son Robert de Leybourne, whose son, the first Sir Roger de
Leybourne, was in arms when quite a youth against King John with
his brother barons, was taken at the siege of Rochester Castle, and
had to pay two hundred and fifty marks for his liberty, which he did
not regain till the following year. The oldest of the buildings of
which the ruins now remain were probably antecedent to that date.
These were on the south side of the *enceinte*, and were on the plan
followed in all manor houses of Norman as well as Saxon times,
and perpetuated in our collegiate buildings. On the western side
we have the remains of the vestibule, with the sewery for the
storing of linen and provisions, and the buttery for wine and beer.
There must have been a drawbridge over the moat, about 20 ft.
to the westward of the vestibule. The vestibule would be connected by a porch with the hall, which was the principal building
of the castle, being used in the day-time as the banqueting-room
and sitting-room, and at night as a dormitory.

Facing the upper end of the hall was the solar, the private room
of the lord of the castle, over the cellar. There are no remains of
the chamber itself, but the building which must have been connected
with it is perfect and interesting. Its lower portion was connected
with the cellar by a square shaft, of which the part is on a level
with the second story, and at the further side, which is the east
end of a vaulted passage, $2\frac{1}{2}$ ft. wide, $7\frac{1}{2}$ ft. high, and if we include
the top of the shaft, 8 ft. in length. There are four apertures, viz.:
the doorway, the shaft, the upper part of a large arched window
over the shaft, and a small aperture on the north side. Above the
shaft, on the north and south sides of it, are cavities in the stone-
work, evidently designed to receive the ends of a beam or roller,
over which a cord might be passed for the purpose of drawing up
things, and possibly persons also, there being no internal staircase
to the solar. It was built close to the moat.

What is now the north wall of this arched passage must have

been part of the inner wall of the solar. It is throughout of good workmanship, and was very strongly built, which has led to its preservation. Immediately below the passage there is an arched doorway, communicating with the chamber below. This lower chamber apparently was the lady's bower. It was nearly circular, 21 ft. in diameter, and it was provided on the south side with a row of seats, formed by carrying up the full thickness of the wall only for a short distance from the ground. At a distance of 38 ft. from the buildings above described, proceeding northward, are the remains of buildings which must have been the chapel, with small rooms for the priest and other officials built to the west of it, and to the south of the passage leading to it. The chapel must have been a remarkably fine building. The entrance to it was by an arched porch and a passage of 20 ft. in length, and the red sandstone jambs of the doorway of the chapel are very perfect.

The greater part of the north wall remains, but it has been much altered, first by the erection of a dwelling-house on the remains of the chapel late in the sixteenth century, and again by the conversion of this into a farmhouse in the early part of the eighteenth century.

The nave, as was usual in chapels of that age, was divided into two stories. The upper chamber was for the lord of the castle, his family and his guests, and the lower one for the other inmates of the castle. They were separated from the sacrarium by screens, and as they were not consecrated they were used for other purposes than those connected with religious worship. The sacrarium, which was consecrated, was the whole height of the building. When the dwelling-house was built the walls of the chapel were raised and a third story was added. Fireplaces were formed for the first and second stories, with a long chimney. The fact of these having been of that later date is shown by distinct lines of demarcation between the masonry of the wall and that of the chimneys. A fireplace was also formed on the third story, which had no chimney; the smoke having been carried into the long chimney by a short lateral flue.

Chambers were formed at this time over the sacrarium. Their comparatively modern date is shown by the fact that the joists of the floor were let into the wall: whereas in the original building these were affixed to beams supported by corbels. When the dwelling-house was converted into a farmhouse, the chapel was turned into a dairy. The two floors which had been placed over the sacrarium were removed, and the space which had been occupied by the fireplace, in the third story, was filled up with stones built into the level of the wall; the chimney place having been built in with them, its face remaining so as to show its position. These details of an ordinary dwelling-house have led to the general belief that the building has never been a chapel; but a careful and minute examination shows this opinion to be entirely erroneous.

The fact that the original building was a chapel, and that it could have been nothing else, may be proved by the following evidence:—First, by its orientation, and by the remains of narrow windows in the eastern wall. Secondly, by its eastern half having been of the full height of the building, while its western half, or nave, was divided into two stories. Thirdly, by the distinctly ecclesiastical character of the north window of the sacrarium, which was evidently divided by a central pillar into window lights, with pointed arches formed of Caen stone; there are no remains of the pillar or of the keystone, but the voussoirs which were on either side of it remain, and on the lower side of each of these there is a triangular incision, showing the position of the tops of the pointed arches, and in one of them a fragment of the Caen stone remains. Fourthly, by the distinct remains of an Easter sepulchre, such as are to be found in a few old English churches, but never before, so far as I have been able to learn, in any castle chapel: it projects about 8 in. from the north wall at its eastern end, and is recessed for 2 ft. into the main wall. A stone which formed part of an ornamental arch has been found by the author of this book among the rubbish; this stone was the segment of an arch with a radius of 2 ft. 6 in., and as the width between the imposts on the two sides of the sepulchre was 4 ft. 8 in., the stone was probably part of a semicircular arch over the sepulchre; it is chamfered on its lower side, and the chamfered portion, as well as its inferior surface throughout, is covered with fine red cement. Fifthly, by a corbel stone, which must have been placed there for the support of a rood beam at its north end; its distance from the east wall is 14 ft. 3 in., and its height above the ground 5 ft. 6 in. Allowing 2 ft. for the length of the wall-piece rising from it, this would make the height of the rood beam 7½ ft. above the floor of the chapel; from the position of the corbels which supported the beams of the floor of the second story of the nave, the rood beam would appear to have been about 2 ft. to the eastward of it, and 8 ft. lower; of the three corbels which supported that floor two remain, and the third was removed when the fireplaces were made. Sixthly, the roof of the building had arches, supported by corbels 24 ft. above the ground. The corbels in the north wall remain at the east and west ends: the intermediate corbel must have been removed for the construction of the fireplace in the third story of the sixteenth-century building. The length of the chapel was 29 ft., and its width 24 ft., and there was probably a Mary chapel, or chantry, on the south side. If there was a partition wall, every vestige of it must have disappeared when the chapel was converted into a dairy.

Sir William de Leybourne, who alienated Leeds Castle to the

Crown—as we shall have occasion to mention—in 1276, returned to
Leybourne; and then, no doubt, he built the gatehouse to the
castle, which was not a usual addition to these buildings before the
reign of Edward I. The gateway was strongly and scientifically
fortified. The towers projecting before the main walls are pierced
with loopholes; besides the portcullis there is a water-cullis, for
throwing down water to prevent the gates from being destroyed by
fire. Provision was made for an ample supply of water for this
means of defence, by an aqueduct communicating with the moat and
passing under the west tower, from which there is a shaft rising
up into the tower. In the west tower, which was washed by the
water of the moat, are a series of foot log-holes, by which a wooden
gallery could be carried across the front and to the west side of it,
serving as an advanced position for the defenders, and a passage
for them to and from the interior.

The author is indebted for the above remarks on Leybourne
Castle to the Rev. J. H. Timins, Vicar of West Malling. They
were read before the Kent Archæological Society in July 1891,
and objection taken to the Easter tomb; but the author himself
discovered the stone mentioned in the above article, and is certain
it could have been part of no other arch but one over such a tomb.

Twice, if not more, this valley in those times must have been
the scene of war; for if it escaped the ravages of William the
Conqueror, probably Rufus marched through these possessions of
Odo, after defeating the proud prelate at Tonbridge, and John,
after his victory at Rochester, if not before, ravaged the lands of
the Kentish barons here who opposed him. Mr. St. John Hope,
of the Society of Antiquaries, is of opinion that the two shafts
led to the places of the *gardes-robe* of the castle, and that there was
undoubtedly a chapel, but that it was on the first story—the two
drum towers, as any one can see, are of much later date. The portcullis arrangements were much praised by him, but he ignored the
water-cullis. It may perhaps be worth while to mention that all
persons who know anything of antiquarian research doubt the
vulgar tradition that there was an underground passage from the
church to the castle. Richard de Tonebridge possessed land, in
the time of the Conqueror, in Aylesford, Halling, Leybourne,
Offham and West Malling.

This old ruin, which stands close to the Rochester road to Malling,
forms with the church a very picturesque scene. The late Sir
Joseph Hawley had this ancient ruin cared for and enclosed in
a fence, so that, though there is no dwelling-place at the castle, it
is kept safe from wanton depredation.

Amongst other relics of this period are the churches of all these
parishes except Halling, which contain more or less Norman architecture. Parker speaks very highly, especially of the piscina of Ryarsh
church in his Glossary, as a specimen of the work of this era.

At this period we have the earliest mention of the names of any parish priests; we accordingly give them, with the accompanying archbishops and bishops of Rochester, from the end of Henry I.'s reign :—

Date.	Archbishops of Canterbury.	Bishops of Rochester.	Rector of Allington.	Vicar of Aylesford.
1132	William Corboil	Ascelin	Robert de Donam	...
1138	Theobald	... ...	... ...	...
1148	... ...	Walter de Canterbury	... ...	Jordan
1162	Thomas à Becket	... ...	... ...	...
1174	Richard	... ...	... ...	...
1182	... ...	Gualeran	... ...	...
1184	Baldwin	... ...	... ...	...
1185	... ...	Gilbert de Glanville	... ...	...
1193	Hubert Walter	... ...	... ...	...
1207	Stephen Langton	... ...	... ...	...
1215	... ...	Benedict de Sansetum	... ...	...

CHAPTER V.

THE times of Henry III., the three Edwards, and Richard II. were stirring times in the history of our valley, owing to the various disputes between the barons and the kings, the Crusades, and perhaps also the Scotch and French wars. In these we find the family of the De Leybournes, whose castle we described in the last chapter, coming to the front as representatives of our valley. The Sir Roger who was freed from Rochester Castle appears to have had a son, another Sir Roger, a veritable knight of those days. We agree with Rev. Lambert Larking in this matter, that Dugdale is wrong in considering that there was only one Roger de Leybourne, for one of the greatest arguments against this is that the rebel baron of 1215 could hardly have been at Evesham fight in 1265. But in Henry III.'s rolls we actually find Roger de Leybourne described as "patrem predicti Rogeri"; it would appear the first Sir Roger died about 1251. The first mention of the Kentish knights and squires in these tumultuous times is that of Hugh de Crescie, who held a manor called Crescie Park, in Trottescliffe; he also possessed the manor of Ryarsh. We find him fighting against the King at Lincoln (1217), but as he died in 1263 we do not know what side he would have taken when the barons were in rebellion. He was succeeded by his brother in 1264, who appears to have died shortly afterwards, when the family became extinct.

Aylesford about this time became the property of Lord Richard de Grey, of Codnor. He does not appear to have been endued with the turbulent spirit of his brother barons, and in 1230 was made Lord Warden of the Cinque Ports and Constable of Dover Castle : he went soon after to the Holy Land, and returning to England founded the first Carmelite priory at Aylesford. Soon after, the first European chapter of this order was held in that village. One of the Carmelite friars of this period who rose into notoriety was John Stock, so called, says Camden, " because he was found in the hollow trunk of a tree near this place " (Aylesford).

In the year 1252, Roger de Leybourne, whose knightly spirit had now reached its most turbulent period, murdered Ernulf de Muntney in revenge for injuries received from him at a former joust at the round table. (This meeting was so named because previous to the joust, that no cause for any malice should arise, the combatants partook of refreshment at a round table, to show that no precedence could be given to any one.) He received pardon from the king for this deed of violence.

We now come to that period of English history which is called the Barons' War, in which the barons tried to prevent the king from giving land to foreign favourites, being quite forgetful that they themselves were but the sons, or grandsons, or in the third and fourth generation of foreign favourites; and we learn, from the pardon obtained next year from the king, that amongst the barons who had opposed the king in the Provisions of Oxford, were Sir Roger de Leybourne and William de Huntingfield, of Paddles-worth. Sir Roger, however, appears from the earliest to have been more or less attached to Prince Edward, and thus, contrary to the wishes of the general body of barons, he persuaded Edward to go on a round of tournaments in France, as they feared that Edward would seize the occasion to hire foreign troops. So different was the feeling of the queen, that she tried to alienate her son from Sir Roger, evidently feeling that he would strive to prevent the prince from acting in this unpopular manner. Whatever was the cause, however, we soon find the two brothers-in-arms separated for a period, and Sir Roger, who was steward of the household, is charged with misappropriating £1000. Being driven to despair, the baron, in company with Sir William de Deptling, appears to have found the means of subsistence by commanding an armed band, with which he was reported to have been making arrangements for an attack upon Dover Castle. This, however, was declared frivolous; nevertheless Sir Roger, with others, marched into Wales and Herefordshire, took Hereford, and made prisoner Peter de Aqua Blanche, the bishop. After this we find him, with Simon de Montfort and others, driving the mercenaries through Kent, and finally expelling them from Dover in July 1263.

At this time, it would appear, the song of the Barons was made in which the hero is thus described :—

> " Et Sire Roger de Leybourne,
> Que sà et la souvent se torne
> Mout a la conquerrant ;
> Assez mist paine de gainer
> Pur ses pertes restorér,
> Que Sire Edward le fit avant."

["And Sir Roger de Leybourne, who often turns him on this side and on that, made great progress conquering, he laboured much to gain and to restore his losses which Lord Edward caused him previously."]

In September, Sir Roger and several other barons, whose quarrel seemed to be over with their monarch as soon as the foreigners were expelled, were pardoned, and joined the king. And thus we find Sir Roger fighting bravely for the king at Lewes and Northampton, and with many another from this valley he defended Rochester Castle against Simon de Montfort in 1264, and was wounded at Evesham while assisting the king against the rebels. In the pardon of Kenilworth were included Lord John de Grey, of Aylesford, who died 1272, and William de Say, of Birling, who died 1293.

In 1265 matters in England had become at last settled, and in the year 1268 we find Sir Roger exchanges with Crecquer, or Creve-queur, Trottescliffe. But this restless baron could not be quiet, though he was made Sheriff of Kent (1264-68), also Lord Warden of the Cinque Ports; still he desired more active life, and thus, amongst those who followed Prince Edward in his expedition to the Holy Land in 1270, we find Roger de Leybourne's name and seal. About a year after, on November 7th, 1271, we find Sir William de Leybourne giving in his homage to Henry III. as son and heir of Roger de Leybourne, deceased; thus we conclude that the grand old knight was dead, but whether before or after the landing of the Crusaders at Acre, which took place at Easter 1271, we cannot tell.

The heart shrine in Leybourne Church, I consider, has been incontestably proved, by the late Mr. Larking, to be the tomb of our great Kentish baron; but for whom the second casket was intended it is difficult to determine. Mr. Larking has evidently shown why his wife, who survived him, had not her heart deposited in the companion niche, she being a second wife, and he being her third husband.

We must be pardoned if, though slightly out of order, we conclude the history of the Leybournes. The son of the great Sir Roger was Sir William, but he, long before his death, gave up his interest in Leybourne; nevertheless, it is to this owner of the castle that this valley was indebted for the only royal pageant it seems ever to have witnessed, when, in 1286, Edward I. visited the castle of his old companion-in-arms, and perhaps went into the church and saw the famous heart shrine. The two crowns which the present rector (Rev. C. C. Hawley) is so anxious about, may possibly have something to do with the votive offerings of himself and his queen at this period. Sir William was a brave soldier: " vaillans homs sans mes e sans si " (a valiant man without buts and ifs), as he is described when present at the siege of Caerlaverock in 1300. He was one of the hundred and four nobles who, headed by Bigod, Earl of Norfolk, and Bohun, Earl of Hereford (when Boniface wrote to Edward I., ordering him to make peace with the Scottish chiefs and nobles, and to send representatives to give evidence of his claim within six months to the papal court, in which the Pope professed to settle such questions), subscribed a document, saying that if Edward proposed to argue the matter before the

Pope they would not allow him to stoop so low. This letter was written January 20th, 1301. He died in 1309, and two years previously his son, Sir Thomas, to whom and his wife he had alienated Leybourne, predeceased him. His other son, Sir Henry, was one of the most violent and restless men of his day, whose name occurs again and again on the *Curia Regis* as a turbulent ruffian ; in 1309 he was one of the knights in the tournament of Stepney, got up by the Earl of Lancaster, and as one of his partisans was taken at Borough Bridge in 1323, having apparently only been released from Scarborough Castle the year before. He was finally outlawed. Sir Thomas * left an only daughter, Juliana, commonly called, from her enormous possessions, the Infanta of Kent. Juliana married John, Lord Hastings, by whom she had a son, Laurence, created Earl of Pembroke, who was succeeded by his son John. This family became extinct in 1389. She afterwards married William de Clinton. She died at Preston-next-Wingham, November 1st, 1367, leaving, it would appear, effects worth £3160 13s. 4d.—an enormous sum for those days. She confirmed her manors in Kent to trustees for her life, with remainder to the king, and for religious uses.

The De Leybournes seem to have been generous benefactors of the Church, especially Sir Roger (perhaps on account of the murder he committed) ; being anxious for his soul, he gave lands from his estate to the priory of Cumbwell, and also certain properties which he had acquired from Ralph Ruffyn, as an endowment to the church at Leybourne, for two capellans. Out of this the latter's wife, Feodina, received her dower. Sir William de Leybourne confirmed this endowment. It appears that Roger Ruffyn trespassed on these lands. We learn from Thorpe's *Registrum Roffeuse* the land was in Leybourne and Caumpes, the meadow called Ruffynsmead being reserved ; to the original grant Sir William de Leybourne gives five marks from the manor of Radleke. The land at Caumpes,† 100 acres, still belongs to the glebe of Leybourne. This grant is interesting, as giving us the names of the first clergymen we know of connected with Leybourne, viz. : Thomas Bacun, who, in order to maintain better the chantry founded by Sir Roger in 1279, was allowed to hold Leybourne and Langley ; Peter, another rector ; and John, the capellan or chaplain. It, moreover, caused several disputes as to boundaries between the family of Leybourne and the clergy, and in

---

* On the death of Sir Thomas we have the first intimation of Malling being considered an assize town, as the Inquisitio post mortem was held here on July 8th, 1307.

† Great Comp Wood and Comp are still where there is a detached portion of Leybourne. All kinds of food except bread and drink are called companage, which Spelman interprets : " quicquid cibi cum pane sumitur." In the manor of Fisker-ton, Notts, some tenants, when they performed their boons, or work days, had three boon loaves with companage. In the time of Mary Wade, widow, there was given one parcel of ground to " fynde a compe in the church of Trottescliffe, also two garden plotts to fynde two compes, one in the church and the other in the chancell."

the fourth year of Edward II. we learn, through these disputes, that Walter de Leghton, or Lecton, was incumbent of Leybourne in 1311, and he was still so in 1314. The manor of Leybourne was, it would appear, in accordance with the Countess' will, given to the abbey of St. Mary Grace's on Tower Hill, upon Sir Simon Burley, who had been Lord Warden of the Cinque Ports (1387), and the king's governor, being attainted for treason. He was executed, together with Tressilian, in the year 1390. We subjoin the pedigree of the De Leybournes :—

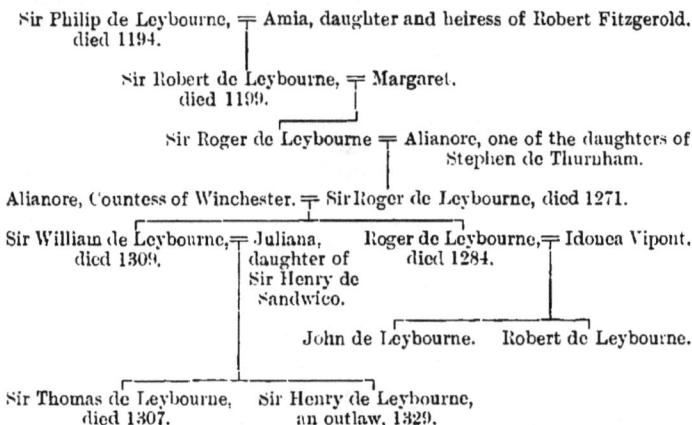

Sir Philip de Leybourne, ⚯ Amia, daughter and heiress of Robert Fitzgerold. died 1194.

Sir Robert de Leybourne, ⚯ Margaret. died 1199.

Sir Roger de Leybourne ⚯ Alianore, one of the daughters of Stephen de Thurnham.

Alianore, Countess of Winchester. ⚯ Sir Roger de Leybourne, died 1271.

Sir William de Leybourne,⚯ Juliana,   Roger de Leybourne,⚯ Idonea Vipont.
died 1309.   daughter of   died 1284.
          Sir Henry de
          Sandwico.

          John de Leybourne.   Robert de Leybourne.

Sir Thomas de Leybourne,   Sir Henry de Leybourne,
died 1307.   an outlaw, 1329.

The late Wykeham Martin, Esq., M.P. for Rochester, of Leeds Castle, claimed to be descended from the De Leybournes. Sir Roger's arms were *d'azure*, six lioncels, *argent*; and Sir William's six lions rampant, *argent*. The granddaughter of Sir Roger, Idonea de Say, was the ancestress of the Lords de Say and Sele, and it is through her the Wykeham Martins claim their descent.

The manor of Cressy Park, in Trottescliffe, which, as we have seen, came to the Crevequeurs in 1248, and for which they obtained full rights from the De Leybournes in 1276, went to the Bishop of Rochester in 1278. In 1292 Trottescliffe was valued at £6 13s. 4d. for ploughlands, £1 5s. for beasts, and 5s. for hens—a total of £8 3s. 4d.; in 1360 it had risen to £15 3s. 10d. The Bishop of Rochester, after this date, is the sole lord of the manors of Trottescliffe.

Birling appears to have continued during this period in the hands of the De Says, of whom Geoffrey, who succeeded his father William, died in the twenty-third year of Edward III.; William his son died twenty-six years afterwards, when the property came to his grandson John, who died in 1384, and left the estates to Geoffrey, Lord de Say, who, towards the end of Richard II.'s reign, alienated

the manors to Richard Fitzallan. Of these Geoffrey and William
were both summoned to Parliament.

In Edward I.'s reign the manor of Ryarsh, the remaining
property of the De Cressie's we have not spoken of, was given to
the great family of the De Mowbrays. John de Mowbray joined
the Lancastrians, and was taken at Borough Bridge and hung at
York (1322); nevertheless his son John was received into favour,
but came to a premature end in the plague (1348). His son
John was slain by the Turks at Constantinople in 1369. His
eldest son, John, was created Earl of Nottingham, and was present
at the coronation of Richard II., but died in 1383, and was
succeeded by his brother Thomas, Lord Marshal and Duke of
Norfolk, the famous peer who quarrelled with Henry of Lancaster,
then Earl of Hereford, afterwards Henry IV. On account of their
disputes Richard banished both the discontented nobles; but our
Kentish hero did not return, like his more successful rival, and died
at Venice. His son Thomas succeeded him.

The castle of Allington, which was surrounded by a moat filled
from the Medway on three sides, and the Medway itself on the
fourth, presents us with a fair opportunity of seeing how these
ancient defences were made, as much of it is very perfect (especially
the gateway), and will repay a visit.

In the end of Henry III.'s reign, Stephen de Penchester, who
was Sheriff of Kent (1269-70), Lord Warden of the Cinque Ports
(1275), and Judge of Common Pleas, having married Margaret,
daughter of Hugh de Burgh, acquired this manor, and obtained
leave from Edward I. to fortify this place, hold a free market,
and have free warren. This would lead us to think that he really
was the founder of Allington Castle; and when we consider that
there is nothing Norman, apparently, about the place, the date 1300
seems a good time to consider as that of its erection. The fair
was held on St. Lawrence's day. He died without children, and was
buried at Penshurst. By his daughter Margaret it passed to the
noble family of the Cobhams, of Cobham Hall, in this county, who
held it till the days of Edward IV. In the year 1307 Stephen
de Cobham was summoned to Parliament, and died 1333, and was
succeeded by John de Cobham.

In the year 1216 King John pardoned William de Cosyngton
for taking the oath of allegiance to Louis. His son Sir Stephen
was at the siege of Caerlaverock, and founded the chapel which stood
on the manor for many years. His son Sir William, who was Sheriff
of Kent 1307, appears to have died here about 1332, and the manor
continued in the family down to the reign of Henry VIII. The
chapel dedicated to St. Michael, Thorpe seems to think he traced;
but Philpott says that even in his days it was crumbled into so
desolate a heap of rubbish that he could hardly trace its form
amongst its ruins. However, the author and a friend saw distinct

traces of there having been a manor house and a chapel here this
year (1892).

The manor of Tottington was held by Richard de Rokesley of
Hamo de Crevequeur in Henry II.'s and Edward I.'s reign. In
Edward II.'s reign, Sir Richard de Rokesley, Governor of Poitou
and Montreuil, in Picardy, held it. In Edward II.'s reign he left, by
his wife, daughter of John de Criol, two daughters, the elder of whom,
Agnes, married Richard de Poynings, whose son Richard, Lord
Poynings, died 1383. He founded a free chapel in honour of St.
Michael, of which Thorpe found traces in his days in an orchard.
This fact may account for a singular entry in the burial register of
Aylesford, 1666. " A travelling man, who sold earthen pots and
other earthenware, being found dead in Thomas Smith's barn, was
buried in the said Thomas Smith's orchard the seventh day of
February." If the orchard was the one in which Tottington Chapel
stood, he may have been buried there because the country folk still
held it as sacred. There is a farmhouse here now, and many blocks
of building stone lying about, besides a number of sarsons, but no
traces can be seen of the ancient manor-house and chapel.

Of the De Greys of Aylesford, John, whom we have mentioned
as being pardoned at Kenilworth, died in 1272, and left Henry his
son as heir ; though he served in Wales, Edward I. brought a trial
to evict him, which, however, failed, and he died 1309. His son
Richard obtained royal favour, and gained a market (1331) on
Tuesday, and a fair on the eve and day of the Ascension, for Aylesford ;
he also added three acres to the Carmelite friars' possessions, and
died 1336. His son John, for his knightly acts, was presented by
the king with a mantlet of white cloth embroidered with men
dancing, and having buttons of pearls ; he was exempted also from
serving in Parliament in the year 1372, and had accoutrements of
Indian silk with the arms of Penchester bestowed upon him ; he
died 1393, and was succeeded by his second son, Richard, his eldest
son, Henry, having predeceased him. Henry de Aylesford, Abbot
of Battle (1281-97), was a native of Aylesford. In 1272 it is
noticeable that a felon fleeing into the church of Elsford (Aylesford)
was declared to have abjured the realm. Aylesford was not subject
to the Constable of the Hundred.

The manor of Addington was in the possession of Roger de
Mandeville and his son Arnold, according to Hasted, in the times
of Henry III. and Edward I. ; the former of these warranted the
manor to Roger de Scaccario, and three years later we find the
said Roger entailed the manor on his wife and son, Laurence, who in
1271 did homage for it. In Edward III.'s reign, according to Hasted,
John de la Chekere held it ; and I am inclined to think with Rev.
T. S. Frampton, that the stones with Lombardic characters in the
south chapel of the church at Addington are his monument. As Mr.
Frampton says, they run : 𝔥𝔈ℜ𝔞ℜ𝔞 : 𝔊𝔍𝔖𝔗 . . . on one stone,

and on the other . . . . ᚱᚲ : ᛃᛞᛖᚨᚱᚲ : ᛞᚲ : ᛚᚨᛊᚲᛚᛃᛊ. The inscription, Mr. Frampton says, may have run: "Sire Johane de Leschekere, gist en ce place "; but though I agree with him as to the one stone being the tomb of Sir John, I fancy that possibly the other stone may be another tomb of the Chekeres. Nicholas de Daggeworth obtained the manor, and alienated it soon after to Hugh de Seagrave, who was Steward of the Household and Keeper of the Great Seal and Treasurer of England; he conveyed it to Richard Charles, who is buried in the church, and whose monumental brass is still nearly perfect. ·The inscription runs: "Hic jacet magister Ricardus Charles, qui obiit in festæ sanctæ. . . . anno domini Millesimo CCCLXXVIII., Cujus animæ propicietur Deus " (Here lies master Richard Charles, who died on . . . of the sacred festival . . . . . in the year of our Lord 1378, on whose soul may God have mercy). His daughter Alice, who survived her brother Robert, married William Snayth, to whom she conveyed this manor. These monuments are the oldest in this neighbourhood.

The manor of Preston, in Aylesford, was held by Sir Thomas Colepepper, judge, in King John's days. His son Sir Thomas had two children—Sir Thomas of Bayhall, in Pembury, and Walter. Sir Thomas of Bayhall, in Pembury, had a son John, who was sheriff in Edward III.'s time; his son Sir Thomas was Sheriff of Kent in 1394 and 1395, and was commissioned with Nicholas atte Crouch to go against the rebels Wat Tyler and Jack Straw, the latter of .whom was said to have been a native of Offham. At any rate, John Sales of Malling took part in the rebellion, and came to Canterbury with a great multitude; but the descendants of the second Sir Thomas are those whom we follow as heirs of Preston Hall. Of this branch of the family his second son Walter was the founder; his eldest son, Sir Thomas, dying without children, the estates passed to Sir Jeffrey, who was Sheriff of Kent in 1366 and 1374.

Another manor in the parish of Aylesford was called Rowe's Place, and was in the possession of Robert Rowe in Edward III.'s reign. There are squared stones in the wall of the garden of what is still called Rowe's Place, at Eccles, kindly pointed out to me by the present owner, that suggest a handsome mansion and probably a chapel: Mr. Shaw also kindly informs me that he has ploughed up a number of sarsons on his property.

Offham was held by William de Offham in Henry III.'s time, one of whose sisters was married to Stephen de Penchester, his other sister, Matilda, claimed a third from Richard de Courtone. Robert de Courtone passed the manor to Ralph de Ditton, whose daughter Isabella, being granted it, carried it in marriage to Thomas de Plumsted: it afterwards went to the Colepeppers. The other manor, Goldwell, passed from Robert de Courtone to John de Melford, by whom it was conveyed to the Brownes of Beechwood Castle.

Ditton was held by William de Ditton (1290), of the Earl of

Gloucester, in the time of Edward I. He also held Brampton and
Sifletone. In the reign of Edward II. (1307) Ralph de Ditton,
who was in 1259 convicted of injustice, and Joan de Leukenoe,
were owners both of Ditton and Brampton; these soon after passed
to the Aldons. Sifletone passed into the hands of Robert de Burgh-
hersh, Constable of Dover Castle and Lord Warden of the Cinque
Ports. In 1305 his son Stephen obtained a charter of free warren
from Edward II. He was succeeded by Bartholomew, Lord Burwash,
who acquired the three manors, from whom they passed in 1347
to Thomas de Aldon. He died in 1362, and in 1392 they passed to
Sir Walter de Pavely, from his family to that of Windsor, whence
they were sold to Lord Clifford of Clifford, in Herefordshire, in 1392.

We have already spoken of Walter de Huntingfield * being one of
those who took part in the Provisions of Oxford. He appears to have
been in possession of Paddlesworth in 1313, when we learn that he
and his son John had a dispute with John de Wereslee concerning
the manor of Paddlesworth, and Dodechirche and Paddlesworth
churches. Two years previously, Philip de Povenasshe had a dispute
with Reginald de Boclonde concerning lands in Snodelonde, Berlyngg,
Paddlesworth and Dodechirche. Now, as the ruin of Dode Church
stands about a quarter of a mile from a farm called Bocland, these
disputes are interesting, as showing us how Dowde about this time
became stitched on to Paddlesworth, and the two parishes virtually
united; since it would appear that, being small, and the lord of the
manor not residing at Buckland, but Paddlesworth, the former was
allowed to fall into decay. Richard Charles, of Addington, possessed
one-third of this manor, which had belonged to Basing in Richard II.'s
reign. The Povenasshes, who appear to have been long connected with
this district, had their name still surviving in Punish Hill till a few
years ago, when the same place got re-named Holly Hill. It may be
from the fact that the parishes of Paddlesworth and Dodecirce having
passed away, Snodland and Birling contested the tithes of this part,
and it thus got called Holy Hill.

As the rest of the valley was at this time in the hands of the
Church, we cannot do better than go on with the Church history of
the part, which gives us a fair insight into the history of the Church
of England generally during this period. In the year 1319 Hamo
de Heth obtained the see of Rochester, and was probably as good a
specimen of the episcopal bench of his day as could be found. He
was loyal to a fault to Edward II., yet did not shrink from inform-
ing him that had he been commanded to preach before him at
Tunbridge he should have chosen Haman for his subject. He was
persecuted for being one of the only four bishops who refused to
agree to Edward II.'s resigning the throne, and would not join in

---

* The arms of the Huntingfields were *or* on a fesse gule, three plates. Robert
de Huntingfield's arms are described as a *d'or* " à la fess des goules, et tres torteux
d'argent en la fesse."

the acclamations of praise to Edward III.; but when he learnt his master had resigned, he joined in the act of the coronation of his son, and in the same year, in Lent, dined with the king.

It is from this bishop's time that the registers of Rochester are well kept, beginning in 1326. At this period the parishes of Snodland and Halling had vineyards in them, showing how much better our fathers knew the use of the sunny southern slopes of the downs than the present holders; for it would appear that they made wine of the grapes, and certain of the bishop's tenants in Halling and Snodland had to gather blackberries to mix with the grapes, to make a beverage which the bishop then thought fit for the board of Edward II. Perhaps this mixture may have been the origin of Port. Soon after this we find the bishop repairing the palaces of Halling and Trottescliffe, and at this period we notice the chapels in these palaces were the scenes of his and his successors' ordinations. His statue, we are told, was outside the palace in Halling, and being blown down about 1720 was presented by Dr. Thorpe to Bishop Atterbury: what became of it after is unknown.

It would appear that all Halling was not under the Bishop of Rochester as Trottescliffe was (where Hamo also restored the palace); but the manor of Langridge belonged to Adam de Bavent, whose son Roger shared the estate with John Melford, lord of the manors of Offham and Goldwell in Offham in 1347. On De Bavent's death, in 1358, John Melford possessed the whole manor, which, after continuing some time in his family, was at last alienated to the Raynvells. The bishop also possessed Snodland, or as it is called in Domesday Book, Esnoiland.

Hamo de Heth, in 1323, built a mill here; possibly where the present mill on Holborough brook is.

The Palmers were tenants here during the period we are now considering. Greenwood tells us of a quaint epitaph once in Snodland Church :—

> " Palmers al our faders were,
> I, a Palmer, lived here ;
> And travylled till worne wythe age.
> I ended this world's pylgremage
>
> On the blest Assention daie,
> In the cheerful month of Maie,
> A thousand wyth four hundred seven,
> And took my journey hence to Heaven."

This epitaph, however, appears to have been gone in Thorpe's time as he does not mention it. This family continued in Snodland down to 1660, as we have in the registers of the parish the following entries :—

1607. Mr. William Palmer was buried the first day of June.
1607. John Handsylde, gent, and Mary Palmer, were married the first of July.

1608. Bridget, the daughter of William Palmer, was baptised the third day of July.
1612. Buried was Joan Palmer, the widow of William Palmer, gent, the 19th day of February.
1631. Buried was Elizabeth Palmer, ye wife of Will Palmer, gent, being Saturday paid for her interring in the Church, July 23rd, 6s. 8d.
1632. William Palmer was buried, and paid for his interringe in ye Church, 6s. 8d., on Easter Monday.
1646. Samuel, the sonne of Thomas Palmer, was baptised the third of April.
1659. Thomas Palmer was buryed April the eighth.

The ancient manor of Veles belonged to the family of Vitulus, or Veles, from whom it borrowed its name in the thirteenth century. In 1209 Ralph Camerarius gave land to the wife of Robert Vitulus, in Sifleton, Aylesford, Ryarsh, and Farleigh : thence Veles was transferred to Blunts, Turvyes, and Harveys in turn.

Holloway Court, also in Snodland, was in the possession of Henry de Holloway in Henry III.'s reign. In Edward I.'s thirtieth year, 1302, William de Holloway had inherited it, and from him it passed to the Tylghmans, of whom we shall speak in the next chapter.

In 1272 Archbishop Boniface confirmed the grants of Hallingas, Snodelonde, Trottesclyve, and Mallingas to the monasteries.

In 1295 Solomon de Rochester was fined and imprisoned by Edward I., with other judges. He was a justice itinerant, and appears to have lived at Snodland, the parson of which parish, Wynand de Dryland, has come down to us as one of the first known clergymen of the place; but with the unenviable notoriety of having poisoned his unfortunate parishioner.

East and West Malling were at this period both under the rule of the abbey, of which the following is an historical sketch:—In the year 1278 the abbess claimed certain liberties granted by Henry III., and a market, weekly, on Wednesday and Saturday—this was held in Dugdale's time, but has now long been discontinued; and though an attempt to resuscitate it was made in this century, it failed. In 1272 we learn that a man was attached in Malling market for debt, and they threatened to carry him to Tonbridge Castle and keep him till the money was paid. West Malling was not subjected, we understand, to the Constable of the Hundred at this period. The abbess also claimed, as granted by King John, to have out-fangtheof and warren on all her lands at Malling; and to hold fairs, as granted by King Henry, time beyond memory, on the eve, day, and morrow of St. Matthew, which would be October 9th, 10th, and 11th; and on the eve and day of St. Leonard, November 16th and 17th; and on St. Peter ad Vincula, August 12th (old style). The fair on November 17th is still maintained, and Messrs. Codlin & Short's puppets, Mrs. Jarley's wax-works, with toys and ginger-bread, worn-out horses, and hot sausages, still attract an admiring crowd. In 1292 Edward I. granted such franchises to the nunnery that Malling became an important place, and losing its old title of Little Malling, received the name of Town Malling instead. The abbess paid annually to the see of Rochester, at this time, a boar and 10 lbs. of wax. In 1321 the bishop proceeded to appoint Agnes de Leybourne abbess, on account of the affairs of the nunnery having been badly managed   The abbess complained of was one of the powerful family of the De Badlesmeres, of whom Bartholomew was the head at this time—his nephew was Bishop of Lincoln. His wife refused Queen Isabella admission to Leeds Castle on October 13th, 1321 : she was imprisoned for this conduct, and De Badlesmere lost, for the time, his land.   Now we learn that the Bishop of Rochester, a great friend of Edward II.'s, visited the abbey, and heard the complaint of the nuns at the king's request ; and though the nunnery was said to be ruined by her bad management, he did not allow them to choose and himself assent to and nominate the new abbess, which he would have done had the complaint come by the ordinary channel of his clergy.   We can thus easily fancy that the " She-wolf of France " really caused the disgrace of the abbess because of the insult offered to her by one who was so nearly related to the abbess as Lady de Badlesmere, whose sister-in-law she most probably was, and at any rate one of a family whose power Isabella feared would upset the plans of herself and Mortimer. In 1324 the bishop, on the death of Agnes de Leybourne, appointed Laura de Retling, but forbade her to give a corrodium * to her maid. He further sequestered the common seal of the abbey, inhibiting its being used without the bishop's permission—the seal, we are told, represented the Virgin crowned, under a Gothic canopy, with the Holy Jesus in her right hand and a sceptre in her left ; in a niche a figure praying.   The motto ran : " Sigillum commune Monasterii Beatæ Mariæ, de West Malling."

In 1339 a dispute arose about the tithes of West Malling, when

---

* A corrodium, or corrody, signifies a sum of money, or an allowance of meat, drink, and clothing, due to the king from an abbey or other house whereof he was the founder, towards the sustentation of such a one of his servants as he thought fit to bestow it upon.   The difference between a corrody and pension seems to be that a corrody was allowed towards the maintenance of any of the king's servants in an abbey, and a pension was given to one of the king's chaplains, for his better maintenance, till he was provided with a benefice, and as to both of these see *Fitz. Nat. Brit.*, fol. 250, where are set down all the corrodies and pensions that our abbeys, when they were standing, were obliged to pay the king.   Corrody is ancient in our laws, and it is mentioned in *Staundf. Prærog.* 44 ; and by the Statute of Westminster, c. 2, s. 25, it is ordained that an assize shall lie for a corrody.   It is also apparent by statute 34 & 35 Hen. VIII., c. 26, that corrodies belonged sometimes to bishops and noblemen from monasteries ; and in the new terms of law it is said that a corrody may be due to a common person, by grant from one to another, or by common right to him ; that is, a founder of a religious house not holden in frankalmoigne, for that tenure was a discharge of all corrodies in itself.   By this book it likewise appears that a corrody is either certain or uncertain, and may not be only for life, or years, but in fee.

Robert de Beulton, the vicar, was granted the lesser tithes, and the personal tithes n Holyrode Street and Tan Street; * the greater tithes and the prebend's house were to belong to the abbey. The vicar, in return for the lesser tithes, was to find everything for the use of the church—bread and wine for the sacraments, processional tapers, lights for the chancel, accustomed minister's rochets, surplices, unconsecrated napkins, vessels, basins, and green rushes to strew the church if necessary. The document recording this speaks of two streets in Malling—Holyrode Street and Tan Street; and of two inhabitants—Thomas atte Shoppe and William Cake.

In 1348 the plague called the black death, which spread from China over Europe, was said to have been introduced into this country by our soldiers from France. It appears to have carried off two abbesses of Malling, as Hamo de Heth instituted two abbesses in one year; in the neighbouring parish of Addington four rectors were appointed; and three Archbishops of Canterbury were also consecrated in the same year. At this time Sir John Lorkyn was perpetual curate of East Malling, and he obtained from the archbishop the augmentation of the living of East Malling, for himself and his successors, by the addition of all oblations and offerings of what kind soever made in the chapel of St. John Newhythe. This chapel still forms a cottage behind Newhythe Street, and has an Early English piscina, which the unlettered peasant who lives there informed the writer was the cupboard where the Romans kept their idols! Sir John Lorkyn also obtained from the revenues of the nunnery at West Malling an addition to the income of himself and his successors from the archbishop, the abbess and convent being desirous of providing a proper support for the vicar :—"It was decreed and ordained that the vicar and his successors should have the mansion belonging to the vicarage, with the gardens of it; and six acres and three roods of arable land; and two acres of meadow; which they used to have in past times, free and discharged from the payment of tithes. Together with the herbage of the cemetery of the church, and the trees growing on it, and the tithes of *silva cedua*, lambs, wool, pigs, geese, ducks, eggs, chickens, calves, cheese and the produce of the dairy, pigeons, hemp and flax, apples, pears, pasture, honey, wax, beans planted in gardens, and of all other seeds whatsoever sown in them. And also the tithes of sheaves arising from orchards or gardens dug with the foot, together with the tithes as well of the cattle of the religious in their manors or lands wheresoever situated within the parish, either bred up, feeding, or lying there; and of all other matters above mentioned being within the said manors and lands, as of the cattle and matters of this sort, of all others whatsoever arising within the parish; and further, that the vicar and his successors ministering in the church should take at all future times all manner of obligations, as well in the parish

* Now Swan Street and Frog Lane.

church as in the chapel of St. John, at Newhythe in this parish, and all other places within it then or in future. And the tithes of businesses of profit, of butchers, of carpenters, brewers, and other artificers and tradesmen whatsoever, to this church in any wise belonging; and likewise the residue of the paschal loaf after the breaking of the same, and legacies then or which afterwards might be left to the high altar, and the rest of the altars or images. And he decreed that only the tithes of the two mills in this parish belonging to the religious, and also the great tithes of sheaves and of hay whatsoever, arising within the parish, should in future belong to the abbess and convent. And he taxed this portion of the vicar at ten marcs sterling, yearly value, according to which he declared that the vicar should pay the tenth whenever the same ought to be paid in the future, and that the vicar for the time being should undergo the burden of officiating in this church, either by himself or some other fit priest, in divine services, and in finding of bread and wine for the celebration of the sacraments, and of the two processional tapers as heretofore; and that he should receive and undergo all other profits and burthens otherwise than as before mentioned."

The vicarage is valued, we are told by Hasted, in the king's books at £10 8s. 4d., and the yearly tenths at £1 0s. 10d.

In 1349 Hamo de Heth, going on a visitation, found the monasteries of Lessness and Malling were so decayed as to be hardly restorable.

It may be worth while to mention that Lessness Abbey, near Belvedere, in Kent, did not last till the Restoration; but Henry VIII. granted it to Cardinal Wolsey (with the sanction of Pope Clement), who gave the abbot a small sum of money and turned out the monks to shift for themselves. We subjoin a list of the abbesses of Malling so far as we have succeeded in finding them during this period :—

Avicia, the first abbess, who was appointed by Gundulf (and took the oath we have mentioned) in Henry I.'s reign.

De Badlesmere : this abbess was the one who was dethroned by Hamo de Heth, or de Hythe, at the desire of Edward II.

Agnes de Leybourne, chosen by Hamo in 1322 ; she died in 1324.

Laura de Retling, chosen as her successor and refused a corrody.

Elizabeth Grapnel, a nun of Malling, so described, and custodian of St. Leonards in 1343, may possibly have been abbess.

Guester de Bonasge, appointed 1344.

Isabella de P(ar)h(a)m was appointed in the year 1349 by Haymo de Hethe—the only one of the three abbesses said to be appointed in this year in William de Dene's *Life of Haymo de Hythe* that I find in the Rochester register.

In addition to the abbesses, the abbey had a dignitary who officiated in the abbey church, named the Prebend of the High Mass;

of those that belonged to this period the following have been preserved :—Thomas de Alkham, 1328; Radulph Roach; John Watson, 1392. also Vicar of West Malling ; John Graunger, 1392 ; Colne de alba Clara ; Thomas Gerard, 1398.

Besides the abbey of Malling, we have at this time various notices of certain chapelries that had sprung up in the neighbourhood, which we subjoin :—Gregorie de Elmham, Vicar of Aylesford, in 1285 was ordered to say prayers at Cosyngton, which chapel is declared to belong to St. Mary's Hospital in Strood. In 1330 John Tredelant was licensed to the Hermitage of Longsole (long pond), near Barming Heath ; he was succeeded by Robert de Kilnardeby, 1337 ; he by Galfridus Hert de Debenham, 1341 ; and he by Stephen Fynamour in 1351. Besides these we have mentioned, John Mold was chaplain in 1391, Reginald Herbe in 1453, and John Rodes in 1462.

In 1422 an inquiry was made as to whether this chapel belonged to Aylesford or Allington. Thorpe tells us that a manor pound of Allington is fixed near it, through which the inhabitants of both parishes go to beat the bounds, and one rut of the road leading from Aylesford to Barming Heath is repaired by Allington, and the other by Aylesford. The lords of Allington Castle were the patrons of this chapel. In Thorpe's days the old chapel was used for a barn ; the walls had great breaches in them, and were damaged by putting in barn doors. The door-case of the chapel, at the west end, was still remaining, and the farmer remembered there was a door like a church door in the framing and finishing. There was also another old door-case still visible inside the barn, in the east wall on the north side of the altar of the chapel. A barn now occupies what remains of the ancient building. This chapel gives a protest against the foolish arrangement of this year, by which Allington has been transferred from the Malling to the Sutton deanery ; a transaction contrary to geographical and all historical and ecclesiastical considerations.

The Carmelites at this time increased considerably at Aylesford. John Ringer desired to give them a priory, but died before doing so ; but the friars petitioned for this and obtained it in 1369. In 1396 Richard of Maidstone was buried in the priory, and in 1404 the king granted them a spring and land (from henceforth, no doubt) called Haly (Holy) Garden in the parish of Burham.

The Friars has been turned into a modern dwelling-house. On the gateway we see the crest of the Sedleys, and the date 1590. The drawing- and dining-rooms have their ceilings beautifully decorated. There is a niche where one of the friars is supposed to have been entombed in the walls. The arches of the cloisters can still be traced. On a wall are two paintings : one of the learned monk, which says :—

" Richard Maidstone, S.T.P., friar in this priory, died 1396 and was buried in this cloister.

Quid cupit hic servire deo nisi semper et esse
Pacificum, lætum, nilque perire bonum,
Sic Fovet Ecclesiam statuens Statuum moderamen
Sternere ne liceat quod statuere patres."

The other is that of the founder, whose inscription runs : "Richard
Lord Grey de Codnor, founded this Carmelite House 1240, on his
return from Jerusalem." In the windows are several shields.

The chapel of St. Laurence at Upper Halling is first mentioned
in 1348, when Thomas Glanville was appointed chaplain. In 1369
Thomas Watton, canon of South Malling, changed with Walter
Dautre the chaplain ; he was followed, it would appear, by John
Bromyng, who in 1397 was succeeded by John Hall.

The chief men of this part of Kent had, during this period, two
grand opportunities of recording their names : one on the occasion of
the payment to knight Henry III.'s son, and the other on the payment to knight the Black Prince. On these memorable occasions the
parishes were thus represented :—

On the first occasion : Allington by Robert Longchamp. Aylesford
by Richard de Grey ; by Richard de Rokesly, representing Tottington for Hamo le Crevecouer, and Eccles for the Count of the Isle ;
and by John Marsh for Preston. Birling by Robert de Engebergh
for William de Say. Ditton by Ralph Schoford for the Earl of
Gloucester, and William de Sifleton and William de Brampton for
Ralph Schoford. Halling, under Reginald Cobham, was paid for by
Adam de Langereche, Peter de Camera, Roger de Bavent, Thomas
le Chivaler and William Martyn. Snodland and Trottescliffe were
paid for, the former by Reginald Harynges, Henry de Pevenseye,
Anselin Lad and Richard Veal ; and the latter by Hugh de Cressye
for the Bishop of Rochester. Ralph Chetwind paid for Paddlesworth, John Malterre for Offham, Roger Mowbray for Ryarsh, and
Walter de Berstede and Roger de Leybourne for the manor of Ewell
in Malling.

To knight the Black Prince the assessments were : Lord John
Grey for Aylesford, 40s. William de Clinton for Leybourne, 10s.
The wife of William Lybaude for Tottington and Eccles, 30s. (in
Aylesford). Thomas de Sifleton in Ditton, 20s. John
of Cobham for Allington, 20s. Nicholas de Dagworth for Addington,
40s. Thomas de Ditton and John Melford for Offham, 40s. Thomas
de Aldon for Brampton and Ditton, 30s. The Parson of Leybourne
for Great Comp in Leybourne, 16s. John de Huntingfield for
Paddlesworth, 16s. Lord Wilfrid de Say for Malling and Ewell,
20s. Richard Povenashe, John Melford, John Lad and Richard
le Veal for Snodland, 20s. John la Doune for Birling, 4s. Roger
Bavent and John Melford for Malling, 30s.

CHAPTER VI.

WARS OF THE ROSES (1399—1509.)

THE eventful Wars of the Roses, as we shall see, made some changes in the families of our part of Kent, though no one of its great battles was fought in the county. In fact, an internecine strife which may be said to have decimated the inhabitants of England, and exterminated some of its proudest families, could not have but left its mark upon men who so stood up for their rights and the constitution of England as the gallant squires of Kent. On account of the Reformation times beginning in the reign of Henry VIII., we have taken this chapter from the accession of Henry IV. to the death of Henry VII.; and really, when we examine history, we think that this period should be always dated thus, because the disturbances that began with Henry IV.'s accession never really ended till the commencement of the reign of Henry VIII., Henry V.'s French war being an attempt by that politic king to call off the thoughts of his nobility from home affairs to foreign ones. During this period the first great historical event that attracts our notice, after the battle of Shrewsbury, in which we do not find any of our families, is the rebellion—if it can be so called, since Henry IV. himself had no right to the throne—of Thomas Mowbray, Earl of Nottingham, lord of the manors of Ryarsh, Richard Scrope, Archbishop of York, the Earl of Northumberland, and Lord Bardolph. By the alertness of Ralph Nevill, Earl of Westmoreland, the insurgents were captured, and Mowbray was executed (1405); but Elizabeth his wife died in possession of Ryarsh. Their son John married Katherine the daughter of the said Ralph Nevill, and serving in France was restored to the family honours and estates by Henry V. in 1416, and died in 1432. His son John died in the first year of Edward IV. His son, who was created Earl Warenne, died fourteen years afterwards (1475), leaving an only daughter, Anne. At her death the vast possessions of the Mowbrays fell into the hands of the Howards and Berkeleys, who were descended from this family, and we soon after find Ryarsh in the hands of the Nevills.

Calais Court, or, more correctly, Carew's Court, in this parish, was a manor held by the Carews of Beddington in the reign of

Henry VI., and obtained its present nomenclature from a corruption of the name of that family.

Addington in the beginning of this century was, as we have already seen, in the hands of William Snayth, who was sheriff in 1408. He died in 1441, and was buried in the Church of Addington. Round his brass, which is that of himself and his wife with a lion, runs the inscription: " Hic jacet Willielmus Snayth, armiger quondam dominus de Adynton, ac vicecomes Kanciæ et Alicia, uxor ejus qui quidam Willielmus, obiit XII.º die Martii, anno domini MCCCCXLI. Quorum animabus propitietur Deus. Amen." Snayth left an only daughter, who married Robert Watton, who succeeded to the manor of Addington,. which continued in his family till the year 1775, when the last of the Wattons died, leaving an only daughter. She carried this in. marriage to the Bartholomews. In the year 1797 the manor went to the Stratfords in the same way. The Wattons who possessed Addington during the period we speak of were Robert, then William, whose will was proved in 1466, who gave directions for his body to be buried in the chapel of the Assumption of the Blessed Mary, in the parish church of Adyngton—his brass has gone ; he was succeeded by his son Robert, whose brass still remains, on which we read : " Hic jacent corpora Roberti Watton, armigeri filii et hereditarii Willelmi Watton et Aliciæ, uxoris predicti Roberti filiæ, Johannis, clerk unius baronum saccarii domini regis qui quidam Robertus istius villæ dominus et hujus ecclesiæ, verus patronus, obiit die Novembris, anno domini MCCCCLXX. Quorum animabus propitietur Deus. Amen " ; Robert was succeeded by his son Edmund,* who stands first on the Watton monument—to his memory are inscribed the words : " Juxta hunc locum jacet sepultum corporis Edmundi Watton, hujusce loci armigeri qui adjunxit sibi Elizabetham filiam Roberti Arnoldi de Gillingham, in comitatu Cantii armigeri, obiit anno domini 1527." None of these Wattons appear to have distinguished themselves.

Hasted tells of Addington that the church here was built at this period, to prove which he quotes the rhyme :—

> " In fourteen hundred and none
> Here was neither stick nor stone ;
> In fourteen hundred and three
> The goodly building which you see."

Had Hasted consulted Domesday, or had he honoured Addington with a visit, the tomb of Mr. Charles would have proved to him that this rhyme could not apply to Addington in Kent; since Mr. Charles' monument is of older date, and Addington Church is mentioned both in Domesday and *Textus Roffensis*, together with the other churches of the neighbourhood.

The castle of Allington was in the hands of Thomas de Cobham,

* See registers of Watton family, pp. 228-9.

a descendant of Henry de Cobham, who obtained it by his marriage
with Margaret de Penchester ; he alienated to Robert Brent, whose
grandson William alienated it to Sir Henry Wyatt, who was a
privy councillor of Henry VII. Sir Henry had been imprisoned
in the Tower by Richard III., and being either purposely or through
neglect deprived of food, his life was wonderfully preserved by a cat,
that used to bring him a pigeon daily ; he was ever afterwards so
fond of cats, that he is always painted with one. The Abbot of
Boxley having privately visited Allington in his absence, Sir Henry
Wyatt's lady had him seized, carried to the gatehouse, and put in
the stocks. He complained of this indignity to the privy council,
whereupon Sir Harry told them that if they acted in like manner
no doubt his wife would treat them the same.

Richard, the son of Richard, Lord Grey of Codnor, who held the
manor of Aylesford, was high in favour with Henry IV. and
Henry V., and died at Argentoin Castle, in Normandy, 1419.
John, his son, died without issue 1431, and Henry, his brother, died
1444, leaving a son, Henry, Lord Grey, who died 1496, leaving a
widow, Lady Catharine, who carried the manor to Sir W. de la
Pole. After her death it devolved on John Zouche, who died in
possession of the manor.

In the time of Henry VI. the Poynings transferred their manors
of Tottington and Eccles to the Palmers, whose history as lords of
the manor of Snodland we have already traced.

In 1426 John Cosyngton died, and was succeeded by his son,
Stephen de Cosyngton ; and his descendant, Thomas, died in the time
of Henry VIII., leaving three daughters, heiresses to the manor of
Cosyngton.

Preston was held by Sir John Colepepper in 1401, who died in
1428—his son John was a Justice of the Common Pleas, and died in
1450. This family about this time began to spread over all this
part of Kent, and hence we find their numerous monuments in many
churches of this district. The manor of Ditton, having passed
through the hands of Windlesor and Sir Lewis Clifford, came to
them, and in 1485 Sir Richard Colepepper (sheriff) of Oxenhoath
died in possession, having come into it in 1472. His three daughters
—Margaret, married to William Cotton of Oxenhoath, Joyce, married
to Lord Edmund Howard, and Elizabeth, married to Henry Barham,
Esq., of Teston—sold Ditton, in Henry VII.'s time, to Thomas Leigh,
of Sibton ; they also possessed Borough Court, in Ditton, by inherit-
ance, which they alienated to Francis Shakerley, and his eldest son,
Richard, inherited it.

Goldwell also passed, as we have already seen, to the Brownes of
Beechwood Castle at this time, one of whom married a Colepepper.
Her son, Sir Michael Browne, alienated to Richard Nortop, alias
Clerk. Offham came to the Colepeppers, and following the same
fate as the rest of Sir Richard Colepepper's manors, came into the

hands of Thomas Leigh; the Preston Hall property, however, remained in the hands of William Colepepper, Sir Richard's brother, who transmitted it to his son Edward.

Fitzalan, Earl of Arundel, Lord Treasurer and High Admiral, left a son Thomas, who succeeded to the estates in Birling, but died without children. His sister Joan, in 1416, married William Beauchamp, who died in 1422; his daughter Elizabeth was married to Sir Edward Nevill in 1436, who was created Lord Burgavenny; * he died 1480, and was succeeded by his son George, who died in 1492. He was present at the fight at Tewkesbury, for which he was knighted.

As this family were the cousins of the king-maker Warwick, they no doubt drew upon their retainers in our valley for men to fight for the White Rose, and hence this part of the county, as the rest of it—

> "Joined with York
> And did the work,
> And made a blest conclusion."

Sir George Nevill's son George was imprisoned for joining in the conspiracy of the Duke of Buckingham, but was afterwards released. The family of the Nevills made their home at Comfort, in Birling, and we shall have occasion to remark upon their family records, which are left us from this time forward in the registers and in the parish church of Birling. Few relics of their old mansion remain, part being pulled down, and part occupied by an old farmhouse; one mediæval room, with a barred window bricked up, still remains, and some handsome squared stones and two archways exist in the surrounding walls. From the Birling register it would appear the Nevills had a chapel here.

During the period we are speaking of Paddlesworth passed from the Huntingfields, who appear to have died out, and went into the families of Bele, Bullock, Diggs, Peckham and Vinely, who alienated it to the Bambergs; from them it came to the Wattons of Boughton Malherbe, while Langridge manor was alienated in 1502 to Robert Watton of Addington.

The famous insurrection of Jack Cade seems to have been nearly confined to Kent, and consequently we are not surprised to notice that some persons from our valley took a share in this curious rebellion, "which required of Henry VI. the dismissal of evil counsellors." Amongst those of this neighbourhood we find Robert Somery of Aylesford, described as a gentleman, William Rowe and Edmund Rowe of the same—possibly two of the Rowes of Rowe's Place in Aylesford, as they belonged to this parish; of West Malling we have John and William Downe, gentlemen of West Malling, Robert Langley of West Malling, gentleman, and Thomas Edolff and

---

* See registers, etc., of the Nevill family, pp. 126, 221.

William Gunne of West Malling, yeomen, besides John Elphy, Richard Welcock, and William Browne of Birling, carpenters. It speaks well for the loyalty of this valley, that as Cade's riot extended to Blackheath, and as he was victorious at Sevenoaks, many more persons were not mentioned as assisting in his rebellion from this neighbourhood.

In the list of the gentry of Kent, we find in the time of Henry VII., in this neighbourhood : Lord Burgavenny, Alexander Colepepper, William Colepepper, James Walsyngham, Richard Cosyngton, Edward Myllys, Robert Watton, Thomas Palmer, and Edmund Watton, the squires of Birling, Preston, Ditton, Cosyngton, Paddlesworth, Snodland and Ditton of these days. James Walsyngham was one of a family at first established at Chislehurst and Ightham Mote ; afterwards we find a branch of them, at the end of the seventeenth century, at Ryarsh.

Of the Millys we have three records :—One a brass in Birling church : " Off your charitie pray for the soule of Walter Myllys, sumtime Reseyvor unto my Lord of Burguevenney, the whyche decesyd the xv. day of March, the yeare of our Lorde God, mccccxxv., on whose soule J'su have mercie. Amen " : there is a figure of a man and four children. The second, a brass of a man in West Malling church, with the inscription : " Orate p āīā Willīm Millys, qui obiit v° die Jañā, a° dm millīo ccccclxxxxvii., cuj's āīe prōet͏r d's. Amē." —in English : " Pray for the soul of William Millys, who died the 5th day of January, in the year of our Lord 1497, on whose soul may God have mercy. Amen." The third mention of the family is the appointment of one of the same name, Rector of Leybourne, to sit on a commission to inquire into the patronage of the Prebend of the High Mass of the High Altar in the monastery of West Malling, A.D. 1493, of which we shall speak presently.

Besides the monuments already mentioned, there are also the following in the churches belonging to this period. In Aylesford church : " Hic jacet Johannes Cosyngton, qui obiit secundo die mensis Aprilis, anno domini millesimo cccc° xxvi°, et Sarra uxor ejus, quorum animabus propitietur deus. Amen " : arms three roses, with same as crest. In Halling church there is an undated brass, with " Orate pro animabus Johannis Colard, nuper unius clericorum regis in Scaccario suo per XXXVIII. annos, et Margerii consortis, ejusdem quorum animabus propitietur deus," and the arms of the Colards, possibly belonging to this period. In Ditton there is a brass with the inscription, " Orate pro anima magistri Ricardi Leggatti, qui obiit anno domini mcccclxxxi, vi° die mensis Junii." In East Malling church there is also one brass of this period, with the figure of an ecclesiastic, and the words, " Orate pro āīā magistri Ricardi Adams, qundam pbdarii magne misse in monasterio de West Malling, ac Vicarii ppetui pōchiæ de Est Mawling, qui obiit sexto die mensis Maii, a° dñi mdxxii quᵉ āīe pproet͏r deus—" Pray for the soul of Master

Richard Adams, formerly Prebend of the Great Mass in the monastery of West Malling, and perpetual Vicar of the Parish of East Malling, who died on the sixth day of the month of May, in the year of our Lord 1522, on whose soul God have mercy." * Snodland possesses two. One to one of its rectors : " Hic jacet Thomas Dalby, quondam rector istius ecclesiæ, qui obiit vi° die Octobr., anno domini mccclxxii., cujus animæ propitietur Deus. Amen "; and another " Hic jacet Johannes, filius Lancastri heraldi armor, qui obiit x° die mensis Junii, anno domini millesimo ccccxli, cujus animæ propitietur deus. Amen." In Trottescliffe church : " Hic jacet Willm̄ Cotton, generosi filius, baccalaur juris civilis et legis p̄itus, ex collega de Grey's Inn, qui obiit xviii.° die March, A° Dn̄i mcccclxxxiiii., et Margeria ux͡r eiˢ, qui obiit die a° dn̄i mcccclxxx., cujus āiābus propitietur deus. Amen " —" Here lies William Cotton, son of a gentleman, bachelor of civil law, and skilled in law, of the college of Grey's Inn, who died the 18th day of March, in the year of our Lord 1483 ; also Marjory, his wife, who died in the year of our Lord 1490, on whose souls God have mercy. Amen." We have also, in Addington parish church, inscribed to one of its incumbents: " Hic jacet dominus Thomas Chaworth, quondam Rector ecclesiæ de Addington et Long Melford, unus clericus dni° regis in cancellariâ suâ, ac cognatus Elizabethæ, uxoris Roberti Watton, armgr̄i quorum animabus propitietur deus." Another in East Malling church ran : " Hic jacent Robertus Selby, olim civis et goldsmith London, ac Johanna uxor sua qui quidam Robertus, obiit xii° die Augusti, anno domini mcccclxvii., quorum animabus propitietur deus. Amen."

Thorpe, in addition to these, tells us of the following brasses in West Malling church, now lost or hidden, that were extant in his time :—One to John Rose, Vicar of West Malling, that ran : " Hic jacet vir sacer, divinæ baccalaureus theologiæ, mensis Octobris X. nono die sepult, anno millieno cccc quat LX. subt octo, dormiat in pace cum Christo semper. Amen "—" Here lies John Rose, a pious man, bachelor of divine theology, buried the nineteenth day of October 1452, may he sleep with Christ for ever. Amen." Also one of the effigy of a man, which was destroyed ; but there remained a heart on which was inscribed " credo quod," and three labels branching out from it, on which were written :—

"Redemptor meus vivit
De Terra Surrecturus Sum
In Carne mea videbo Deum Salvatorem Meum."

There are certain gifts of land at this time which introduce us to a few of the old inhabitants' names. Thus the grant of John Marchant, of Birling, speaks of Peter Fisher, William Marchant, Andrew

* Another in East Malling church runs : " Hic jacet Robertus Ereby, olim civis et auri faber de London, Joanna et Joanna uxores sui qui quidam Robertus, obiit 15 Augusti, anno domini 1477. Hic jacet Thomas Ereby et Isodia uxor ejus, qui obiit 1 Sept., 1478."

Chapman, William Smyth, John Luke, Simon Spayn, John King, and John Peckham; while another grant tells of Thomas Bennett, Edward Pekerynge, John Dobby, John Beanley, Richard Palmer, and several other persons belonging to Snodland, in addition to Thomas Dalby, the parson. We have now gone through all the eventful period included in this chapter, except that which refers to the parishes under ecclesiastical surveillance: out of these, of Leybourne we learn nothing, as it was under the control, as we have already stated, of the Abbey of St. Mary Graces.

Halling, Trottescliffe, and Snodland were under the Bishop of Rochester, and in the two former were his country residences and palaces: indeed, in this century his ordinations were held in the chapels of these parishes. It is said that the famous Hamo de Heth was so fond of these country retirements that he neglected his duties, and was reproached for this by Archbishop Meopham, to whom, however, he showed no animosity, as he attended his funeral, which took place shortly after.

Thomas Brinton, appointed to Rochester 1372, gave his secretary, Bartholomew Waryn, the living of Snodland in 1401, which, however, he exchanged for Hadstocke in Middlesex. Bishop William Wellys, the initial letter of whose register is quite a picture, died at Trottescliffe on February 24th, 1443. His successor, Bishop Lowe, amongst other matters, heard a process against John Pure, or Purrs, Vicar of Malling, who was charged with having used divers incantations over the bread in the Lord's Supper, and with administering it to persons suffering from fever. He admitted that he had taken the wafers (but not consecrated ones), scored them quarterwise with a knife, using the words, " Petrus autem jacebat super Petruno "—" but Peter was laying upon Peter "—and having observed this form with six wafers he gave them to the diseased, who were to eat one of them a day. Some of his patients informed him they were healed by them, but of this fact he professed ignorance. He owned that he had received money for them which he had spent on the church. He concluded his defence by declaring that Bishop Langdon ratified and commended the work, and desired his servants might be instructed in the art. Bishop Lowe died in his chair at Halling on the last day of September 1464, and was buried at Rochester in the cathedral.

During the prelacy of John Russell, 1476-80, nine persons of Snodland and Halling, being convicted of playing tennis on Thursday in the week of Pentecost, and confessing their guilt, those who belonged to Snodland were ordered to walk barefoot after the procession on the Lord's Day, each bearing a halfpenny taper, which they were to offer at the Holy Cross; while the parishioners of Halling were to do the like, only they were to make their offering, which was to consist of two tapers, at the high altar.

Thomas Savage, bishop, 1493-96, instituted an inquiry into the patronage value and circumstances of the Prebend of the High

MALLING ABBEY.

Mass in the Abbey of Malling.   The Palmers continued tenants of the manor of Snodland, but Holloway Court changed into the hands of the Tylghmans, whose family lived into the seventeenth century; we give an extract of their registers from the parish church of Snodland :—

1559. Richard Tylghman and Margaret Valentine were married 10th July.
1563. Joane, ye wife of William Tylghman, was buried 20th September.
1567. William Tylghman and Dorothy Reynolds were married 11th August.
1572. Dorothy, ye wife of William Tylghman, gent, was buried November 21st.
1576. Whetenhall Tylghman, ye sonne of William, was baptised September 4th.
1578. Dorothy Tylghman, daughter of William, was baptised January 10th.
1581. Oswald Tylghman, sonne of William, was baptised October 11th.
1582. Charles Tylghman, sonne of William, was baptised October 18th.
1584. Lambert Tylghman, sonne of William, was baptised April 12th.
1586. Lambert Tylghman, the son of William Tylghman. gent, was buried
       21st November.
1586. Lambard, ye sonne of William Tylghman, gent, was baptised August 18.
1587. Gervise Olyver, ye sonne of John Tylghman, was baptised 31st January.
1592. Armigill and John, ye sonns of Christopher Tylghman, were baptised
       23rd April.
1592. John, the son of Christopher Tylghman, the 3rd of June.
1593. William Tylghman was buried the 24th of February.
1607. Whetenhall Tylghman and Eleanor Kemsing were married the 4th day
       of February.
1608. Mary, the daughter of Whetenhall Tylghman, was baptised 11th day of
       December.
1608. Charles Tylghman was buried the 25th of May.
1611. Buried was Edward Tylghman, gent, the 23rd day of the same
       (December).
1613. Buried was Mrs. Tylghman, the wife of Edward Tylghman, gent, the
       23rd day of October.
1625. Joseph Tylghman, son of Whetenhall Tylghman, was baptised the 2nd
       day of January.
1627. Baptised was James, the son of Whetenhall Tylghman, March 2nd.
1632. The thirtieth of December Mrs. Ellen Tylghman, wife of Mr. Whetenhall
       Tylghman, was buried.
1633. Benjamin Tylghman, son of Whetenhall Tylghman, was baptised
       Jany. 25th.
1643. Elynour, the daughter of Isaacke Tylghman, was baptised the 16th day
       of June.
1644. Isaac Tylghman dyed December ye 21st, and is buried under the great
       Chancel window, at the East End, in Snodland Churchyard, noaw
       if we can make any guess, his father, Whetenhall Tylghman of this
       parish, gent, is deposited.
1645. Elizabeth, the daughter of Isaacke Tylghman, baptised 14th day of
       July.

The parishes of East and West Malling continued during this period under the quiet sway of the nuns.   By means of the records of Rochester, we learn that in 1425 Cecilia Batisford was abbess; and in 1479 Catharine Skefton.   In the year 1493, when Joane Moone was abbess, Thomas Savage, then Bishop of Rochester on the death of Thomas Cook, having appointed John Whitmore as his successor, issued an inquiry into the patronage, value, and other circumstances of the Prebend of the Great Mass of the High Altar in the monastery of West Malling, anno domini 1493. Twelve persons are said to have been on this court of inquiry, and are represented to the Bishop as worthy of faith. The clerics. described domini, are William Spayne de Offham, Thomas Hundbache de Nettilstede, John Punkar de Barmynge, William Millys de Leybourne, Master Thomas Revell, rector, Nicholas None (Vicar de Faldynge), and the rest laity. The inquiry determined that John Perot finds the place vacant, and appoints John Whitmore. Those high clerical dignitaries, styled Prebends of the High Mass, are given us pretty regularly during this period in the records of Rochester : they are Thomas Wall, 1402 ; Thomas Gloucester, 1426 ; Thomas Compton, on death of last, 1444 ; Richard Stone, 1447 ; Daniel Everard, 1455 ; Thomas Cook, died 1493 ; John Whitmore, 1493 ; Thomas Nevill, 1499 ; Richard Adams, died 1522 ; Robert Dokett, 1522 ; John Bamborough, 1522 ; Henry Fletcher, 1524.

Of other chapels we find little mention during this period, except that in 1406 John Chesterfield succeeds John Gold at the hermitage of Longsole. In 1453 Reginald Herbe is licensed to it, and John Rodes to the same in 1462, after which there is no more record of this chapel having a minister.

Of the chaplains of St. Laurence, in Halling, we have John Perot, who was Rector of Snodland in 1453, chaplain previous to Robert Sharpe, who was appointed in 1518 ; this last was succeeded by Richard Back, who in 1531 was succeeded by Robert Truelove, who was appointed rector of Snodland ; after him there are no more chaplains of St. Laurence.

CHAPTER VII.

TUDOR TIMES (1509—1603).

THE period of the Reformation, which caused the dissolution of monasteries in 1539, could not be without interest in a district so greatly under ecclesiastical control as Malling Valley, since so many churches and properties found in this way new masters. Moreover, from getting hold of the fair lands of the Church, Bluff King Hal, who was one who had always an eye to his own interest, learnt that many people had properties in various parts of the country superior to the royal demesnes, and therefore did not hesitate to deprive them of such, either by giving an inferior property for them, or, if the whim seized him, he prevented any further opposition by executing them. We shall speak of the various properties in this valley, and their owners, under the different divisions : of gentlemen fortunate enough to pass through this stormy period altogether free ; then of those who, though they escaped the rapacity of Henry VIII., could not, like the Vicar of Bray, change their religion at pleasure, or had to bow to the greediness of England's merrie monarch ; and lastly we shall show the various hands into which our ecclesiastical properties got, and also the different changes that took place in our churches during this stormy century. Of those who appear to have been amongst the most fortunate, in that they retained their properties all the way through, we find in this district the Wattons of Addington, the Colepeppers of Preston Hall, and the Nevills of Birling. The Wattons continued throughout this period serving their country, as justices of the peace and commissioners of the sewers, according to Hasted ; but they have left no record of either obtaining even a knighthood from the sovereign, or possessing amongst their family a single man that was noted either for his position in the ecclesiastical military, or literary world. Quietly their beautiful lands descended from father to son, and quietly were their bodies buried in the pretty church. Edmund Watton was the squire of Addington at the beginning of this period, and died in 1527. George Watton gave away the living in 1533, but he is not mentioned in the Watton monument, though Thomas Watton, son of Edmund, is, who died in 1580, and was succeeded by his son Thomas. To these two last the

Watton monument pays the following tribute: "In hac ecclesia etiam jacet corpus Thomæ Watton, armigeri filii predicti Edmundi Watton, qui sibi conjugem habuit Eleanoram, filiam Edmundi domini, Sheffield, obiit anno domini 1580, sepultus 26 Julii. Hoc sepulcro clauditur Thomas Watton, armiger filius prædicti Thomæ Watton, qui uxorem habuit Martham, filiam Thomæ Roper de Eltham, in comitatu Cantii, armigeri qui ex Vita hac emigravit 16° Septembris, anno domini 1622."

Of the lords of Comfort, George Nevill, having been released from imprisonment, died in the year 1536: he was present at the Field of the Cloth of Gold, and saved Kent from joining the Cornish rebels, and was buried at Birling. His son Henry was summoned to Parliament in 1551, and got into trouble for striking the Earl of Oxford, but was pardoned; he, with his brother Sir Thomas, collected an army to oppose the Isleys, who were marching from Sundridge and Sevenoaks to join Sir Thomas Wyatt at Rochester, and encountered them in Blacksole field, in Wrotham, and defeated them. Blacksole (black pond) field, the Rector of Wrotham kindly informs me, is still to be found (being a part of his glebe), not far from Wrotham Church; in it relics of the combat—as swords and skull-caps—have been found. Sir Henry died 1587, when his cousin Edward claimed the titles and estates of the Lords of Burgavenny in opposition to his daughter Frances; the matter was not settled till the first year of James I., when it was determined that Edward should be held Lord Abergavenny, the title and the estates connected with it being declared in tail male; while, in order to indemnify the Lady Frances, she was given the title of Baroness le Despencer (a title originally created in 1264), which was to descend to female heirs if of nearer kin than the next male heir. In this way the late Lady Falmouth succeeded to the title of Le Despencer; her son, who is Lord Falmouth in right of his father, succeeded to the title of Baron le Despencer in right of his mother. The manor of Ryarsh followed of course the succession of Birling, having now become a property of the Nevills.

At Preston, Edward Colepepper was succeeded by John,—who contributed to Henry VIII.'s loan in 1542,—he by his son Thomas, and he by his son Sir Thomas in the year 1587. Sir Thomas, in the year 1598, also acquired the manor of Aylesford, which had been alienated by Sir R. Southwell to Edward Randolph and Richard Argoll.

The Isleys possessed the manor of Bradbourne in East Malling till 1540, when it passed into the hands of the Manninghams. John Leigh exchanged Offham with Henry VIII., who granted it to William Welford, J. Bennet, and G. Briggs, who conveyed it to John Tufton, Esq., of Hothfield, who was sheriff in 1561 and 1575: this family still possess it. In the year 1881 Henry James Tufton, Esq., was created Baron Hothfield on the recommendation of Mr.

Gladstone. In the year 1529 Sir John Zouche died, and the king granted Aylesford to Thomas Cromwell. On his death, in 1539, it passed into the hands of the Wyatts. Paddlesworth continued in the family of the Lords Watton, of Boughton Malherbe, till 1572, when Lord Watton settled it on his daughter Katharine, who carried it in marriage to Lord Stanhope. In 1547 Bavent, in Halling, was alienated to Sir William Whorne, who was Lord Mayor of London, and who built Whorne's Place, in Cuxton, which is still a solid building at the angle where the road to Upper Halling and Trottescliffe leaves the main Rochester road. The estate passed from him to Vane, and thence to Barnewell, from whom it came to Nicholas Levison, Sheriff of London.

As the other events connected with this valley group themselves with the ordinary circumstances of English history, it will be perhaps best to give them in the proper order. The first of these is the suppression of the monasteries in 1539. We left off the history of Malling Abbey when Joan Moon was abbess : she appears to have been succeeded by Elizabeth Daniel, 1524, who is next mentioned in the records of Rochester. In 1531 Elizabeth Rede was abbess; she is mentioned again in the *Valor Ecclesiasticus* of Henry VIII.—we may notice that St. Blaise's Chapel is valued at £2 in this document. The last abbess of Malling mentioned is Margaret Vernon, who, with her eleven nuns, Felix Cocks, Arminal Bere, Rosa Morton, Margaret Gyles, Joan Randall, Letitia Duk(e), Beatrice Williams, Juliana Whitnall (Whetenhall), Joan Hall, Elizabeth Pimpe, and Agnes West(e), surrendered the abbey to the king. It was valued at £218 4s. 2½d., and, according to Dugdale, the abbess received a pension of £40 yearly, and the nuns from £2 13s. 4d. to £3 each. According to another authority, the abbess received £10, the first four nuns £3 6s. 8d., and the others £2 13s. 4d. per annum. This seems more likely, because we find that in the year 1553 there was paid an annuity of £10 to Margaret Vernon, and £2 13s. 4d. each to Agnes White, Elizabeth Pimpe, Johanna Hall, Joan Randulph, Juliana Whetenhall and Lettice Buck. This is the last we hear of the nuns and the abbess of Malling ; the abbey itself, with its temporalities, was granted by the king to the Archbishop of Canterbury, all except the vicarage of East Malling which he kept to himself, in the year 1541.

It appears from the contribution to Henry VIII., that George Pierrepoint paid to this collection as comptroller to my Lord Archbishop of Canterbury. Malling Abbey, it is said, was granted to him ; possibly he was only receiver for the Archbishop of his rents here, and lived in the abbey as tenant. Hasted tells us he was buried in Malling Church. There is a brass half remaining of Elizabeth Pierrepoint, daughter of Sir Anthony Babington—she was also mother of Lady Brett. The motto is wrongly written above the shield of arms (they are those of Babington quartering Dethick), and

appear as " Foy est tout " ; * they should be " Foyes Toute." This is,
first quarter, ten torteaux (4, 3, 2, and 1) and a chief label of three
points ; second quarter, a fesse varry between three water bougets.
The other shields are lost. One, however, was in existence in
Thorpe's day—the quarters were : first and fourth, a lion rampant ;
second, six annulets ; third, three hedgehogs. The Pierpont shield
bore : *argent* a lion rampant *sable*, a dexter baton *or*. Thorpe,
in his *Church Antiquities of the Rochester Diocese*, makes this figure,
wrongly, to be part of that of a man.

The manor and castle of Leybourne, which was given up by
the Abbey of St. Mary Graces, was granted by Henry VIII. to
Cranmer, Lord Archbishop of Canterbury; but it was demanded
back by him, and granted to Sir Edward North.

At Aylesford in the 27th year of Henry VIII., the Carmelite priory
was dissolved, and given by that monarch to the accomplished Sir
Thomas Wyatt, who was lord of Allington Castle, where he was born
in the year 1503. He commenced his education at Cambridge, and
finished at Oxford. He won the favour of Bluff King Hal as a wit and
a poet, and was in consequence knighted by that monarch. He was
employed on several diplomatic missions, was Sheriff of Kent, 1537,
and died at Sherborne in 1542. His poetical works, amongst which
was the original rendering in English of the "Town and Country
Mouse," consist of love songs, odes, satires, and a metrical version of
the psalms. He got into disgrace with Henry VIII. on account of
his attachment to Queen Anne Boleyn, but managed to regain that
monarch's confidence, which speaks volumes for his powers of address.
He was sent as ambassador to Charles V., Emperor of Germany.
As an example of his poetry, we quote "The Lover's Appeal."

> " And wilt thou leave me thus ?
> Say, nay ! say, nay I for shame,
> To save thee from the blame
> Of all my grief and græme ;
> And wilt thou leave me thus?
> Say, nay ! say, nay !
> " And wilt thou leave me thus,
> That hath loved thee so long,
> In wealth and woe among ?
> And is thy heart so strong,
> As for to leave me thus ?
> Say, nay ! say, nay !
> " And wilt thou leave me thus,
> That hath given to thee my heart
> Never for to depart,
> Neither for pain nor smart ?
> And wilt thou leave me thus ?
> Say, nay ! say, nay !"

* The motto of the family of Babington, of Rothley Temple, Leicestershire ;
descended from the Babingtons of Nottinghamshire, of whom Thomas Babington, M.P. for Nottingham (1450), married Isabel, daughter and heiress of Robert
Dethick, of Dethick, in Derbyshire. I am indebted for the above remarks to
the Rev. Scott Robertson, Vicar of Throwley.

A story told of Sir Thomas Wyatt* the elder is that he brought up and made playmates of, at the castle, an Irish greyhound and a lion's cub; so tame were they that they used to wait for his coming home, with great delight, at the hall door. But the lion's whelp at last became violent, ran roaring at his master, and must have destroyed him but that the greyhound leaped on him and pulled him down, when Sir Thomas drew his sword and slew the lion. When Henry VIII. heard this, he observed, "Oh, he can tame lions." Leland's praise of Wyatt is fulsome :—

> " Bella suum merito jactent Florentia Dantem,
> Regia Petrarcæ carmina Roma probat;
> His non inferior patrio sermone Viattus,
> Eloquii secum qui decus omne tulit."

Translated into his native tongue, it is :—

> " Let fair Florence her Dante rightly boast,
> Let royal Rome her Petrarch's songs applaud ;
> Than these no worse in his own country's tongue
> Can Wyatt all the grace of speech afford."

Wyatt was succeeded at Allington Castle by his son, the famous Sir Thomas Wyatt, whose rebellion against Queen Mary is perhaps one of the most noble insurrections that we read of in any history of the various nations of the world, and certainly the grandest in the history of England. He was no claimant to the throne, nor even did he want high honour, nor draw his sword for any other claimant ;— he had been bred up in the Church of England, and he disliked the idea of a foreign king. Situated as he was at Allington, he must have remembered the expulsion of the nuns of Malling, and the Carmelite Friars of Aylesford, in his father's lifetime—he had been Sheriff of Kent under Edward VI. in 1551. Not only had he seen the destruction of the Carmelite Friars, but the Newhythe and Halling chapels were suppressed, as well as those of Cossington and Tottington, and that of Longsole on Barming Heath on his own property, in the suppression of 1545 and 1547. Thus we are not only not surprised that Sir Thomas Wyatt led a rebellion against Queen Mary, but are more astonished that some great movement was not set on foot from several counties to oppose the Queen, since the parliament had been dismissed at the end of the previous year, and had declared its dislike to the match with Spain. Sir Thomas Wyatt, we are told, confederated with Sir Peter Carew, Sir William Pickering, Sir Nicholas Arnold, Sir Nicholas Throckmorton, Sir James Crofts and others, in November 1553, to hinder the marriage. The Duke of Suffolk and his sons tried to raise troops in Leicester, but were unsuccessful, and were lodged in the Tower. Carew and Crofts failed in Devonshire and Wales.

---

* It is interesting to find this Wyatt spent his time at Allington in hunting and hawking. He may possibly have hunted the herds of "savage swine" (wild boars) which infested the Mereworth woods in Elizabeth's days.

The confession of Anthony Norton, of Trottescliffe, is curious, as showing the feelings of Wyatt and others of the time. He tells us that he was sent for, before the trouble began on the Monday, to Allington Castle, where he found Mr. Wyatt in his parlour sitting by the fire. Mr. Wyatt said,—

"I am sure you have hearde of ye comynge of ye Kynge of Spayne, how shal be oure kynge, to ye undoynge of thys realm ; for at the sprynge of ye yer such gentylmen as I, with other, shall be sent into Franssc, with a gret powre of Ingeglysmen to inlarge hys kuntry'sther, and in ye menetyme, unther pretensse of fryndshippe. he shall strenkten ye reme with hys owne nasyon, to ye subver-tynge of oure own nasyon and losse of thys realme." Axynge ye sayde Anthony what he koulde do in ye defensse thereof, ye said Anthony aunsuryd yt hys dwellynge wasse nere unto ye Lord a Burgoyne * and not far from Mr. Southwell, so yt was not in hym to do anythyng. "Well," said Mr. Wyatt, "yf suche as ye are wyll not consyder youre sa(f)te, I can do no more ; but, as one may do, yf ye worste cum, I can go into other partts wher I shall be hartelly wellcum and joyfully reseyvd," and so pawsed. In the menetime in came Mr. Redstone, Mr. Feycher. with one howme I know not. Mr. Wyat sayde unto them yt my Lord Kobam had sent hym word yt hys iii. suns shoulde go with him and shoulde have hys ayde. Ansure wasse made yt suche sendynge wasse ye kastynge away of ye Duke, and sayde yt there lywys wer as dere unto them as my Lord's wasse unto hym. Wherefor, sayde they, let hym go hymselfe and set hys fote by ours. "Well," sayde Mr. Wyat, "how thynke you by Mr. Sowthwell. I wolde spende a thousand pounds yt we kowlde have hym reformeabull"; how hathe kept a worsypfull ho, and by hys gentyl intrety of the kountry he had the hartts of ye parttys, sayinge yf he by cny means mowthe be reformed, they wolde not dowte ye optaynynge ye Lord a Burgoyne, sayynge yt he wolde wrytte a a letter unto hym, forgyvynge frome hys hartte all mattars paste, with as myche fryendshyppe as he kould by ye sayd letter declare. Ansure was made yt yt was best to cawsse sum man to opun ye matter by mowthe, and not to wrythe, and yffe fryndshyppe mythe yt way take plase, then to wrytte. Mr. Wyatt sayde he woulde send for William Iden, how showlde open ye matter to Mr. Sowthewell yffe he kould get hym so to do. Mr. Wyatt sayde further, saynge, "Mr. Sowthewell hath ye lowe and hartts of men in yt partts, if hys worshyppe wolde not prokure men to resyste and cause bloudeshede in yt quarrell, yt he went in and sayde he was owt of dowte yt men wolde not fytte agaynste him ye matter yt he had taken in hand." As far as ye saide Anthony peseyvd, they had hoppe of ayde of ye Quean's shyps and ye questyon was axed Mr. Wyat yffe yt were not best yt men were landed in Shepe (Sheppey), wher mythe be gothen horse-harnes, with other artyllery for warse. Mr. Wyat ansured yt ye howsse of my Lorde Wardensse to have myche quyne and tresure, at the sythe whereof ye men would not abstayne ther hands from robbery wyche he wolde not have commytted, and sayde, "When my Lorde shall cum downe to take up men, he shall perseve ye halffe of hys owne men to be agaynst hym, wych when he perseyvthe he wyll undoubtedly kepe hys Iland and not stere." I heard Mr. Redstone say that "Ye Quean wolde gyve awaye ye supremasy, and ye Byshoppe of Roume shou have hys powre in Yngland as he had before tyme, wyche to thynke on grevyd hym" ; and so began to talke in secret to themselves. Whereupon ye sayde Anthony went into ye hawll, and was desyred by ye parsonn to drynk a kouppe of beare, and so departyed. beynge brouthe on hys way by ye sayde parsun. Alexander Fyssher wylled ye sayde Anthony to speke unto Tylden, ye drapar, yt he showlde repayre unto suche yomen as he knew downewarde to make them prewe to ye bysyness.

This arch traitor, Norton, appears to have belonged to a family of

* Lord Burgavenny=Abergavenny at Birling.

some notoriety in Trottescliffe. I do not know whether it was his son, but Gylles Norton is entered in the baptismal register there for 1560, and the Rev. C. W. Shepherd informed me that his father had been told that this latter man led a band of rustics to London; possibly the names have become interchanged, and this Norton, who saved his neck by betraying his comrades, was the man meant. As regards this insurrection, it only remains to speak of the direction taken, the parties concerned, and their fate. Wyatt got together about 2000 men in Rochester, in spite of having the force of Sir Henry Isley and the two Knyvetts cut off by Lord Abergavenny and the Sheriff (Sir Robert Southwell), aided by Thomas Henley, Walter Taylor, John Raynoldes, John Lambe, Anthony Weldon, Thomas Chapman, Heughe Cartwright, and George Clarke, who marched from Malling and defeated the insurgents, as we have already seen, at Wrotham. We cannot quite understand how some of these men opposed this movement against the Papist King of Spain and Queen Mary, inasmuch as Anthony Weldon, and Heughe Cartwright, it would appear, were of the Church of England. That the Duke of Norfolk should oppose in London, we are not surprised; nor are we astonished that Sir Alexander Brett and the Londoners went over, crying, " We are all Englishmen."

The attempt to enter the city from Southwark having failed, Wyatt crossed the Thames at Kingston and marched on London, repulsing the attack of Sir John Gage at Charing Cross. Failing to get into the city by Ludgate, which was defended by Lord William Howard, he gave himself up to Sir Maurice Berkely on being exhorted to merit the queen's pardon by preventing bloodshed. For all this the unfortunate Queen Jane, her father (the Duke of Suffolk), her uncle, and Sir Thomas Wyatt were executed. It is true that the queen's clemency has been praised on this occasion as the only act of mercy in her reign, but so far from being so, as she had, by her herald, held out Sir Thomas Wyatt to surrender on conditions of mercy, her execution of him was only another proof of the deceitfulness and cruelty so plainly distinguishable in her character.

Perhaps England never lost a greater opportunity to be free than it did in not joining in this brave revolt against civil and religious slavery. The failure of Wyatt was the cause of sorrow which lasted till 1600, or forty-six years. It lighted the fires of Smithfield for those opposed to Rome in the reign of Mary, it brought about the executions for religion in Elizabeth's reign, and though the destruction of the Spanish Armada prevented England from becoming a mere province of Spain, nevertheless it was not till the end of the century that peace was re-established between the countries which the pride of Mary had interrupted, by giving Philip the idea of becoming sovereign of two powerful countries.

We are told that Sir Thomas Wyatt intended to go abroad on the death of King Edward VI., but was prevented by his wife,

who besought him not to forsake his wife and child. Upon their
coming to see him at the Tower on the day of his execution, he said
to them as he was being led out, " See what my love for you has
brought me to."

In Aylesford church we have the following memorial : " Here
lieth John Savill, Gentleman, some tyme servant to Sir Thomas
Wyatt, Knight which deceased the 30th day of March, A. Dmn!. 1545."

In the year 1554 the married clergy were expelled from their
various parishes. We do not find so many changes in this district
as we might expect during these stormy times, but in 1545 we find
Hugh Woodward, Rector of Leybourne, instituted, who is emphatically
declared to be of the *Church of England* in the episcopal register
of Rochester ; and in this eventful year, 1554, we find George Attke
succeeding Nicholas Archebolde at Ditton ; Robert Salsberry,
Bartholomew Bone at Trottescliffe ; and Launcelot Gylhawke,
Thomas Bedlowe at Halling. Of these the latter is distinctly said
to have been deprived ; but as Walter Hait, in 1567, is said to have
obtained the living of Halling pro Thomas Bedlowe, undoubtedly
Launcelot Gylhawke was one of the Roman priests who obtained
a living in Mary's reign, and who afterwards, refusing to assent to
the alterations proposed in the beginning of Elizabeth's, was deprived,
and the former persecuted incumbent restored. As regards Salsberry,
he continued, and we find him incumbent of Addington, and afterwards instituted to Ryarsh, in 1572 ; while George Attke remained
at Ditton till 1565.

We omitted to mention in its proper place, owing to our following
the history of the Wyatts, that in the year 1536 parish registers were
ordered to be kept. Many of the early ones have been destroyed ;
but the register of Oftham now commences in 1538, that of
Trottescliffe in 1540, that of Birling in 1558, that of Ryarsh in
1559, that of Leybourne in 1560, that of Snodland in 1560, that
of Addington in 1562, and that of East Malling in 1570. The rest
of the registers commence later, owing to their having been destroyed.
It is a matter of regret that these registers should be left to moulder
in the vestries, as they contain, as we shall show, much valuable
information, which is yearly becoming more and more illegible ;
in some places the registers have been allowed to be wantonly
destroyed—I myself was actually offered in sale part of the registers
of a parish. The episcopal registers of Rochester, instead of being
carefully re-bound and placed in the cathedral library, are left in a
cupboard in the office of the lawyer for the diocese (Mr. Knight),
by whose courtesy I have been allowed to inspect them, but who
has himself told me he would prefer them placed elsewhere.

Sir Robert Southwell, who was so active against Wyatt, obtained
much land in Kent during this century ; those that he gained by his
marriage with Margaret, daughter and heiress of Sir Thomas Nevill,
fourth son of Lord George Burgavenny, he sold in 1544 to Sir

Edmund Walsingham. Thomas Leigh exchanged Ditton with King Henry VIII., who granted it to Thomas, Lord Wriothesley (1554), whom he made Lord Chancellor in room of Lord Audley. He was grandson of John Wriothesley, Garter at Arms to Edward IV. and Henry VII., and nephew of Thomas Wriothesley, who held the same post at his father's decease—he was created Lord Southampton in 1547. He did not hold Ditton long, as it passed to Sir Robert Southwell in 1554, who, in 1555, conveyed it to Sir Thomas Pope. Sir Thomas was Clerk of the Star Chamber, Treasurer of the Court of Augmentations, Master of the Jewels, and Warden of the Mint ; to him was also intrusted the care of the Princess (afterwards Queen) Elizabeth. He founded Trinity College, in Oxford—the first college founded in either University after the Reformation—March 18th, 1556 ; he died in January 1559, and was buried in the chapel of his college, where his monument may still be seen. As natives of all places where he had lands were to be benefited by his foundation, persons whose sons were born in Ditton, amongst others, could claim preference to the emoluments of the college : his representatives alienated Ditton to Wiseman in 1600. Richard Nortop, alias Clerk, obtained Goldwell in Offham from the Brownes ; his son, George Clerk, was one of those who marched against the Isleys with Lord Abergavenny. Of the others there mentioned, Hugh Cartwright was, after Gervase Pierrepoint, the holder of Malling Abbey. The manor of Leybourne passed from Sir Edward North, to whom Henry VIII. had granted it, to Robert and John Gosnold, from whom it passed in Queen Elizabeth's time to Robert Godden : * whose family appears to have originally sprung from Trottescliffe parish—indeed this family seems to have been the only population of Trottescliffe at one time.

The Grange, in Leybourne parish, belonged to a family called Quintain in the time of Elizabeth. One of this family, Thomas Quintain, having always styled himself son of Oliver, the family finally took the name of Oliver :† we find their family very frequently mentioned, in the registers of Leybourne. The oldest of the Leybourne church bells is dated 1581, and on it is inscribed, besides, " Thomas Goddin, Gentleman, and Robard Oliver, Youman."

This brings to our notice how thoroughly in the Reformation times they appear to have done the work of stripping the churches. I am inclined to believe that the priests carried off much when they were deprived, as well as the lay people, where licence was given to spoil ; certes we notice that none of the better furniture of the Church in this valley belongs to a date previous to the Reformation. The oldest ornaments of the churches we possess in this valley are the cup and paten of Offham, dated 1572 ; then the cup of Trottescliffe, dated 1576 ; then the West Malling stoup, or jug, made of Delft ware, mounted, with foot, neck, handle, and body-straps of silver-gilt, hall-marked " London, 1581," which does not

* See Godden Registers, p. 233.   † See Oliver Registers, p. 230.

look very ecclesiastical, and was probably given in the seventeenth century;—I here quote Canon Scot Robertson on Kent Church Plate. We then come to the Leybourne bell already mentioned. The third bell of Snodland has on it, " by me, Gylles Reeve, 1589"; the first bell of Offham, " by me, Gylles Reeve, 1590"; and the cup of Allington is of 1599. Owing to Sir Thomas Wyatt's rebellion his lands were forfeited to the crown; and thus Allington Castle and the Friars at Aylesford came into the hands of new owners. The castle appears to have been some time in the hands of Southwell, but it was granted by Queen Elizabeth to John Astley, master of her jewels, in 1584.

In 1554 the queen also gave the Carmelite Friars to Sir John Sedley; he bequeathed it to William, made baronet May 22nd, 1611, who resided at the Friars, and left a son, Sir William Sedley, married to Elizabeth, daughter of Henry, Lord Abergavenny.

About 1553 Halling was let to the Denes by the Bishop. The heiress of these Denes was Sylvester, whose curious brass in Halling church cannot but attract notice : it is that of a lady in bed, with two young persons on one side of it and two on the other side; near the latter is a cradle with two children. The inscription on the brass runs : "Gemilliparæ positum" (put up to the lady who had twins), "Sylvester, the daughter of Robert Dene, gent, and Margaret Whyte, his wife, was born the 18th December, 1554, marryed to William Dalyson, Esquier, the 29th June, 1573. After that married to William Lambarde, gent, the 28th October, 1583; and died the 1 September, 1587, leavyng on lyve, by William Dalyson, Sylvester a daughter and Maximilian a son, and by William Lambarde, Multon a son and Margaret a daughter, and Gore and Fane sonnes and twynnes.

> Non illa reverentior ulla deorum." *

This lady's second husband, William Lambarde, was the author of "Perambulations of Kent," and said to be the handsomest man of his day.

In the year 1581 a presentment was made to the jury of the Manor of Preston, in Aylesford, against Thomas Huet and Peter Hartropp, for setting up a pair of butts without licence on the common called Stroude, alias Preston Green ; also a presentation against Robert Palmer and Richard Hartropp as trespassing with cattle on it in 1583.

In June 1584 inquiry was made as to the persons, within the Lathe of Sutton-at-Hone, charged by Lord Abergavenny with furnishing demi-lances and light horse at the musters at Malling Heath.

This was the period of those Papist plots that culminated in the Babington Conspiracy, and it would appear that the move-

---

* No one of the gods was more reverend.

ments we have just mentioned had something to do with this. That the disaffection was spread in this part of Kent is very clear, since we learn, Christopher Dunne, of Addington, in Kent, is examined as being implicated with his son Henry in the conspiracy, when he pleaded ignorance.

In the year 1600, Gervase Pierrepoint—who had held Malling Abbey, and whose wife's brass, showing she was a Babington, appears in the Church, as we have said—with others, had his lands reduced for his share in this rebellion. When Pierrepoint gave up the abbey is not very clear; but as Hugh Cartwright held the manor of East Malling, and obtained a grant of Newhythe chapel * in Edward VI.'s time, and marched from Malling to assist in putting down Wyatt's rebellion, it would appear that it must have been previous to Elizabeth's reign. On the death of the said Hugh Cartwright, his widow, Mrs. Jane Cartwright, became entitled to all the lands that once belonged to Malling Abbey, and carried her interest to her second husband, Sir James Fitzjames. On the north wall of the chancel of West Malling church is her monument, which runs: " Here resteth the body of Dame Jane Fitzjames, widow, first married to Hughe Cartwright, Esq., and afterwards to Sir James Fitzjames, Knighte, one of the 17 daughters of Sir John Newton, Knighte, and of Dame Margaret, his wife, which Dame Jane dyed the xx^{th} day of February 1594, and the 37th of Quene Elizabeth's reigne, and in the sixty-seventh yeare of Hir owne age." The property appears to have come then into the hands of Humphrey Delind, whom Harris styles " a man furnished with a good stock both of Divine and human learning." After him it passed into the hands of the Brookes of Cobham.

In the year 1581 the people of the Lathe of Aylesford were assessed at twopence in the pound towards training the soldiery; no doubt with a view to opposing the Spanish invasion.

In the year 1561, amongst those who are mentioned as paying towards Rochester bridge, who had property in this neighbourhood, we have Sir Thomas Fynch, Mr. Wotton, and Mr. Tufton. This bridge, as is widely known, was ordered to be repaired by different parishes from time immemorial; and as its properties are now not only sufficient to pay for its repairs, but large sums are expended for the benefit of persons in certain parishes, it were well to mention those of the parishes we are writing about that had thus to contribute. Halling, Trottescliffe, and Malling repaired the third pier; Aylesford, the fourth pier; Leybourne, Offham, and Ditton, the fifth pier; and Snodland, Birling, Paddlesworth, and the men in that valley, the ninth pier—these latter words were no doubt intended to signify the inhabitants of Ryarsh, and perhaps Addington.

In the reign of Henry VIII. the Warcups obtained the manors of Eccles and Tottington, and Rowe's Place, which continued in their

* The chapel was valued at 11s. annual value.

family till the end of Elizabeth's or the beginning of James I.'s reign, when Henry Warcup alienated Tottington to Maddox, and Rowe's Place and Eccles partly to the Sedleys, partly to the Goldings, and partly to the Bests, of Chatham. In Henry VIII.'s reign the family of Cosyngton was represented by three daughters, one of whom married Thomas Duke, who possessed this manor and handed it down in his family.

On the January 25th, 1583, we have the inventory of the goods of William Dalyson given us in the *Archæologia Cantiana;* which William Dalyson married the Sylvester Dene of whom we have already spoken. Through him the lease of Halling descended to Sir Maximilian Dalison, of whom we shall speak more in the next chapter.

In the year 1556 we have the will of John Hodsoll proved, in which he desires to be buried in the churchyard of West Malling : his family granted a lease for two thousand years of land in Ightham in 1788. In 1791 the administration of effects of Frances Hodsoll, formerly Tassell, of Town Malling, spinster, granted 'to William Hodsoll, her lawful husband, was sworn under £5. Such is a brief notice of a family connected with this neighbourhood.

It is worthy of remark, that the digging out of chalk as an industry commenced at Halling in 1538, though cement-making was not then established. A few curious memorials in one or two churches which survived the era of the Reformation may be fitly mentioned here. Thorpe tells of two strange windows in Leybourne church (now gone): in the north-west window was a George and Dragon, St. Christopher with the Saviour, and the label, " Xtofore ora p . . . "; while in the east window was the label, " Ave gratiâ plena domina tecum." In Trottescliffe church a window with an idolatrous representation of the Trinity still remains. In Halling church, until a very few years ago, existed the uncouth painting over the chancel arch of which we give a print; it has been variously deciphered, as the tale of an unfaithful wife and the emblematic representation of the seven deadly sins. On an old mantelpiece in Malling Manor House are the words, above, " Ætatis Suæ 43," then " S.P."; on a shield underneath, " 1566." Tradition says that this house belonged to one of Queen Elizabeth's admirals, and these are his initials; unfortunately, one cannot find an admiral of the name, but the oak stairs and the room where the mantelshelf is, as well as the mantelpiece, point to a date in this century that renders the inscription a genuine one if the tradition be not correct.

We cannot do better than continue this chapter with a few remarks on the parish registers that are commenced in this century. The earliest of these registers is Offham, which begins in 1538. The commencing words are,—

The Register book of the Parish of Offham, beginning 1538.

And the first entry is,—

1538. Mary Whiffins was baptised the 17th day of February.

Besides these we have the following specimens of the way the Burial Register was kept :—

1539. Robert Tresse was buryed the xvi[th] day of June.
1543. The third day of January were two children of a stranger buryed.

In the years 1553, 1554, 1555, 1556, 1566, 1568, 1573, 1574, 1578, 1583, 1588 and 1589 there are no entries of burials at Offham.

In 1585 we have,—

A little gyrle of the pshe of Wrotham was buryed ye first of May.
Grante, a nurse childe, was buried the xvii[th] daye of January.

In 1558 we have,—

Henry Kerbye and Katherine Edmunde were married the vii[th] day of May.

1539, 1540, 1542-49, 1557, 1558, 1562, 1567, 1568, 1570, 1574, 1579, 1585-89 and 1596 are without entries in the marriage registers of Offham.

The next is that of Trottescliffe, which commences,—

The Register Book of Trottescliffe, of all the Christenings, Weddings, and Burials in the year of our Lord 1599.

The entries here, however, begin 1540, for we read :—

1540. Roger Barre, the son of Thomas Barre, was baptised the xix[th] day of April, ut sup.
1540. John Hills and Anne Fullor was wedded the 26th of June.

We have in this register the following Latin entries of this century :—

1576. Thomas Sanctilis, filius Anthonii de Sanctilis, nobilis Majoricensis, captanei Brabantiorum, mense July xvi[th] baptisatus fuit.*
1591. John Wood, filia Nicolai Wood, was baptised ye xxv[th] day of April.
1550. Gylles Norton was baptised the 24th day of April, ut supra.†

In the Marriage Register of Trottescliffe we find :—

1540. John Hills and Anne Fullor was wedded the 26th of June.
1592. Richard Baggace was married the xxv[th] of August.

From this date up to 1599 the man's name only is mentioned in the Marriage Register of this parish.

The Birling Parish Register commences with the words,—

1558. The Church Book containing Christenings and Burials.
Marian Westre, the daughter of John Westre, of Birling, was baptised the xxiii[rd] day of February ye second year.
Such as were burials of the said Parish of Birling are following :—
1588. Richard Harte buried the second of January.

---

* Thomas Sanctilis, son of Anthony de Sanctilis, a noble of Majorca, captain of the men of Brabant, was baptised in the month of July xvi.
† This is the man who is supposed to have been the turbulent character we have mentioned.

There are several entries of the Nevills during this century, which will be mentioned at the end of the book; the following one, however, requires a special notice :—

> 1588. Mr. Edward Nevill was baptised the 4th day of June in the Chapell of Comfort.

This entry is interesting, as it not only tells us of the old place of the Nevills at Birling, but is the only existing record of a chapel there that we find. Dr. Harris speaks of Comfort as the place of the Nevills.

The Parish Register of Ryarsh begins,—

> "The Register book of the Parish of Ryarsh, within ye countie of Kent and diocese of Rochester, containing all those names that have been christened, married, or buryed within ye said Parish ; ye said Register book beginning the xx. day of November, in the yeare of our Lord God 1559. Copied out of ye original by John Parker, Vicar of Ryarsh.
> First and foremost. ye xxii^d day of January, anno domini 15$\frac{6}{8}$ was christened Joane Oliffe, ye daughter of George Oliffe.
> Monday ye xiiith day of March, in that aforenamed yeare, was christened John Sanrock, anno 1560 from Christ's Incarnation.
> 1579.   Elizabeth Yonge, filia Edward Yonge, baptizata fuit secundo die Martii anno supradicto.
> Marriages in the Parish church of Ryarsh, from the xx. day of November in ye year of our Lord God 1559.
> First and foremost, on Monday the xxth day of November, and in the aforesaid yeare, anno 1559, Thomas Godderd was married to Elizabeth Littell.
> Monday the 22nd day of Januarie, in the year of our Lord one thousand five hundred and sixty and one, was married Thomas Clegat, a weaver, to Mildred Tyler, the daughter of William Tyler.
> Robert Godden, ye sonne of Thomas Godden of Calliscourt (Carewe's Court), in the parish of Ryarsh, marryed in Seal church to Jane French of ye said parish, on Monday the xxii. day of January 1598.
> Gyles Symons of Ryarshe, and Marie Crispage of Stone, were maried on St. George's day 1599, by a licence granted out of the office at Rochester.
> The names of all such as be buried in the parish of Ryarsh, from the vith day of April, in the yeare of our Lord God 1560.
> The sixth day of Aprill, in the yeare of our Lorde God 1560, was buried Jane Boorman, the wife of John Boorman.
> 1562. The xxvth day of March there was buried Alice Drodger, servant to John Walsingham.
> The third day of April, 1562, was buried William Byshop Clarke of this said parish of Ryarsh, and the said third day of April was Maundy Thursday.
> xxiiird day of Januarie, in the yeare from our Lord's incarnation 1562. was buried Richard Turley, Servant to Thomas Brissenden.
> 1562. The third day of March was buried one Thomas Philpott, a poore man. a stranger, which same late out of Yorks, that died in Richard Boorman's house.
> The xxvith day of Aprill, in the yeare of our Lorde 1573, there was buried Mother Wyborne.
> 1576. The xxvith day of September, in the year of our Lord 1576, there was buried Ann Murphy, servant to Goodman Golden, of Carew's Court.

The titles " stranger," " poor man," or " woman," " goodman," "goodie," "mother," are not unfrequent in the registers of all parishes down to quite recent times.

LEYBOURNE.

The register of the Christenings of Laiborne, beginning Anno Dom. 1560. Annoque Dominæ nostræ Elizabethae secundo. [Then follows, in the old book] :
Anno Dom. Bapt. 1560, June 5, Jane King.
Buried, July 3, William Wheler.
Anno Dom. 1562, married, March 13, Thomas Petley and Elizabeth Oliver.
May 17, Bur: William Oliver.
[Here the boke was very imperfect, and therefore I leve out the yere.]
: George Bredham, Gent, and Mary Godden.
[Here the boke was imperfect and therefore I leave the yeare and day unwritten.]
1594. Ann Morris, widow of Jasper Morris, was buried upon Ascension Day.

It will be very plainly seen that one of the incumbents took the trouble to recopy the registers of Leybourne, for which he cannot be too much praised. The respected Rector of Trottescliffe (Rev. C. W. Shepherd) has set an example to present incumbents by doing the same.

THE REGISTER BOKE OF SNODLAND.

1560. Gyles Andrew, sonne of Frances, was baptised ye 6th of March.

There are many entries after the above fashion, leaving out the father's name altogether, and giving only the Christian name of the mother.

1564. Elizabeth Leûse (Lewis), ye daughter of John of Hamyll, was bapt. 23rd April.

This proves that the proper name of Ham Hill is Ham Mill.

1575. Samuel Godden, ye son of Thomas of Paddlesworth, bap. 17 Sept.
1585. Sarah, filia Johannis Swonei, Rectoris hujus ecclesiæ de Snodland, baptizata fuit. Obiit eadem Sarah 12 die Februarii anno predicto ; sepulta jacet in cœmiterio de Addington, quoniam infra limites parochiae enutriebatur, viz., in ædibus Thomae Whiteing. [Sarah daughter of John Swone, Rector of this church of Snodland, was baptized. The same Sarah died on the twelfth day of February in the year aforesaid ; she lies buried in the cemetery of Addington, since she was being nursed within the limits of the parish, namely. in the house of Thomas Whiting.]

It appears, from this, that children were put out to nurse by people in good position in those days.

Also ye daughter of John Aynsworth, pedlar, was baptised 23 January.
1559. The Booke of Maredges:—
    Thomas Woodroffe and Margery Fielde were married 11 March.
Burials 1559 :—
    George Erpingfield was buried 26th March.
1561. John Usher, of Hoborowe, was buried 16th October (Holborough).
    .. James Clare, a nurse child, was buried 10th December.

1563. From the entries this year we learn that there was a great mortality in Snodland.

1563. July 5, a waterman or sayler, whose name was unknown, who had a wife, as he said, and children at Rye, was buried.

1563. Cuthbert Ersh, a young man of London, was buried 3 September.
1564. Richard Bambridge, a goldsmith of ye town of Rye. was buried 3 March.

We find at this time several nurse children mentioned as being buried, not only in this register but also in the others.

1568. One Cornelius, a poor labouring man, was buried 7 January.
1569. Mary, ye daughter of one Austin, of London, was buried March 3.
  „   William Carr, a taylor, was buried 21st August.
1570. Robert Crane, a singleman, was buried 18th March.
  „   An olde maide, called Phillip, was buried 4 October.
1573. Also Sparrowe, an old wydowe, was buried 7th August.
1571. Sir William Hall, * pson of this parish, was buried 22nd June.
1574. Sir William Apsley * was buried 14 April.
  „   Theobald Hammon, a french. dwelling in London, was buried here June 15.
1575. Peter, a nurse child, was buried 17 November.
1578. A child yt was borne at Swynborn's house of a woman yt askt for lodginge, was buried 20 May.
1583. Father Borden was buried 3rd Februari.
1585. John Leeds, clothier, was buried ye 16 Aprill.
1585. Jane Swinborne, alias Downe, widow, was buried ye 27th October.
1586. Agnes Houlton, single woman, was buried 3rd February.

We have often, in all the registers, such entries as " singleman " or " single woman."

1588. Deborah, the daughter of John Powlter, miller. which was drowned in the mill pond at Holborough, was buried 2 June.
    An infant of Robert Wellingham. born before his time, was buried 9th October.
1589. An infant of John Hammons, which died unbaptised, was buried 28th January.
1591. A man child of William Rice's, which departed as soon as it was born, was buried 1st July.
1592. Nem Tiksall, being drowned the day before at the mill, by goodman Leedes his house, was buried 22 June.

The entries of the parish of Addington commence 1562 :—

1562. Thomas Boorman the third day of March was baptised.
1563. John Brown the last of November was buried.
1565. Joane Godden, the servant to Mr. Tilden, the fourth day of November was buried.
1566. John Martin, a Frameyer, the seventh day of March was buried.
1568. Robert Stilt, alias Round, the fourth day of February was buried.
1582. The eleventh of March, a gentlewoman, of freebirth, from Mr. Dan's house, was buried.
1585. Alice Taylor, widow, the xth day of July was buried.
1568. John Dolton and Joane Taylor, widow, the ix. day of May was married.

No marriages in 1578-1581, 1588 and 1599.
The registers of East Malling commence thus :—

Burials, Marriages, and Christenings, in the parish of Est Malling.

1570. Baptised was John, the sonne of John Lane, the 18th day of April.
  „   Married was William Bannister and Joan Bathersly, 4th day of May.

    * Parsons were styled " Sir " till long after this.

1570. Buried was Marye, the daughter of George Warbutton, the 8th day of April.
  „ 9th of January, baptised was Robert, the sonne of John Thomas, without any swathing.
1571. Buried was Johen Pyne, the daughter of one, Pyne, of London.
1581. Baptised was Sarah, ye daughter of one Margaret, a traveller.
  „ Buried was ye saide Sarah, the 12th of October.
1590. Buried was one Joy, a stranger, the 2nd day of February.
1595. Buried was a child that was nursed in this parish, 5th of April.
1597. Buried was one who said he was born in Hartford sheere, the 9th of August.
1598. Buried was Thomas Foster, a Sussex man, who died soon after he was released out of prison, 28 February.
1599. Buried was Thomas Eves, gentleman, the 13th day of August.

A relic of ancient times of this date is the quintain on Offham Green, still standing. It is reported by some that the lord of the manor is bound to repair it, and by others that the house opposite is bound to do so, but this is denied; however, it is still kept up, and as it is said to be the only one in England it is to be hoped that it will continue to stand. It is a plain, upright post, with one cross-piece at the top, and looks at a distance, to all appearance, like a signpost to a bye lane. When we get near it, however, we notice little holes in the broad end, and on the narrower end a hook, on which, we are told, used to be hung a weight: a man rode at it full tilt, and charging it with his lance, bowed his head to escape the weight—if he did so he was considered to have been successful; but this, as may be seen, was no easy task. It shows us one amongst the games of our ancestors which were known before cricket came into fashion.

There are a few more monuments that we have not yet mentioned belonging to this period. Two in Ditton; one to one of the Shakerlys, who owned the manor for a short period:—

"Here lyeth the body of Rowland Shakerly, gent, sonne and heyre of Francis Shakerly, of Brooke Court, within this parish of Ditton, Esquyer; which Rowlande, beynge Fellow of Gray's Inn, deceased the xxii^d day of June in the yeare of our Lord 1576; and had this memorial of his death made by a young gentlewoman, as an argument of her unseparable good meaning towards him."

The other to one of the Brewers, another family who held a position for some time in this parish, which runs:—

"Here lieth the body of Millicent, the second wife of William Brewer, gent, daughter of Robert Tyas, of London, Esq., who died the xi^th September, 1556 Si moram fecerit expecta."

In West Malling church a brass, whose inscription runs:—

"Of yo' Charitye pry for the soule of Master Wyllīm Skott, Gētylman, which deceasyd the xxiii. day of January, a° dm mdxxxii., & whose soule Jhū have mercy."—"Of your Charity pray for the soul of Master William Scott, gentleman, the which deceased the 23rd day of January, in the year of our Lord 1532, on whose soul Jesus have mercy."

5

Also another that Thorpe tells of:

" Here lyeth George Cattlen, Esquire. deceased the·VIIIth day of February, 1590.

And in Snodland:

" Pray for the soules of William Tilghman the elder, and Isabell and Joan his wives, which William decessyd the XXVII. day of August, anno domini MCCCCCXII., on whose soules Jesu have mercy.

As you are so was I, and as I am so shalt you be."

Also a brass:

" Jesu mercy, Jesu mercy, mercy Jesu. Orate pro anima Rogeri Perot, qui obiit xvii. die mensis Septembris, anno domini MCCCCLXXXVI., cujus animæ propitietur Deus. Amen ":
this seems to have been a relation of the then rector, John Perot.

Also:

" Orate pro animabus Edwarde Bischoptre and Margaret uxoris ejus, qui quidam Edward, obiit primo die mensis Septembris, et domina Margareta, obiit xvii. die mensis Novembris, anno domini MCCCCLXXXXII. Quorum animabus propitietur Deus. Amen."

Old Mural Painting formerly over the chancel arch of Halling Church. Height 7 feet, breadth 6 feet; (from a copy in possession of H. Raven, Esq., The Cedars, Halling).

CHAPTER VIII.

THE first year of the reign of King James, Henry Cobham, alias Brooke, brother to Lord Cobham, was implicated in the conspiracy of Sir Walter Raleigh, Lord Grey, his own brother and others, to place Lady Arabella Stuart on the throne (called the Main). Some see two plots in the matter ; one to place her on the throne, the other to surprise and imprison the king (called the Bye, or Surprise). If we consider it is as two separate plots, it was in the second half that he appears to have been implicated. Thus two of the holders of the abbey were amongst those who planned plots in those dark times, and we might speculate, in our imagination, upon the rooms where their schemes were devised. The treason of Brooke led to the abbey and lands getting into the hands of Sir Robert Brett, of Somersetshire, who bore for his arms a lion rampant, *gules*, with an orle of cross crosslets, *fitchet*, of the second. He was buried in an ornamental tomb on the south side of the altar within the rails ; from the inscription we learn : " Here lyeth Sir Robert Brett, of the ancient family of the Bretts, in ye county of Somerset, with Dame Frances his wife, ye sole daughter of Sir Thomas Fane and Dame Mary, Baroness Le Despencer, his wife ; they were religiously and charitably disposed, as appeareth by legacy given to charitable uses to this towne of West Malling and East Malling ; they had between them one only sonne, Henry Brett, who also lyeth here, interred with them. Sir Robert Brett dyed 1° Septembris, 1620 : The Lady Brett dyed 27° Augusti, 1617 : Henry Brett dyed 12° Octobris 1609.

> In obitum clarissimi viri Roberti Brett, militis
> Hoc elegante qui sepulchro clauditur
> Ipse et Sepulchrum et elegans nuper fuit
> Sed elegantionis hospita domus
> Illa avolabit et manebit et manet
> Æternitatis consecrata annalibus
> Deo Politæ mentis eminens nota
> Patriæ et propingius cognita pietas viri."

This Sir Robert Brett gave by will ten shillings to be distributed to

twenty poor people, and ten shillings for a preacher on market day,
which was formerly Saturday. The sermon is still continued weekly,
the alteration having only been made from Saturday to Friday, and
from Friday morning to Friday evening, the market having been
discontinued in Malling for some period—an attempt made to revive
it in this century did not succeed. The money was to be paid out
of his estate of £80 at Tewkesbury, in Gloucestershire. After this
King James granted the manors to John Rayney, and this grant
was confirmed to his son Sir John Rayney, Baronet of Nova Scotia,
in the second year of Charles I.

Having traced the Malling manors down to this date, let us go
on with the rest after the same fashion till we come to the civil war,
when there will be several things to be found worthy of notice in
our valley. At Addington, as we have already seen, Thomas Watton
continued the Squire, when he was succeeded by William ; during his
lifetime the third and fourth bells were placed in the church steeple.
On them is printed, " John Wilmar made me, 1623," and also the
second bell, which has " J. W. 1635." " Juxta hoc monumentum
sepulturæ traditur corpus Guilielmi Watton, armigeri hujus,
manerii domini, filii predicti Thomae Watton, qui duxit uxorem
Elizabetham, filiam Johannis Simonds, in comitatu Essexiæ, generosi,
per quam prolem habuit filium unicum et tres filias, viz., Guilielmum
Elizabetham, Margeritam et Annam, obiit 28° Octobris, anno domini
1651.   Hujus mortem deflexerunt piisimi liberi et maestissima,
conjux Elizabetha quae in memoriam, tam carissimi mariti hoc
monumentum posuit."—" Near this monument is also placed the
body of William Watton, Esquire, in its burial place, lord of this
manor, son of the aforesaid Thomas Watton, who married Elizabeth,
daughter of John Simonds, in the county of Essex, gentleman, by
whom he had issue one son and three daughters, namely, William,
Elizabeth, Margaret, and Anna. He died the 28th of October, 1651."
As this Watton died within thirty years of his father, just at the
eve of the civil war, this is probably the reason why we do not find
the name of Watton, associated with one or the other of the two
great parties that rent every family in England asunder at these
times, as the next Watton (William) was probably a minor ; and
thus this family passed through a second period of civil war still
keeping their lands. We shall remark further on this when we come
to the Addington registers, which throw a great deal of light upon
those times, when parties through their bitterness upset the rules of
order and religion.

Allington, which had been granted by Queen Elizabeth to John
Astley, master of her jewels, was bequeathed to his kinsman, Jacob
Astley, in 1639.

The Friars at Aylesford, as we have seen, passed into the hands
of the Sedleys. Sir John Sedley was created a baronet by James I.,
and in 1622 was Sheriff of Kent ; in 1639 Sir Charles Sedley was

born at Aylesford, of whom more hereafter. Preston and Aylesford continued in the hands of the Colepeppers. Sir Thomas Colepepper died 1604, and is buried in Aylesford Church, where his monument still may be seen ; on it is inscribed : " Here lyeth intombed Sir Thomas Colepepper, knt., by bloode and desarte descended of many worthy ancestors, in his lifetime for his worth and desarte beloved of all men, and in his death as much lamented as bemoned; he had by dame Marie, his only wife, at the time of his decease, three sonnes and two daughters, which dame Marie, to perform her last duty in remembrance of her faithful love to her deceased husband, at her own cost erected this gratefull monument, under which he resteth, and by his lively faith hopeth a joyful resurrection. He died the 12th of October, 1604." He was succeeded by his son Sir William, who was Sheriff of Kent in 1637.

About this period the other manors of Eccles, Tottington, and Rowe's Place passed to the Sedleys from the Warcups. The Aylesford cup and paten are of this date, being of the year 1628, and so were possibly the gift of a Sedley or a Colepepper.

We have seen that Sir Edward Nevill obtained the earldom of Abergavenny by the decision of the court, but he appears to have died before the decision, in 1589, and his son, Lord Edward Nevill, in 1602 succeeded to the honours his father had thus won. He was succeeded at his death in 1622 by Lord Henry Nevill, his son, who was followed at his death by his son Lord Thomas, who died in 1658, his two children having predeceased him.

The old cup of Birling is dated 1617, and was perhaps a gift of the Nevill family ; the fourth, fifth, and sixth bells of Birling also belong to this period, and are dated 1631. On the fourth and fifth are, " Joseph Hatch made me," and on the sixth is, " Josephus Hatch me fecit, 1631."

Ditton passed through the hands of the Wisemans to Sir Oliver Boteler, of Teston, knight. He died in 1632, and his son John dying without children, his brother William Boteler obtained the properties, and was created a baronet in 1640.

Boro' Court came into the possession of the Shakerlys, one of whose monuments we have already spoken of ; another runs as follows : " Here lyeth the body of Elizabeth Shakerley, late wife of Richard Shakerley, of Ditton, Esquior, who died the 17th of February, anno 1626," which is no longer extant. Her granddaughter is also buried in this church, as appears by the monument : " Here lieth the body of Elizabeth Bewley, the daughter of Peter Bewley, gent, and Mary his wife, who departed this life the 14th day of October, 1638."

Another monument of the same date is to the incumbent : " Hic jacet corpus Guilielmi Prewe, in artibus magistri, hujus que ecclesiæ rectoris fidelissimi, qui obiit Septembr 28, anno domini 1638.

Nicholas Levison passed Bavent in Halling at this time to his

son Thomas, whose son Richard was knighted by Charles I.
Maximilian Dalison, the son of the lady whose brass we described
in the last chapter, was knighted by James I. He had eleven
children, and died in St. John Street, Clerkenwell, and was buried
in the parish church. He was Sheriff of Kent in 1613: in 1649
his name is found as Secretary for the Committee of Kent. His
half-brother, Multon Lambarde, was knighted in 1607. The Dalison
family still hold property in Halling, but the present family of
Dalison took the name about fifty years ago.

Bradbourne, in East Malling, continued in the hands of the
Manninghams. Richard Manningham's monument in East Malling
church has the following inscription in Latin :—

> Richardus Mannyngham,
> Honesta natus familia mercaturam.
> Juvenis exercuit satis copiosam,
> Ætate provectiori ruri vacavit.
> Literis et valetudini in studiis tam,
> Divinis quam humanis eruditus ; Latine.
> Gallice, Belgice dixit : scripsit.
> Eleganter et proprie : nec alieni Appe-
> Tens nec profusus sui ; amicos habuit
> Fideliter et benigne pauperes : for-
> Tunis suis sublevavit affines et
> Consanguineos auxit : animi cando-
> Re vultus suavitate et gravitate
> Conspicuus sobrie prudens et sin-
> Cere pius, languido tandem confec-
> Tus morbo expiravit 25 to Die Aprilis.
> A° Salutis 1611, et ætatis suæ 72.
> Desideratus suis maxime Johanni
> Mannyngham heredi qui monumentum
> Hoc memor mærens que posuit."

[Richard Mannyngham, born of an honourable family, as a young man carried
on a sufficiently plentiful trade. In age he gave up merchandise for
the country, for letters and for health, being skilled in studies divine
as well as human. He spoken in Latin, French, and Flemish. He
wrote elegantly and correctly ; neither greedy of other men's property
nor wasteful of his own, he held faithfully and kindly to poor friends :
with his own fortune he relieved his neighbours and assisted his
relations; by the candour of his mind, the sweetness and gravity of his
countenance, conspicuous ; soberly provident, and sincerely pious. At
length, worn out by languishing disease, he expired on the twenty-fifth
day of April, in the year of our salvation 1611, and of his own age 72.
Longed for by his friends, especially by John Mannyngham, his heir,
who raised this monument in memory and in grief.]

Hasted tells us (wrongly) that this was the Manningham who sold
the estate to Sir Thomas Twisden ; if so that judge must have shown
legal powers of a peculiar kind, inasmuch as his monument, which is
exactly opposite Richard Manningham's in East Malling church,
tells us that he was not born till January 8th, 1602, and consequently,
according to that author, the judge at nine years old acquired manorial

rights—"Credat Judæus Apella." It is very possible that the East Malling paten, which is dated 1609, was the gift of the Manninghams.

The East Malling Register supplies a few entries of the family :—

1598. Buried was Jane Manningham, the daughter of Richard Manningham, the 30th day of January.
1600. Married was Richard Manningham, Esq., and Mildred Jane Manningham, widow, of Maidstone, the 26th of February.
1607. Buried was Jane Manningham, the daughter of John Manningham, gentleman, the 1st of April.
1607. Baptised was Amy Manningham, the daughter of John Manningham, the 3rd day of April.
1611. Baptised was John Manningham, the son of Mr. John Manningham, the 16th day of May.
1612. Buried was Richard Manningham, Esquire, the 27th day of April.
1612. Baptised was Elizabeth Manningham, the daughter of Mr. John Manningham, the 23rd day of February.
1618. Baptised was John Manningham, the son of John Manningham, Esquire, the 24th day of September.
1619. Buried was Anne Manningham, the daughter of John Manningham, Esq., the third day of October (twice entered).
1622. Buried was John Manningham, Esq., the 25th of November.
1637. Baptised, December the 7th, was Anne, the daughter of Mr. Richard Manningham and Bridget his wife.

It is evident that this last is the Mr. Richard Manningham of whom Sir Thomas Twisden bought the estate, and not the one whose monument is in East Malling church, who, it seems to me, was probably the grandfather of the man last mentioned in this register. Since he was born in 1539, his great-granddaughter would probably be born about 1637, and his grandson would be of the same age, or thereabouts, as Sir Thomas.

It would appear that the manors of East and West Malling, together with the abbey, went to John Rayney, created a baronet in 1641. His son passed the manor of East Malling, by sale, to Thomas Twisden, Esq., who had already acquired Bradbourne ; and the abbey, also by sale, to Edward Honeywood, Esq.

In 1637 the fourth, sixth, and seventh bells were hung in Town Malling church tower, as is testified by the date upon them. On them, also, is the name John Wilnar ; but who presented them we cannot tell. The monuments in this church down to this date not already mentioned are :—

"Here lyeth buried Elizabeth, the late wife to William Manley, of the ancient house of the Manleys of Manley Hall, in Cheshire, she died the 4th of Januarie, 1622."

Thorpe also tells us of another monument :—

"Ætatis suæ 51. Here lieth interred the body of John Baron, gent, of this parish, who dyed . . . . the 14th, anno domini 1630, religiose pacificus."

The Goddens passed Leybourne Castle and Manor to Sir John Levison ; his representative, Sir Richard Levison, passed it in the time of James I. to Henry Clerke, Recorder of Rochester, who seems to have died 1649.

Offham remained in the hands of the Tuftons, Ryarsh in the hands of the Nevills, and Holloway Court continued to belong to the Tilghmans.

Paddlesworth about this time fell into the hands of Catharine, Lady Stanhope, from whose family it passed to the Marshams, of whom we shall speak hereafter. The last recorded institution to the parish of Paddlesworth-cum-Dode is that of Edward Aldey, in the year 1623; after this we hear nothing of this ancient parish. This institution is the second in the century, Robert Chambers having been instituted in 1600; while in 1599 we read in the register: "Take care that no one institute or induct to the rectory or the parish church of Paddlesworth, Rochester diocese."

During this period the parish register of Allington commences— in the year 1630. Perhaps the earlier records were destroyed at the time of the civil war, as may have been also the case with Aylesford, Ditton, Town Malling, and Halling; but these latter two parishes commence their registers much later than the others, and therefore these valuable diaries must have been lost by some one of those wanton destructions of which we have numerous records, amongst which is the well-known story that the next pages were torn out to use to baste a goose. Which was the greatest goose, the one that was cooked or the one that ate it ? is a question to be solved; but we are certain that the man had not a much deeper head than the bird, if it was as deep.

We now come to the times of the civil war. In the conflict between king and parliament the men of our valley played a considerable part. Sir H. Tufton, the lord of the manor of Offham, who was M.P. for Maidstone, is said to have been warm for the Parliament in the Committee of Kent, which also numbered amongst its members Maximilian Dalison, one of the Manninghams, and one of the Wattous. It was, however, in 1648 that the Kentish Royalists rose under Lord Goring; and after Husbands had defeated some of them at Northfleet, he was ordered to join Fairfax at Malling, whither he had marched from Maidstone, where Sir John Rayney and Sir W. Brockman were stationed with 1000 horse. We are told that at this time from 6000 to 7000 proscribed persons assembled on the high ground at Bluebell, and viewed the valley with longing and sorrowful eyes; but they had not, it would appear, the hearts of the men of Kent of other days, or Fairfax would not have left the valley so easily as he did. We find James Gosling, in 1651, was spoken of as riding in a troop of horse with sword and pistols at Town Malling; he was probably one of the many who did not relish the turn things took.

But of all the people who took part in the scenes of those times, perhaps none were more to the front than two persons who lie in the quiet old church of East Malling—we mean Colonel Tomlinson and Judge Twisden. Colonel Tomlinson, whose picture is in the Bradbourne collection, was brother-in-law to Judge Twisden, fortunately for himself. He was a colonel of horse in the army of the Parliament, and conducted Charles I. daily to and from Westminster. He turned approver against the regicides, and said he rebuked the people for taking tobacco before the king, and for refusing to take their hats off before him. He attended Charles I. till Hackett took charge of him on the scaffold, and appears to have been friendly with his unfortunate prisoner, as the monarch gave him a gold toothpick in a case before his execution, and confided his George and seals to him, which the colonel forwarded by his sister, Lady Twisden, to Charles II. On a plain stone in the chancel of East Malling we read, "Matthew Thomlinson, Esq., obiit ye 5th of November 1681, and Pembrook Thomlinson ye 10th of June 1683. In the register we read, "1681. Buried 9th of November was Matthew Thomlinson, Esquire." His wife's name is not in the register; she was the eldest of four daughters, and co-heiress of William Cobham (who died 1668), the son of George Brooke, Esq., who, with his brother, Henry, Lord Cobham, was attainted for his share in the Raleigh Conspiracy. Sir Thomas Twisden was the second son of Sir William and Lady Twysden, of Roydon Hall, East Peckham, where he was born in 1601-2. His father, Sir William Twysden, was the son of Roger Twysden, Esq., who was Sheriff of Kent, and Captain of the Lathe of Aylesford Light Horse at the camp of Tilbury : he married Ann, the daughter of the famous Sir Thomas Wyatt. Roger Twysden lived at Wye Court; but on that being burnt he removed to Chelmington, and thence to Roydon Hall, which he inherited from his mother, Elizabeth Roydon—this lady's father lost both his sons while they were bathing in the Medway at Brandbridges; he strongly opposed Wyatt's rebellion—her picture, as well as those of the Wyatts and Roger, hangs at Bradbourne.

The elder brother of the judge was the famous Sir Roger Twysden, born in 1597. He was an antiquary and historian, the friend of Dugdale, D'Ewes, Kenelm Digby, Junius, Biondi, Philpott, Selden, and Somner; he was confined in prison because he was an ardent Royalist and friend of the Church; he wrote, amongst other works, *A Historical Defence of the Church of England, The Laws of Henry I.,* and *Treatise on the Government of England.* He was J.P. for Kent, and died of apoplexy in the Malling Woods when on his way to the Petty Sessions in 1672.

Another of the family was Dr. John Twysden, born in 1607, who died in 1688. The sisters of the judge were Ann, Lady Yelverton, and Elizabeth, Lady Cholmley. Sir Thomas's aunt married Henry Fane, of Hadlow, who was related to the Earls of Westmoreland of that name ; her son altered his name to Vane, and is well known as Sir Harry Vane the elder, at one time Lord Treasurer to Charles I.,

but who afterwards took part against him ; his son was Sir Harry
Vane the younger, who sat in the Rump Parliament, and is so well
known because of Cromwell's exclamation, " The Lord defend me
from Sir Harry Vane." Thus father and son were first cousin and
first cousin once removed to the judge.

Sir William Twysden, the judge's eldest nephew, was born in
December 1635, and died in 1697. He served in several Parliaments,
and was M.P. for Kent in James II.'s first year ; he spoke very
strongly against a standing army, and more especially a popish
standing army, officered by men who were enemies to the constitution. He had seventeen children, of whom the seventh was a
contributor to the *Tatler*, and was killed at the siege of Mons ; the
eighth, John, was a lieutenant in the Royal Navy, and was one of
the unfortunate persons drowned with Sir Cloudesley Shovel off the
Scilly Isles in 1707 ; Josias, the ninth, was killed by a musket shot
while fighting in Flanders ; and the fifth son, William, who became
baronet, married Eleanor, the granddaughter of the judge, to whom
we now return. He was a member of the Inner Temple at the early
age of seventeen, and was called to the bar when twenty-four ; he
was made a bencher in 1646, and a serjeant in 1654 ; Cromwell sent
him to the Tower for defending the rights of the City of London, on
account of which the grateful city still preserves his portrait in the
Guildhall. He bought Bradbourne from Sir John Rayney it would
appear in 1642, and in 1653 his name first appears in the East
Malling register : * " 1653. September 3, baptised Isabella, daughter
of Thomas Twisden, Esq." Having founded a new family, he changed
his name from Twysden to Twisden. He married Jane, daughter of
John Thomlinson, Esq., of Whitby, sister, as we have already said,
to the famous Colonel Thomlinson. On the Restoration, in 1660, his
loyalty was rewarded, and he was knighted and made a Judge of the
King's Bench, in which official capacity he tried the regicides. He
was made a baronet, and died on January 2nd, 1683. His picture,
with those of his family we have mentioned, hangs at Bradbourne—
he is painted as serjeant-at-law, in a scarlet robe and black coif.
His monumental tablet in East Malling church runs :—

D. Thomas Twisden, Eq. Aur. et Baronetus,

Guilielmi Twisden de East Peckham, Eq. Aur. et Bar., fil. secundus Rogeri
Twisden de East Peckham. armigeri nepos natus. die VIII°. Januarii 1602-3.
Legum studiis in Interiori Templo, operam dedit et ob eximiam juris-
prudentiam, a serenissimo rege Carolo II., eo qui rediit anno ; Justitiarius
ascriptus est in Banco Regis, et cum per annos octodecem officio integerrime
functus, est A° Ætatis 77, senio et dysuria laborans, Judicis honore in termi-
num vitæ continuato officii veniam impetravit uxorem duxit Janam filiam
Johannis Thomlinson, de Whitby, in com. Ebor armg. e qua, suscepti sunt
Rogerus, Thomas, Guilielmus def., Heneagius def., Franciscus Jana, ux Jo.
Sympson, Eq. Aur., Anna def., Marg ux. Tho. Style, Bart., Elizabetha, Elnora,

---

* See Twisden Registers, p. 224.

ux Felicis Wylde, Bart., Isabella def. ; mortuus est 2ᵈᵒ die Januarii, Aᵒ Ætatis LXXXI., Dⁿⁱ MDCLXXX$\frac{\text{II}}{\text{m}}$. Venerabili parentis memoriæ posuit D. Rogerus Twisden, Baronetus qui, uxorem habet Margaritam, filiam Johannis Marsham, de Whorne's Place, Eq. Aur. ct Bar.

In English :—

> [Sir THOMAS TWISDEN, Knight and Baronet, second son of Sir William Twisden of East Peckham, Knight and Baronet, grandson of Roger Twisden of East Peckham, Esq., born the 8th day of January, 1602. He paid attention to the studies of the law in the Inner Temple, and on account of his excellent skill in law was appointed Judge of the King's Bench by the most serene King Charles II., in the year that he returned ; and when for eighteen years he discharged his duties with all integrity, suffering from old age and dysentery,* in the 77th year of his age, his honour of judge being continued to the end of his life, he obtained excuse from his office. He married Jane, daughter of John Thomlinson, of Whitby, in the county of York, Esquire, by whom he had Roger, Thomas, William, deceased, Heneage, deceased, Frank, Jane, wife of Sir John Kympson, Knt., Anna, deceased, Margaret, wife of Sir Thos. Style, Bart. ; Elizabeth, Eleanor, wife of Sir Felix Wylde, Bart.; Isabella, deceased ; he died on the 2nd day of January, in the 81st year of his age, A.D. 1682. To the venerated memory of his parent, Sir Roger Twisden, Baronet, who married Margaret, daughter of Sir John Marsham, of Whorne's Place, Knight and Baronet, erected this monument.]

This same Sir Roger succeeded his father in the estate, and died in 1703, when he was succeeded by his son Sir Thomas. On the family tablets down to this date we read :—

Dame Jane Twisden died September 24, 1702, aged 91. Sir Roger Twisden, Bt., eldest son of Judge Twisden, died Feby. 28, 1703, aged 62 years. Dame Margaret, wife of Sir Roger Twisden, and daughter of Sir John Marsham, died Jany. 30, 1687. They had issue Thomas, Roger, William, Heneage, Francis John, Jane, and Elizabeth, married to Sir Richard Newdigate."

In the year 1695, the second, third, fourth, fifth, and sixth bells were hung in East Malling church. The second has on it, "This bell was added by Benefactors, J. B. me fecit"; the third, fourth, and fifth, "James Bartlett me fecit"; the sixth, "John Grosse, Vicar, 1695, Abraham Walter, Thomas Hobert, Churchwardens. James Bartlett, me fecit."

During the period of the Commonwealth, Addington still continued in the hands of the Wattons, and remained with them during the rest of the time of the Stuart dynasty. William Watton the younger succeeded his father William, and died in 1703, leaving an only daughter, Elizabeth, who was married, first to Leonard Bartholomew, Esq., and afterwards to Sir Roger Twisden, Baronet. On the rest of the Watton monument we read :—

Hic etiam conduntur reliquiæ Guilielmi Watton, filii supradicti Guilielmi, in uxorem duxit Margaretam Moreland, quæ 7 liberos peperit, Elizabetham, Guilielmum, Edmundum, Robertum, Thomam, Francescam et Martham, necnon filii ejus Guclielmi, qui obiit sine prole superstite A.D. 1703, et Mariæ, uxoris

---

* "Dysuria," probably a mistake of the stonemason for "dysentura."

ejusdem Guilielmi, filiæ Roberti Fane, quæ obiit A D. 1695. Sub hoc marmore etiam depositæ sunt exuviæ Edmundi Watton, Fratris et Heredis Guilelmi Watton, et Saræ uxoris ejus, quibus natæ fuerunt quattuor filiæ, Elizabetha, Maria, Margareta. Anna. e quibus Elizabetha tantæ stirpis et tot progenitorum unica hæres parentis sola superstitil.

[Here also are hidden the remains of William Watton. son of the above-named William Watton. He married Margaret Moreland, who had 7 children—Elizabeth, William, Edmund, Robert, Thomas, Frances and Martha ; moreover of his son William, who died without surviving offspring. A.D. 1703, and Maria, the wife of the same William, daughter of Robert Fane, who died. A.D. 1695. Under this marble also are laid the remains of Edmund Watton, Brother and Heir of William Watton, and Sarah his wife, to whom were born four daughters—Elizabeth, Marie. Margaret, Anne, from whom Elizabeth alone survived, the sole heiress of a parent of so great a race and so many ancestors.]

The oldest cup and paten of Addington belong to this period. The castle of Allington was bequeathed by John Astley to his kinsman Jacob, who was Governor of Oxford and Reading, and Lieutenant-General of Worcester, Stafford, Hereford and Salop during the civil war, for the king. He was made a baron, and dying in 1651 was buried at Maidstone ; his son Isaac, Lord Astley, died in 1662, and was also buried at Maidstone, as was his son Jacob, who died in 1688, but who had previously alienated the property to the Marshams, whose descendants, the Earls of Romney, still possess it. The Astleys seem to have let Allington Castle to the Bests, as the registers of Allington, which commence in 1630, give us several entries of this family, as follows :—

Elizabeth Best, one of the daughters of John Best the younger, of Allington Castle, gent. was christened upon the twentieth day of May, Anno Domini 1631.

Mary Best. one of the daughters of John Best the younger, and being borne upon the eighth day of April last, was christened upon the twentieth day of this instant April. Anno Dni. 1632.

John Best, the son of John Best the younger, Esq., was baptised the second day of June, 1633.

John Best. son of John Best, Esq.. was buried the third day of June, 1633.

Humphrey Best, the sonne of John Best. Esq., was baptised the third day of April, Anno Dni. 1636.

Anna Best. the daughter of George Best, citizen and grocer of London, was baptised the fourteenth day of August. Anno Dni. 1636.

Alice Best. the daughter of John Best. Esq., was baptised the 8th day of July. 1637.

Will Best. the sonne of John Best, Esq.. was baptised the 25th of November, 1638.

Humphrey Best, son of John Best. Esq.. was buried the fourth of April, 1638.

Alice. the daughter of John Best, Esq., and Elizabeth his wife, was buried the 18th of September, 1639.

There were a few memorials to this family in Allington Church. They were, on one stone :—

John Best. an infant, died 2nd June. 1633.

Humfre Best.

On another :—

Here lieth the body of Anne Best, the wife of John Best of Allington, Gent, and one of the daughters of Gore Tucker, of Milton-next-Gravesend, Esquire, who deceased the 19th day of December, 1626.

A shete may hide her face, not her good name,
For fame findes never tomb t'enclose the same.

The ancient church of Allington has been removed, and with it we are sad to say its old memorials—all but three. Even so late as this they are in the porch :—

Here lyeth buried John Maplesdon, Maidstone, gentleman, and Ellen his wife, who died the second day of . . . . . the age, 1644.

Here are deposited the remains of the two children of William and Anne Stevenson, of Maidstone, goldsmith. Sarah died 26 October, 1666, aged 2 years and one month, John died 17th March, 1760 ; and also the remains of William Stevenson, died 24th October, 1769, aged 70 years, and Elizabeth, wife of Samuel Stevenson, died 2nd Sept., 1775, aged 70 years.

Here lyeth the body of John Thatcher, who died 2nd June, 1698, aged 40 years, also Ruth his wife, who died ye 18 Jan., 1743, aged 84 years ; likewise these children : Thomas, 25 May, 1694, aged 3 years. Joseph, 8 Nov., 1698. John died July 1738, aged 13 years. Charles died 14 Nov., 1731, 28 years.

This family of Best is probably the same as that of Thomas Best, of Chatham, who obtained part of Eccles Manor in Aylesford about this time, and whose grandson possessed the same in the time of Hasted.

The Church bell of Allington is of this period, and is marked " J. M., 1653."

The manor of Cosyngton at this time was in the hands of the Dukes, of whom Edward Duke was knighted in the time of James I. ; he was succeeded by his son, George Duke, who was knighted in the time of Charles I. There are several of this family in the Aylesford register :—

1656. George, ye sonne of George Duke, Esq., and Frances his wife, was born the xxvi[th] of June.

1664. Frances, the daughter of Edward Duke, gentleman, and of Mary his wife, was baptised the 26th day of June, 1664.

1655. Frances, ye daughter of George Duke, Esq., and Frances his wife, was buried ye ix[th] day of October.

1658. George, the sonne of George Duke, Esq., and of Frances his wife, was buried the xx[th] day of June.

1666. Catharine, the daughter of Edward J. (D)uke, gentleman, and of Mary his wife, was baptised the xx[th] day of September.

1669. Anne, ye daughter of Edward Duke, gent, and of Mary his wife, was baptised ye seven and twentieth day of June.

1669. Frances, the wife of George Duke, Esq., was buried the 11th day of October.

1691. George Duke, Esq., was buried Feby. 24.

In Aylesford Church we find this monument to one of the Dukes :—

Here lieth Thomas Finch, only sonn of Anthony Finch, of Coptree. gent, by Anne his wife, daughter of Thomas Duke, of Cosington, Esq., *hee* died 29 of August, 1629.

He left one daughter, Mary, who died 1696 without children, and left the property to S. White, Esq., who passed it on to the Staceys. In the register we find :—

1756. Mrs. Mary Duke was buried from Maidstone, May 26.

George Duke was a justice of the peace for the county, and thus we find his name appended in the Aylesford register :—

John Birchall, of Aylesford, in the county of Kent, Taylor, being chosen by the parishioners to be register of the said parish of Aylesford, was this day sworne before me, one of the Justices of the Peace in the said county, and doe approve him to be Parish Register, according to an act of the late parliament entitled an act touching marriages and the registering thereof, and also touching births and burials ordered to be printed the 24th of August last past. Witnesse my hand hereunto set the day and yeare first above written : GEO. DUKE, 23 Januarie, 1653.

The act above referred to was passed in Oliver Cromwell's administration, in the year 1653, by which it was ordered that marriages might be solemnised by justices of the peace and the banns should be published in the Market Place. A *Jeu d'esprit* was made by Fleckno upon this law :—

"Now just as 'twas in Saturn's reign,
The Golden Age is returned again ;
And Astræa again from Heaven is come
When all on earth by justice is done.
Amongst the rest we have cause to be glad
Now Marriages are in Market made,
Since Justice we hope will take order there
We may not be cousened no more in our ware.
Besides each thing would fall out right,
And that old Proverb be verified by't:
That Marriage and Hanging be both together,
When Justice shall have the disposing of either.
Let Parson and Vicar then say what they will,
The Custom is good, God continue it still:
For Marriage being now a trafique and trade,
Pray where but in markets should it be made?
'Twas well ordained they should be no more,
In Churches nor Chappels then as before,
Since for it in Scripture we have example
How buyers and sellers were driven out of the Temple.
Meantime, God bless the Parliament
In making this Act so honestly meant.
Of these Good Marriages God bless the breed,
And God bless us all, for never was more need."

Some examples of these marriages during the Commonwealth are recorded in the registers about here. In the parish register of Addington we read :—

1654. John King and Margaret Sladden were married, their Banns being first published three several daies, the 8th day of June.
1656. Thomas Hatch and Margaret Hatch were married by Justice Madden, of Boxley, and by the minister of Addington, the twenty-fourth day of September.

John Kindon, of Stanstead, and Mary Woollett, of Meopham, were married
the 15th day of September, 1657, having three several market days
their banns published in Rochester.

William Shileren (Children) and Helen Stimpson, and both of Tonbridge, were
married the 31st day of January, 1658, their Banns being three
several daies published in the market of Tunbridge by the Jtce. of
the Peace.

John Dennis, of Boxley, in Kent, and Mary Boorman, of Ryarsh, in the same
county, their intended marriage being three several market days
published in Maidstone, and no exception against them, were married
the 7th day of April, 1659.

John King and Susannah Prior, both of Wrotham, having had their proposed
marriage three several Lord's days in the congregation of Wrotham
published, were solemnly married at Addington the 13th of October,
1659.

In the register of Trottescliffe we read :—

According to an act touching marriages, and ye registering thereof, also
touching births and burials, bearing date Aug. 24th, 1658, Robert Hills being
chosen registrar by the parishioners of Trottescliffe, was sworn before me,
Justice of ye Peace, to be registrar of ye said parish, October ye 5th, 1654.
W. JAMES.

A purpose of marriage between Richard Daniell, of ye parish of Luddesdown,
and Joan Miller, of ye Parish of Trottescliffe, hath been published three Lord's
days in ye parish Church of Trottescliffe, and nothing was by any objected to
hinder or forbid the same, for testimony whereof I hereunto subscribe my hand,
Oct. ye fifth, 1654.

Upon ye aforesaid Certificate, and also another from ye Registrar of the
Parish of Ludsdown of ye like purpot, the marriage between ye said Richard
Daniel and Joane Miller was solemnised before me upon ye 5th of October,
1654, in ye presence of Robert Hilles and John Granger.   W. JAMES.

Entries of this sort take place for the next three years in this
parish register, in one of which the dates of Banns are given. The
marriages resume their usual form in 1658.

In the registers of Aylesford we read :—

23rd Januaire, 1653, John French, of the Parish of Aylesford, Bootmaker,
and Katharine Knight, also of this parish, Spinster, according to the act abovesaid were married by George Duke, Esqr·, ene of the Justices of the county.
Witness his hand : GEORGE DUKE.

November 24th, 1656, John Wood, of Wouldham, husbandman, and Anne
Baldock, of ye same, spinster, were married at Cossington, in the presence of
Michael Maylam and William Booth, Registrar of Wouldham, by George Duke,
Esq., one of the Justices within the county. Witness his name : GEORGE DUKE.

These cases we have chosen from the different registers will suffice
to show the peculiarities of the Act previously mentioned. No
doubt stringencies were placed on marriages, and this probably
explains why certain places where they were a little lax became
favourite places for marriages, though Mr. Burns, in his book on
parish registers, has said that Aylesford was the favourite place for
marriages because of the passage boats from Rochester and Chatham,
and I was told, at the meeting of the Kent Archæological Society, that
some favourite justice of the peace was chosen. I must call attention

to the fact that the retired parish church of Allington has a wonderful number of marriages during this period—about the time of the Act. This little parish never appears to have contained many more than sixty people at a much later date than this, yet here we have twelve marriages in 1648, eighteen in 1649, thirty-two in 1650, thirty-eight in 1651, twenty-nine in 1652, and twenty-five in 1653; while in 1654, 1656, 1657, and 1659 we have no marriages at all entered.   It is very clear, we think, from this, that the Act of Cromwell put a stop to a number of illegal marriages, the more especially as most of these people described themselves widows or widowers, and were strangers. From 1661 to 1668 the number of weddings, and especially of widowers, is also remarkable at Aylesford; but during times of civil war this might be explained, and the fact of Mr. Duke being a favourite justice, before whom these ceremonies were performed, would decide the matter as regards Aylesford.   But the marriages of Allington being previous to the Act, make me fancy that Gretna Greens were found in these early days nearer than Scotland, and that Cromwell's Act did good service in preventing marriages that ought not to take place, and which had become common in out-of-the-way parish churches, but which, nevertheless, were not far from some well-known means of transit.

On account of Mr. Duke's signature occurring in some of these registers we have given them here.   We find, in the register of Offham, that he also approved the registrar of that parish, for in that book we read :—

These are to certify I do approve of the choice of Andrew Dunning to be register for the parish of Offham, he having taken his Corporal oath before me the 27th day of June, 1657.   GEORGE DUKE.

Of the other manors of Aylesford, Tottington was alienated by Henry Warcup to Maddox, who, in the early part of the eighteenth century, alienated it to Mr. Thomas Golding, of Ryarsh.   The Friars and Rowe's Place continued the property of the Sedleys.   Sir John Sedley, who was Sheriff in 1622, left a son, Sir Henry, who died without children, and passed these properties to his brother William, who sold them to Sir Peter Rycaut.   The brother of Sir Henry and Sir William was the famous Sir Charles Sedley, one of the wits of Charles II.'s witty court, and a poet of no mean pretensions: he was born at Aylesford in 1639, and lived till 1701.   The following is a specimen of his poetry :—

> " Thyrsis, unjustly you complain,
> And tax my tender heart
> With want of pity for your pain,
> By sense of your desert.
> By secret and mysterious springs,
> Alas ! our passions move,
> We women are fantastic things,
> That like before we love.

You may be handsome and have wit,
Be secret and well bred ;
The person love must to us fit,
He only can succeed.
Some die, yet never are beloved,
Others we trust too soon,
Helping ourselves to be deceived,
And proud to be undone."

There are only two entries of the Sedley family in the registers of this neighbourhood. One, at Trottescliffe :—

1667. John the son of Sir Isaac Sedley, Knt., and Dame Cicely his wife, was baptised the 23rd of January.

The other, at Aylesford :—

1658. George, the son of John Sedley, gent, and Dorothy his wife, was buried the sixth of October.

The Hospital of the Holy Trinity, in Aylesford, was founded by Sir William Sedley, Knt., who was heir and sole executor to his brother, John Sedley, of Aylesford, who died in 1605, and who left by his will that :—

" A convenient house " was " to be built for six poore, aged, and impotent Persons, in the streete of Aylesford where my tenements be, if a convenient quantitie of the land adjoining may be purchased for that purpose, or in such other place in that Parish where my executor shall think fit, and that there be bought in Lands and Tenements, to be enjoyed by the saide poore persons for theire mayntenance, threescore pounds by the yeare, to be continued for ever ; my said Brother, William Sedley, and his Heires, placing therein from tyme to tyme such poore persons as they shall think, and always providing that one of the sixe shall be able to reade Prayers to the residue daylie, morning and evening."

Sir William bought a piece of ground and built a stone house to be an Hospital, or *Maison de Dieu*, in 1607, and also bought an acre of ground contiguous to the Hospital. He endowed it with two messuages, and lands of one hundred and eighty-four acres, which he purchased in the parish of Frittenden, of the yearly value of £76 ; and he placed in the Hospital six poor persons, four men and two women, and left directions that the inmates of the Hospital should always consist of a warden, who was to be the head of the poor and impotent persons dwelling there, and these poor people were not to exceed six in number. The deed of foundation and incorporation was signed and sealed by Sir William Sedley, October 2nd, 1617 : thus did the Sedleys leave a lasting memorial to Aylesford of their having once held property here.

Sir Peter Rycaut, Knt., alienated the manors to the Banks. There are two entries of his family in the register of Aylesford :—

1654. Petra ye daughter of Peter Rycaut, Esq., was buried the VIIIth day of October.
1700. Sir Paul Rycaut, Knight, was buried Nov. 27.

6

Sir Paul appears to have been a native of Aylesford, and his monument shows us that Aylesford has cause to be proud of him : it runs :—

Here lieth the body, of Sir Paul Rycaut, Knight, the tenth and youngest son of Sir Peter Rycaut, Knight. by dame Mary his wife, without the interposition of a daughter ; who after many years' travels in foreign parts in Asia, Africa, and Europe, and after several publicque offices performed by him. as secretary to the Earl of Winchelsea ; ambassador extraordinary from King Charles the Second to Sultan Mahomet Chan the Fourth, in which and in two voyages from Constantinople to London and back again, one of which was performed by land, through Hungary, and where he remained some time in the Turkish camp, with the great and famous Vizier, Kupriogly, for publick affairs of the English nation, in which he passed seven years ; after which he was made consul for the English nation at Smyrna, where having exercised that office the space of about eleven years, to the great and entire satisfaction of the Turkey company, he obtained a licence at his own motion and desire to return to England, where having lived the space of seven years, in honour and good esteem. as also in peace and plenty, he was, in the reign of King James the Second, called by the Earl of Clarendon, lord lieutenant of Ireland, to be his principall secretary for the provinces of Leinster and Connaught, also by the said King James to be one of his privy council for Ireland, and judge of the high court of admiralty, in which he remained until the great revolutions in England and Ireland, at which time he was employed by King William the Third in quality of his resident with the Hans Towns in Lower Saxony, namely, Hamburgh. Lubeck, and Bremen, where, having continued the space of more than ten years, to the satisfaction of all that knew him, as well to the senators of those republicks as also to the government thereof, and to the company of English merchants residing there, and having written severall books which are now extant.   He dyed the 16th of November, 1700, aged 72, and according to his desire lies interred near the body of his father and mother.

Requiescat in pace. Amen.

At his death, Caleb Banks, Esq., left these properties to his son John, who was created a baronet 1661, and who married Elizabeth, daughter of Sir John Dethick. He died in 1699, and as both of his sons predeceased him, and two of his daughters had no children, the property descended to his other daughter, Elizabeth. Sir John's monument in Aylesford church reads :—

Memoriæ sacrum.   Hinc felicem expectant resurrectionem Johannes Banks, de Aylesford, in comitatu Cantii, Baronet ; uxor etiam ejus Elizabetha, Johannis Dethick, militis comitatu Norfolciæ, olim prætoris Londinensis. filia : necnon filius utrius que communis Caleb Banks, maritus quidem sed liberis orbatus. His præterea nati sunt liberi quatuor, Martha, Elizabetha, Maria, et Johannes. Filias tantum duas superstites relinquerunt Elizabetham et Mariam, Martha et Johanne extinctis, quarum altera nempe Elizabetha, nupta fuit Heneagio Finch, Heneagii comitis Nottinghamiæ, Summi Angliæ Cancelarii, filio natu secundo auspiciis serenissimæ reginæ Annæ, Baroni de Guernsey.   Maria vero Johanni Savill, Johannis de Methley, in comitatu Eboracensi, armigeri. Filio primogenito exuvias deposuerunt.

Caleb Banks, Septbris 13°, An° 1696. Ætatis 37.

Elizabetha, Octbris 21°, An° 1696, Ætatis 59.

Johannes, Oct bris 18, An° 1699, Ætatis 72.

[Sacred to their memory.   From this, John Banks, baronet, of Aylesford, in the county of Kent, and also his wife Elizabeth, daughter of Sir

John Dethick, in the county of Norfolk, knight, formerly an officer in London, await a happy resurrection. Moreover Caleb Banks, the son of them both, who was married indeed but deprived of children. They had four children besides, Martha, Elizabeth, Maria, and John. They left surviving only two daughters, Elizabeth and Maria—Martha and John being dead—one of which, namely, Elizabeth, married Heneage Finch, second son of Heneage, Earl of Nottingham, Lord High Chancellor by the pleasure of the most serene Queen Anne. Lord of Guernsey ; but Maria to John Savill, son of John Savill of Methley, in the county of York, Esq. They buried them near their eldest son.

> Caleb Banks, the 13th September, 1696, aged 37.
> Elizabeth, October the 21st, 1696, aged 59.
> John, October the 18th, 1699, aged 72.

The grandfather, Caleb Banks, was M.P. for Kent 1686; the father, Sir John, M.P. for Kent 1690; and the son, Caleb Banks, Esq., M.P. for Kent 1695. Of those mentioned on this tomb the following are found in the parish register:—

1669. John, ye son of Sir John Banks, Baronet, and of Dame Elizabeth his wife, was buried ye one and thirtieth day of May, anno domini 1669.

1676. Mrs. Martha, the daughter of Sir John Banks, Baronet, was buried 2nd of Sept.

1696. Caleb Banks, ye son of Sir John Banks, Baronet, was buried Sept. 21st.

1696. Dame Elizabeth, ye wife of Sir John Banks, Baronet, was buried Nov. 2.

1699. The Right Worshipful Sir John Banks was buried.

From these monuments, and from these registers, it will be seen that Elizabeth, daughter of Sir John Banks, carried the estates in Aylesford in marriage to Heneage Finch, who was M.P. for Oxford, afterwards created Lord Guernsey, March 15th, 1703, and finally Earl of Aylesford, Oct. 19th, 1714. He died in 1719.

Sir William Colepepper, of Preston Hall, left a son, Sir Richard, who died in 1660, and passed it on to his son Sir Thomas, and he, it would appear, to his son Sir Thomas, who was sheriff in 1704 and died 1723.

The fourth and fifth bells of Aylesford church were hung in 1652, and the seventh and eighth bells in 1661 and 1666; that of Allington in 1653, and the first bell of Ditton in 1656. None of the families we have spoken of have any record of their doings mentioned upon them ; we, however, give their inscriptions. The fourth and fifth bells of Aylesford, " Michael Darbie made me, 1652. T. Macelgin, J. Bogherst, churchwardens "; the seventh bell has on it, " Robert Kemsley, Phillip Grange, churchwardens. God save King Charles ye 2nd, 1661 "; the eighth, " Anthony Bartlett made me, 1666. Capt. Ward, Liveftenant Long, two of His Majsties' Hoymen : Thomas Cossington xxs· to buying a treble bell for Aylesford ": this inscription is interesting as pointing to the royalties of the Medway—we believe sturgeon caught in the Medway are still royal fish.   The name of Cossington appears for the last time in 1708.   The sixth bell was placed in Aylesford, which was simply inscribed, "Thomas Goodman, John Taylor, junior, Churchwardens, 1708."   There is a flagon in Aylesford church inscribed, " Ex sumptu parochiæ de Aylesford, et Thomæ Tilson, Vicarii, conjunctim, A.D. 1711 " (at the expense of the parish of Aylesford, and Thomas Tilson, the vicar, conjointly, A.D. 1711).   The paten has on it, "Tuum est domine tibi reddo, T. Tilson, Vic. Aylesford, 1724-5" (Thine, O Lord, it is I restore to Thee.—THOS. TILSON).   The Alfington bell is only inscribed, " J. M., 1653," and the first bell of Ditton, "Tm. Cw. Wh., 1656."

The church cup of Ditton is also of this period, and dates from 1689.   Ditton Manor continued in the hands of the Botelers of Teston, but Borough Court was carried by Mary Bewley, in marriage. to Mr. Basse, of Suffolk, who in the reign of Charles II. alienated it to Sir Thomas Twisden, of Bradbourne.   Ditton Place was the seat of the family of Brewer.   There are a few entries of this family in the register :—

1663. Dec. 23, William Brewer, son of William Brewer, Esq., was buried.
1666. Aug. 21. William Brewer. gent, was buried.
1675. July 14. Elizabeth Brewer, daughter of William Brewer. Esq., was buried.
1675. July 22. Mrs. Dorothy Lyng, the wife of Thomas Lyng, Esq., the daughter of William Brewer of this parish, gent, was buried.
1691. Dec. 11th. Mrs. Mary Brewer, widow of Richard Brewer, Esq., was buried.
1692. July 30th, William, the son of Mr. Brewer, was buried.
1695. Elizabeth, the daughter of Mr. Thomas Brewer, and Isabella his wife, was baptised January 1st.

In the beginning of the last century it came into the possession of Mr. Thomas Golding, who gave it by will to his nephew of the same name, of Ryarsh ; whence it again came into the possession of the Brewers in the person of John Brewer, whose niece Mrs. Carney, however, in 1735, re-conveyed it to Mr. Thomas Golding.   Birling continued during this period in the hands of the Nevills, who still possess it.

In 1662, John, Lord Abergavenny, the brother of Sir Thomas, died, and was succeeded by his brother George, who died in 1666, and he by his son George, who died in 1699, who was the first baron belonging to the Anglican Church, and who died in 1723.   The Birling cup is 7 inches high, 4½ inches deep, and 3⅞ inches wide; engraved " I.H.S.," with cross and nails *en soleil*; near mouth " P. D.," with two mullets above and one below ; initials in a shield dated 1685-6.

The ancient palace of the bishops of Rochester, at Halling, was already in ruins in Harris' day (1692), who, with Philpott (1656), flourished towards the latter part of this century.

Langridge, which had been part of the property of Sir Richard

Levison, was sold by him to the Barbers, who alienated it to the Goldings.

All the bells of Halling belong to this date. The first three have on them, "John Hodson made me, 1675. Henry Acorte, churchwarden"; the fourth has, "John Hodson, Christopher Hodson made me. Henry Acorte, churchwarden"; and the fifth, "John Hodson, Christopher Hodson made me, 1675. Henry Acorte, churchwarden."

Leybourne, which had also become the property of the Levisons, was alienated by them to Henry Clerk, Recorder of Rochester. His son and heir, Francis, possessed it in Philpott's day. Francis Clerk lived at Restoration House, Rochester—in which Charles II. dined and slept as his guest the evening before the memorable 29th of May, 1660, for which the king showed his gratitude by knighting him that evening in that house—from whom the castle descended to Gilbert Clerk, of Derbyshire, and from whom it came into the possession of Captain Saxby, of the Grange. This last property he inherited from his father, who acquired it from his marriage with Elizabeth, daughter of Edward Covert, who came into the property in right of his wife Julian, who succeeded to her father's (Mr. Thomas Olyver) estates at his death in 1678.

The chalice of Leybourne has on it, "The gift of Henry Ullock, D.D., Deane of Rochester and Rector of Leybourne in Kent, 1691."

The abbey of West Malling was sold by Sir John Rayney to Edward Honeywood, from whom it passed into the hands of his son. His son and heir, Frazer Honeywood, a banker in London, on succeeding, pulled down the old house, then occupied by a fellmonger named Seager—probably Richard Seager, senior, buried June 2nd, 1737, at West Malling. With the materials, we are told by Thorpe, he built the present house. The third bell of West Malling has on it, "John and Christopher Hodson made me, 1677," and the fifth, "Abraham Mason, John Fleete, Churchwardens, 1698; John Weekly, junior, gent."

Offham continued in the hands of the Tuftons, of whom Nicholas, in 1628, was advanced to Lord Tufton and Lord Thanet. His son Nicholas was third earl, having succeeded in 1664; he died in 1679, when his brother John, the fourth earl, succeeded him. He having also died without issue, was succeeded by his brother Richard, who died unmarried, when Thomas, the sixth earl, succeeded, in 1684. Of this family there is only one entry in the Offham register :—

1625. Richard Tufton, ye son of Richard Tufton, Esq., was buried ye second day of July.

An entry in the parish register of Offham, as it is the earliest record of hopping in this valley mentioned, we think worthy of notice :—

1666. Thomas, ye son of Thomas and Mary Clark, strangers, and came a hopping, baptised 4th September.

This not only shows hopping to have been an industry in these parts two hundred and thirty years ago, but also shows that even then the persons in the valley were not sufficient to gather in that harvest.

The second bell of Offham has on it, " John Hodson made me, 1674. C. H. Robert Lurinden, churchwarden," and the third, "John Wilnar made me, 1633." The Offham paten was given in 1675 by the Rev. William Polhill, who was also rector of Addington. In the registers this gentleman's name is frequently written Polly by the clerk, but when he writes his own name, as he does two or three times in the Oftham register, he always writes it Polhill. He was, I have little doubt, one of the old family of the Polhills of Sundridge.

Ryarsh, as already stated, came into the possession of the Nevills. Carews' Court in this parish, which had, as already stated, been in the hands of the Carews of Beddington, was sold about 1670 to Thomas Watton, who passed it to William, who sold it in the days of Charles II. to Edward Walsingham,* whose family had lived in Ryarsh, according to the registers, from 1572, but seem to have died out in 1724. The Walsinghams were probably an offshoot of the Walsinghams of Ightham Mote, and so of those of Chislehurst. Ryarsh second bell has on it, " Joseph Hatch made me, 1616."

The Tilghmans of Holloway Court alienated half their estate to Sir John Marsham, of Whorne's Place, in Cuxton, Clerk in Chancery in the time of Charles I., who was also M.P. for Rochester in 1660, and was knighted and made a baronet in 1685. He passed the property on to Sir John, who was the second baronet, and he to Sir John the third baronet, in 1696. It then passed to Sir Robert, fourth baronet, who was M.P. for Maidstone in 1681, and died in 1703, when Sir Robert, fifth baronet, came into the property, who was created Baron Romney, 1716, by George I. The other part passed to the Clotworthys, and was bequeathed to Thomas Williams, whose family occurs occasionally in the Snodland register :—

1652. Thomas, the son of Thomas Williams, gent, of Holboro, was baptised April 3rd.
1668. Elizabeth, the daughter of Mr. Thomas Williams, baptised August 16th.
1671. Grace, the wife of Thomas Williams, gent, was buryed May 2.

Thomas Williams alienated the property to Richard Manley, Esq. He and his wife were both buried at Snodland : the entries are :—

April 4th, 1682. Madam Manley, ye wife of Richard Manley, Esq., was buried in linen April ye 4th, 1682 the just forfeiture, according to ye act, was paid to ye poore of ye parish ye weeke following.
1684. Richard Manley, Esq., was buried in woollen, May ye 2nd.

Whether the other two entries belong to the family or not seems to be doubtful :—

* See Walsingham Registers, p. 231.

1792. William Southgate and Frances Manley, married Feby. 29.
1819. Amy, daughter of Stephen and Margaret Manley, Farmer, baptised
Feby. 17.

Also, whether the stone found in West Malling church belongs to
them we cannot tell :—

Here lyeth buried Elizabeth, the late wife to William Manley, of the ancient
house of the Manleys of Manley Hall, Cheshire, she died the 4th of Januarie,
1622.

The monuments of Richard Manley are to be found in Snodland
church : that to him runs :—

Richard Manley, armiger, qui obiit vigessimo nono die Aprilis, anno
salutis 1684.

That to his wife is as follows :—

Here lyeth the body of Martha, eldest daughter of John Baynard, late of
Shorn, in the county of Kent, gent. First married to Bonham Faunce, of St.
Margaret's, Rochester, gent, by whom she had issue which survived her, two
daughters, viz., Mary, her eldest, married to William Man, of the city of
Canterbury, Esq., and Martha, her second daughter, married to John Cropley,
of St. Margaret's, Rochester, Esq. She was afterwards married to Richard
Manley, of Holloway Court, in this parish, Esqʳᵉ., by whom also she left issue
surviving her, viz., one son, Charles, and one daughter, Frances. She died
the 29th day of March, anno domini 1682, and in the 58th year of her age.

Hoc monumentum amoris et pietatis ergo idem generi posuere.

Charles Manley sold his property to John Conwy, and his son,
Robert Conwy, M.D., sold it to Thomas Pearce. It then passed into
the hands of Thomas Best, and afterwards to the Vincents, but
none of these have left any memorials of their holding land here.

The fifth bell of Snodland has inscribed on it, "J. W. made
me, 1636."

Trottescliffe was first rented from the bishops of Rochester by
the Whittakers in the beginning of the eighteenth century. The
paten of this parish church is dated 1699. The church bell has on
it, " William Hatch made me. I.G., I.D., C.W., 1639."
The remaining registers commence during this period.

Allington parish register commences with the words, " The registers
of the Christenings, Marriages, and Burials in the parish of
Allington, near Maidstone, in the county of Kent, began in Anno
Dni. 1630."

Aylesford register simply commences, " Joseph Jackson, Minister
of Aylesford, 1654." The choosing of the registrar has already been
alluded to.

Ditton parish register has as its opening words, " The Register
book of Ditton beginning Anno Dom. 1663. William Jole, rector,
inducted Rector of Ditton, 1st August, Anno Dom. 1663."

Also at the commencement we read (August 1, 1711) that every
acre of woodland in the parish of Ditton by immemorial custom
pays tithe to the rector.

Greatly to the honour of the clergy and parishioners of Ditton, the parish registers have been rebound.

West Malling parish registers commence with two entries of the children of the vicar, Rev. T. Pyke, while the baptismal register begins some months before the others; the date of them is the beginning of 1700. In this register we read : "1704. Sept. 11, William Briggs, Dr. of Physick." Hasted informs us that this gentleman was physician to King William III., and to St. Thomas' Hospital; he was also a great traveller. He was buried, we are told, in the church.

There was a Confirmation in the parish church of West Malling on Friday, November 7th, 1712, the Bishop of Killaloe and the Archdeacon of Rochester being present with the bishop : there were between two and three thousand people confirmed.

This seems strange to our modern notions, but bishops formerly held a Confirmation for large parts of their dioceses, and churches were filled once or twice with the candidates.

The village chronicler of this period further tells us : "There happened a terrible and great tempest of thunder and lightning, and set on fire the spire, and broke down through the roof and ceiling of the body of the Church, and through the Belfry-doore ; broke down the pendulum of the Clock, melted the bottom of the pendulum, went through the head of the Chancesell, and did a great deal of other damage, especially to the spire; on Munday morning, about six o'clock, the seventeenth day of November, 1712."

The parish register of Halling commences very abruptly, as late as 1705, evidently showing that the former registers have been lost or destroyed.

I have now only to record, to finish the history of this period, the various monuments that have been found in the different churches, which have not as yet been mentioned, down to the time of the accession of George I.

In Allington were the following memorials, according to Thorpe :—

Here lyeth Mary Fletcher, daughter of Thomas Fletcher the younger, buried the 57th of July, Ano. Dom. 1651.

This one was strange, owing to the day in July having been wrongly cut by the mason, and never remedied.

A brass, mutilated, had :—

Here lies the body of Sir Gyfford Thornhurst, Baronet, who died the 15th of December, 1627. He had issue one sonne, deceased. and two daughters, now living, Frances and Barbara, by Dame Susann, th. . . only daughter of Sir Alex.

Here lyeth John Wykes, who died the 10th of October, 1677. aged 68 years.

Here lieth the body of Susannah, wife of John . . . . , who died Aug. 1709, aged 46 years.

This last is in the porch.

In Aylesford, besides the monuments already mentioned, we have the following :—

Here lyeth the body of Patrick Savage, sometime cook to the Right Worshipful Sir William Sedley, Knt. and Baronet, deceased, who hath given to the poor of this parish LX. pounds, to be employed at the discretion of the Minister and Churchwardenes of this pshe for the time being, for a poore stock for the most comfortable relief of the poore here, and also has given towards the reparations of this Church 10 poundes, towards the inlaying of the Communion Cup of this parish xx. shillings, and towards a Communion Cloth vi. shillings and viii. pence ; he was born in Ireland, and died here at Aylesford the second day of Maye, Anno Domini 1625, Anno Ætatis 57.

A.D. MDCCII. Lapidem hunc marmoreum posuit Thomas Tilson, hujus Ecclesiæ, Annos triginta sex Vicarius ; Joannæ piam in memoriam conjugis, suæ merito dilectæ, ex qua septem liberos genuit, Quinque eorum sunt superstites, Thomas trium filiorum unicus, Sarah, Elisabetha, Maria, Martha. Obijt charissima Aug. 28, 1680. Obijt ipse July 24, 1702. Ætat 61.

[In the year of our Lord 1702, Thomas Tilson, for thirty-six years Vicar of this church, placed this marble monument to the pious memory of his deservedly beloved wife, by whom he had seven children ; five of them are surviving, of his three sons Thomas alone, Sarah, Elizabeth, Mary, and Martha. This most loved one died Aug. 28, 1680. He died July 24, 1702. Æt. 61.]

Here lieth interred the body of Thomas Ward, late of Mill Hall in this parish, who departed this 10 Feb., 1714, in the 64th year of his age : by his first wife, Mary, daughter of John Fowle, of Chart Sutton, he had issue six sons and one daughter, three of which survived, viz., John, Daniel and Samuel. He afterwards married Catherine, daughter of Edward Goodman, and widow of John Lake of this parish, by whom he had one son, who died an infant. The abovesaid Catherine survived him.

In the Aylesford Belfry is the inscription ; " P. PALLMER, R.M., CHURCHWARDENS, 1687, T. PACK."

Besides the memorials to the Abergavenny family in the parish church of Birling, of which we shall speak hereafter, there is this inscription :—

Here lyeth interred the body of Michael Rabbett, Vicar of this parish the space of 32 years, who departed this life the twenty-fifth day of March, in the yeare of our Lorde 1692, ætatis suæ 84.

In Ditton we have a monument to one incumbent :—

Hic jacet Guilielmus Jole, in artibus mʳ hujus ecclesiæ rector, qui obiit Septembris decimo nono, anno domini 1678.

And to his wife :—

Here lyeth the body of Katherine, the wife of William Jole, rector of this parish, the daughter of Henry [Adye], of West Malling, gent, who died the 12th of September, 1677.

The letters in brackets are now illegible. Below this is one to his son :—

Here lieth the son of William Jole, Rector of Ditton, who died in the date 1675.

Also these inscriptions to the Brewer family belong to this period :—

> Here lyeth the body of Richard Brewer, gent, who deceased 8th of April, 1616.
> Here lieth the body of Martha Brewer, late wife of Richard Brewer, of Ditton, gent, deceased, and daughter of William Hamon, of Acrise, Esquiar, who died the 24th of December, anno domini 1629. Mors mihi lucrum (*Death to me is gain*).
> Here lieth the body of Dorothy, the wife of William Brewer, gent, and daughter of John Haward, of Feaversham, gent, who died the [10th] day of January, 1638. Exp^c^tans expectavi (*Waiting, I have waited*).

The figures in brackets are illegible.

> Here lieth the body of Millecent, the 2nd wife of William Brewer, gent, daughter of Robert Tyas, of London, who died ye xi. day of September, 1656. Si moram fecerit expecta (*If he has made delay, Await*).
> Here lyeth the body of William Tyas, Esq., who died the 24th day of Feb. 1654.
> Hic jacet corpus Guilielmi Brewer, de Gray's Inn, armiger, filius Guilielmi Brewer, hujus parochiæ gen., qui obiit 6th die Decembris 1657. Statutum omnibus semel mori.
> Here lieth the body of Martha, the daughter of Richard Brewer, of this parish, gent, and the late wife of Thomas Godden, of Hadlowe, gent, who was heere interred the third day of December, anno domini 1662.
> Here lyeth the body of William, the grandson of William Brewer, of Ditton gent, and sonne of William Brewer, of Gray's Inn, Esqr., by his wife, Ann, the daughter of William Watton, of Addington, Esq., who died the 17th of December, 1663, anno ætatis suæ 8°. Nascentes moriamur.
> Here lyeth the body of Richard Brewer, gent, sonne of William Brewer, of Ditton and grandsonne of the said Richard, [who] dyed February the 20th day, 1672, aged 38 years.
> Hic jacet corpus Guilielmi Brewer, gen., filius supradicti Ricardi Brewer, hujus parochiæ, qui obiit 14th Augusti, anno domini 1666, ætatis suæ 68. Aliorum majore damno quam suo.

> > [Here lies the body of William Brewer, gentleman, son of the above-named Richard Brewer, of this parish, who died 14th August, in the year of our Lord 1666, and of his own age 68. A greater loss for others than his own.]

> Richardus Brewer, filius Guilielmi Brewer, generosi, obiit Feb. 20, 1672, duplici conjugio felix Rachelam, quippe uxorem duxit Thomæ Deacon, mercatoris Londinensis, filiam ex qua quatuor liberos genuit :

> > | Guilielmum et Richardum | defunctos. | Rachelam Thomamque | superstites. |

> Mariam deinde altero matrimonio sibi conjuxit Adriana Evans, Londinensis, armigeri natam. Illa quidem sine prole moriens, Dec. 2, anno domini 1691. Privignum suum Thomam heredem reliquit. Lapidem hunc marmoreum, lubens merito posuit Thomas Brewer, parentum in memoriam et grati animi monumentum.

> > [Richard Brewer, son of William Brewer, gentleman, died Feb. 20, 1672, happy in two marriages, since he married Rachel, daughter of Thomas Deacon, a London merchant, by whom he had four children—William and Richard, who are dead, Rachel and Thomas, who survive. After that he contracted a second marriage, with Mary, daughter of Adrian Evans, of London, Esquire. She indeed dying without children on the second of December, in the year of our Lord 1691. He left

Thomas, his firstborn son, his heir. Thomas Brewer, with pleasure, placed this marble as a monument well-deserved to the memory of his parents, and of his own grateful feelings.]

Here lieth the body of Mary, the second wife of Richard Brewer, gent, and daughter of Adrian Evans, of London, Esquire, she died December 2, 1691.

Here lieth the body of Richard Brewer, sonne of Richard Brewer, of West Malling, gent, who died the 26th day of September, anno domini 1699.

Besides the monuments to the Twisdens, we have one or two of this period in East Malling church to different people :—

D. Jana Sympson, D. Thomæ Twisden, Equitis Aur. et Bar. filia natu maxima D. Johannis Sympson, Equitis aurati, servientis ad legem Caroli Secundi, nuper regis Angliæ, vidua filios habuit tres, Thomam, Edward def., Johannem def., filiam vero unicam Janam, def., Obiit 7° Decembris, A° Dni. 1690. Hoc monumentum charissimæ matris memoriæ posuit Thomas Sympson, de Interiori Templo, London, Armiger.

[Dame Jane Sympson, eldest daughter of Sir Thomas Twisden, Knt. and Bart., the widow of Sir John Sympson, Knight, serjeant-at-law to Charles the Second, late king of England, had three sons, Thomas, Edward who is dead, John who is dead, and one daughter, Jane, who is dead, she died December 7th, A.D. 1691. Thomas Sympson, Esqre· of the Inner Temple, in London, erected this monument to the memory of a most dear mother.]

Reliquiæ Roberti Whittle, nuper hujus ecclesiæ vicarii, hic placide quiescant. Vixit annos LXXXI., M.I., D.III. Obiit XIII. Julii, MDCLXXIX. In spe Domini resurgam.

Here lieth the Body of Edward, the son of Edward Belcher, by Sarah his wife, who departed this life the 27th of April, 1701, aged four years.

Here lyeth ye body of Roger Tomlyn, of East Malling, Gent, who departed this life September ye 6th, Anno Dom. 1704, in ye 54th year of His Age. He left issue 5 sons, Thomas, James, John, Edward, Roger.

Also :—

Here lieth ye body of Mary, ye wife of Richard Bruse, Citizen of London, Daughter of Mr. James Fletcher of this Parish, who departed this life on the 16th of November, 1663, at the age of 19 years.

> Youth, beauty, vertue here enclosed do lye,
> Fate nere could boast so dear a victory ;
> Twas heaven, not death, thus ravished her away,
> For such perfection never could decay.
> Her ashes in this monument must rest,
> Her liveing tomb is in her husband's breast.

This is, perhaps, one of the most exquisite poems that have been left to languish on the desert air.

Besides, we have :—

Hic dormit Thomas Furner, de East Malling, dum vixit de multis bene meritis, expiravit secundo die mensis Maii, anno domini 1674, ætatis suæ 55, in cujus memoriam hoc posuit monumentum uxor ejus mœrens Maria Furner. Nescis Lector quam cito sequaris.

In English :—

[Here sleeps Thomas Furner, of East Malling ; whilst he lived of many deserved well. He died the second day of the month of May, in the year of our Lord 1674, in the 55th year of his age, to whose memory his sorrowing wife placed this monument. You know not, reader, how soon you may follow.]

Also we have :—

Hic adjacet marito suo Maria Furner, quae obiit 25 die mensis Septembris, anno Ætatis suæ Sexagesimo Septimo, Annoque Domini Millesimo Sexcen-tesimo Septuagesimo Octavo. Superstitibus adhuc duobus filiis Rogero et Francisco unaque filia, Martha, uxor frugi indulgens mater.

[Here lies, near her husband, Maria Furner, who died the 25th of September, in the 67th year of her age, and in the year of our Lord 1678. Two sons, Roger and Francis, and one daughter, Martha, a kindly wife and indulgent mother, still survive.]

In West Malling we find, from Thorpe, and from monuments still in existence, that a number of the family of Chambers were buried. There still remain these inscriptions :—

Thomas Chambers dyed the 29th of November, 1660.
Dorothy Chambers, the wife of Thomas Chambers, dyed the 19th of July, 1664.
Here lieth the body of Dorothy, the wife of Robert Chambers, who was buried the 16th day of February, 1675.

Besides the above, the following inscriptions have perished already:—

Francis Chambers, the sonne of Robert Chambers, and Dorothy his wife, was buried the 1st day of March, 1675.
Ann Chambers, the second wife of Robert Chambers, was buried the 24th day of April, 1684.
Robert, the eldest sonne of Robert Chambers, and Anne his wife, was buried the 30th day of September, 1680.
Robert, the second sonne of Robert Chambers, and Anne his wife. was buried the 11th day of April, 1681.
Elizabeth, the daughter of Robert Chambers, and Anne his wife, was buried the 5th day of June, 1683.
Here lyeth also the body of Robert Chambers, late of this parish, gent, who departed this life the 8th day of April, anno domini 1707.

The next three are to the Chapmans, a family of importance in Kent, whose property has become the subject of a lawsuit. One of this family we have already mentioned :—

Here lyeth the body of Robert Chapman, gent, who departed this life September ye 5th, 1703, in the 49th year of his age. Also the body of Lydia, wife of Robert Chapman, who departed this life Decem^r 26th, 1726. in the 76th year of her age. To whose memory their children have erected this monument.
Here lyeth enterred ye body of Mary Chapman, the wife of John Chapman ; she is not dead but [sleepeth]. Obiit December23rd, Anno Domini . . . . . Anno ætatis suæ 35.

Thorpe tells us the word sleepeth was in the monument in his day, but the date was gone.

There still remains part of another memorial to this family, half of which is covered by the font. The end of each line beyond our division is the part now not able to be seen :—

Here lies the body   of William
Chapman and Ma   ry his wife, late of
this parish, she died   the XXIV. day
of January, 1679,   in the LX. year.
He died the 12th day of   August,
1694, in the 77th year   of his
                          age.

The next inscription still remains, though it is not found in Thorpe :—

Here lyeth the body of Thomas Ducke, sonne of Thomas Ducke, late of the parish of Wrootham, who died the 23rd day of December, anno domini 1674, aged 21 yeares.

This stone tells us that the neighbouring parish of Wrotham was spelt as pronounced. An inscription now gone read :—

Here lyeth the body of Margaret, wife of George Baynard, and daughter of Jarvice Maplesden, of Shorne, who departed this life September the 18th, 1682, aged 42.

We restore the parts of the two following inscriptions that are in brackets from Thorpe :—

Thomas Kidwell, [who] died November 28, 1684. Ætatis [suæ 81]. Here lieth Mary, the wife of Thomas Kidwell, [of West Mal]ling, Daughter [to Nicholas] Higgins, yeoman [of East] Peckham, [died] September 14, ætatis suæ 60, anno domini 1671.

Here lyeth the Body of Peeter Boorman, son of Nicholas Boorman, of *Raysh*, who departed this life [the 19th] day of May, 1708, in the Eightieth year of his age. [Here] lyeth the Body of Susanna, wife of Peter [Boorman], of this Parish, [daughter of] John Bennet, [of Swinbridge, in] the county [of Glo'ster], who departed [this life the 27th] of *Desember*, 1684.

The spelling on this last stone is unique.

There is one to the incumbent's baby in Latin :—

Maria Pyke, infans suavissima, post menses II. et dies XXVI., abrepta est ad Christi oscula, March 25, 1700.

[Mary Pike, sweetest child, after two months and twenty-six days was snatched away to Christ's embraces, March 25, 1700.]

We have also in this church :—

Here lyeth ye Body of Anne, Eldest Daughter of William and Anne Maynard, late of Cranbrook, in ye county of Kent, gent, who departed this life ye 23rd day of August, Anno dom. 1705, aged 19 years.

Here also lyeth the body of Ann, only child of Isaac and Elizabeth Hawkins, and granddaughter to William and Ann Maynard above named, who departed this life the 28th day of May, Anno Dom. 1721, in the fifth year of her age.

Ann, mother of the above, who departed this life Jany. 1755, aged 92 years.

Abrahamus Mason, chirurgeus, et civis Londinensis, qui ætate matura animam suam Christo Redemptori resignavit, anno domini 1712.

[Abraham Mason, surgeon, and citizen of London, who in mature age resigned his soul to Christ our Redeemer, in the year of our Lord 1712.]

In Leybourne there is :—

Here lyeth enterred the Body of James Walter, the son of James Walter, who departed this life the 4th of October, 1697, aged 29 years.

Also here lyeth enterred the body of Elizabeth Walter, late wife of James Walter, who departed this life the 12th day of November, 1699, aged 61 years.

Also these monuments, now gone. [On an achievement *gules*, on a chevron *argent*, three roses of a field impaling ; impaling *argent* on a pile *azure*, three lions' heads erased *or*; the crest, a pelican in her nest feeding her young ones, *proper*] :—

" Blessed are the dead which die in the Lord."—Rev. xiv. 13.

" The dust shall return to the earth as it was, and the spirit shall return to God who gave it."

Here lyeth interred the body of Henry Ullock, Doctor of Divinity, Dean of the Cathedral Church of Rochester, and rector of the parish of Leybourne, who died the 20ᵗʰ of June, in the year of our Lord 1706, and of his age 67.

"The sting of death is sin, and the *strength of* sin is the law."—
1 Cor. xv. 56, 57, 58.

The body of Meric Head, Esq., eldest son of Sir Richard Head, baronet, doctor in divinity, rector of Leybourne and Ulcombe, lyes here interred. He marryed Elizabeth, daughter of the learned and truly pious doctor Robert Dixon, prebendary of Rochester, and had issue by her Eliza and Sarah, twins. Sarah lyes buried beside him.

This learned man died March 6, 1686, aged 42, was lamented by all, especially the poor, whom he cherished and defended.

These two monuments appear to have been lost or hidden when Leybourne church was restored.

In Halling church we have :—

Here lyeth the body of William Dibley, of this parish, ... this life April 10, 1694, aged about . . . . . . leaving 3 sons and 3 daughters.

Here lyeth the body of John Dibley, son of William Dibley, of this parish, gent, and Frances his wife, who departed this life Oct. 22, 1720, aged 39 years.

Also :—

Here lyeth the Body of James Taylor, son of John Taylor, gent, and Frances his wife, who departed this life [November ye 7, 1727].

The date in brackets is from the burial register, the date being hidden under the altar steps.

In Offham :—

[Here lyes buryed] Mrs. Rebecca Omer, daughter [of John de] Critz, of London, Esq., who [was wife and] widow of Laurence Omer, [of Staple, in Ke]nt, Gent, by whom she [had two chil]dren, whereof one lyes [buried here with] her, the other surviveth. [She dyed December] the 16ᵗʰ, anno dom. 1663.

[Here lyes al]soe with her [Frances de Critz,] sister to the [said Rebecca Omer.]

[Here's wife] and mother now at rest,
[With husband] once and children blest.
[He slept, and she] did sigh and weep
[Soe sore for him] she fell asleep.
[Daughter and sister], too, in bed
[Here, under] this cold coverlied.
[But sons she's left], since she's gone hence,
[The care of friends] and Providence.
[Reader, I could] tell a story
[Of her grace, and of] her glory ;
[But all's husht] till His powerful charme,
[Whose trump] shall sound the world's alarme,
[When goe thy wayes] and hence prepare
[To meet with] the Lord [i' th' ayre].

Thessal. yc 4th, 16 *and* 17 *vers.*

Also :—

Here lieth the Body of Frances De Critz, daughter of John De Critz, Esq., and Grace His wife, of St. Martin's in the Fielde. She ended this life the 23rd of February, 1660, aged 32 yeares.

Also :—

Here lieth the body of Miss Mary Omer, the daughter of Laurence Omer, gent, and Rebecca his wife, she ended this life the 28th day of June, 1659, aged a year and 22 days.

In Ryarsh :—

Here lieth the body of Margaret, wife of Richard Coosens, gent, of Parrock, in the parish of Milton, near Gravesend, daughter of Edward and Joan Walsingham, gent, of this parish, who died November 11, 1713, aged 38 years.

Here lieth the body of Jane Walsingham, daughter of Edward Walsingham, gent, and Felix his wife, died July the 17th, 1708, in the 25th year of her age. [Arms, a lion rampant, impaling between three cinquefoils].

Here lieth the body of Mr. Edward Walsingham, gent, of this parish, who departed this life the 6th of March, 1713, aged 69 years.

Also :—

Here lieth the body of Mary, sole daughter of Will'm Addison and Mary his wife, who departed this life the 20th of June, 1711, aged 4 years.

In Snodland we have two monuments of this date. The first is :—

Here lyeth the body of Isaac Tilghman, son of Whetenhall Tilghman, of Snodland, gent, who dyed the 21st day of December, 1647, aged 36 years, and Lisbona his wife, who dyed the 10th day of September, 1678, aged 58 years, and of their two daughters, Elizabeth and Eleanor.

We have dwelt upon this ancient family in a previous chapter. The other monument is to a former Rector of Snodland.

Here lyeth the body of Mr. John Walwyn, rector of this parish 31 years, who departed this life the 8th day of January, 1712, aged 59 years, and Mary his wife, who died the 15th of September, 1712, aged 55 years.

We must include in this chapter the two following extracts from Snodland parish register, with some explanatory remarks :—

Memorandum that, whereas there has formerly been a difference between the Parishioners of Snodland and those of Berling concerning Groves House, it was unanimously agreed upon by both parties, having putte the business to arbitration, that the said house should hereafter be ever accompted and acknowledged to stand entirely in the Parish of Snodland, in consideration of four pounds and ten shillings to be given by the ministers and parishioners thereof to the inhabitants of Berling, and that the said sum was actually given and received accordingly ye 21st day of March, 1698, in testimony whereof they have hereto set their names :—

THEO. BECK, Vicar de Birling.
The mark of JOHN CARRALL.
The mark of WILLIAM FLOOD.   JOHN WALWYN, Rector of Snod'and.
WILLIAM PAIN.
RICHARD KNOWLTON.
The mark of JOHN KNOWLES.
    Witness of sealing hereof : EDWARD WALSINGHAM.

The other memorandum which we shall give is interesting, as it introduces us to the family who probably were the last worshippers at Paddlesworth, and who, in this generation, having come to Snodland to worship, obtained a pew in the following way :—

Memorandum that I. John Walwyn, Rector of the Parish of Snodland. for ye better ornament and Beauty of God's house, did. in ye year of our Lord 1706, permit George Wray, of Paddlesworth, to erect a wainscote pew in the Chancell for his own and his family's use during mine own times, and 'tis presumed and to be hoped that my successors will have such a true sense of honour and ingenuity in them as to suffer Him and His, quietly and peaceably to enjoy the same ever after, by reason he built it at his own proper cost and charge, and since it was also upon soe laudable an account that I permitted him to doe it. In Testimony whereof I have hereto set my hand this 18th day of December, 1709 : JOHN WALWYN.

And I have the stronger presumption that they will grant this favour to ye said George Wray and his family, in consideration he has voluntarily set his hand to the following certificate, and is ready to justify ye same upon oath whenever occasion shall serve.

These are to certify whom it may concern, that ye bounds of Snodland Parish, both in mine own time and my father's before me. did extend to a White Thorn Tree in West Beacon Field upon the Hill, by which there formerly stood a stock style. so that John Green was the first man that pretended to remove them from thence to a certain oak. and did actually mark upon it, tho' it is about 5 or 6 rods distant from the White Thorn Tree aforesaid. and I told him at the same time of the wrong he did to Snodland. Nor can it be supposed that I have any sinister or ill design in makeing the affidavit. since I oblige myself by it to pay the tythes of about one or two acres, more or less, to the Rector of Snodland, which should be wholly exempted in case the bounds went no farther than ye said oak. for I pay a certain rate to Birling but Tithes in kind to Snodland, in testimony whereof I have hereto set my hand this 12th day of October, 1712 : GEORGE WRAY. Witness MARY WALWYN, CATHERINE WALWYN. The mark of ┬ JAMES WIBOURN.

We have the following entries in the burial register of Snodland :—

Jany. 18, 1676. Hammon [W]ray, of Paddlesworth, was buried.
March ye 6, 1698. David, ye son of George [W]ray, of Paddlesworth, was buried in woollen.
May 24, 1698. John Wray, of Paddlesworth, was buried in woollen.
July 12th, 1709. Thomas, ye son of George Wray, of Paddlesworth, was buried in woollen.
April 17, 1712. George, ye son of George Wray, of Paddlesworth, was buried in woollen.
January 15th, 1722. Elizabeth Wray buried.
June 9th, 1729. Elizabeth, the wife of Nicholas Wray, of Paddlesworth, was buried.
March 17, 1735. Buried Mrs. Jane Wray ; received an affidavit for being buried in woollen.

After this date we lose sight of the Wrays of Paddlesworth.

In the year 1695 we learn that the patrons of the various livings, and the value, were entered as follows in the register of the diocese of Rochester :—

			£	s.	d.
Addington ...	...	W. Watton, Esq.	6	6	0
Allington ...	...	Lord Aylesford ...	6	16	0
Aylesford ...	...	Chapter of Rochester	10	0	0
Birling ...	...	Lord Abergavenny	6	9	4½
Ditton ...	...	Royal	11	15	0
Halling ...	...	Chapter of Rochester	7	13	9
Leybourne...	...	Sir F. Clark	17	13	4
West Malling	...	Fitzjames	10	0	0
Offham ...	...	Royal	6	0	0
Paddlesworth	...	Sir J. Watton ...		Nil	
Ryarsh ...	...	W. Watton, Esq.	6	6	0
Snodland ...	...	Bishop of Rochester	20	0	0
Trottescliffe	...	Bishop of Rochester	10	2	11

East Malling is not given, not being in the diocese, having always been a peculiar of the archbishop, and counted to the Rural Deanery of Shoreham.

CHAPTER IX.

THE period we now enter being free from the many internal disturbances of previous centuries, we have less to record than in former times.

By the death of Edmund Watton, without sons, the manor of Addington was taken into another family by his only surviving daughter, just as it had come to them ten generations before by an only daughter. Elizabeth Watton married first, Leonard, son of Leonard Bartholomew, Esq., and secondly, Sir Roger Twisden, of Bradbourne, in East Malling. She died in 1775 and was buried at East Malling, her son Leonard, by her first husband, succeeding to the estate. He died in 1810, leaving an only daughter, and thus the manor of Addington soon passed from the name of Bartholomew. On a monument in Addington church we read this history. It runs :—

> Sacred to the memory of Leonard Bartholomew, Esq., second son of Leonard Bartholomew, Esq., of Rochester, and Elizabeth, sister and sole heiress of Sir Borlace Miller, Bart., of Oxenhoath, in this county. He married Elizabeth, daughter and heiress of Edmund Watton, Esq., of this place, and by her he had three children, Elizabeth, born 1727, died an infant; Leonard, born 1728; Edmund, born 1729, died 1743. He died A.D. 1730, leaving his wife surviving, who in 1737 married Roger Twisden, Esq., brother and successor to Sir Thomas Twisden, Bart., of Bradbourne, in this county. She died in 1775, and was buried at East Malling.
>
> Also of Leonard Bartholomew, Esq., son and heir of the said Leonard, who married Frances, daughter and co-heiress of Isaac Wildash, Esq., of Chatham, widow of George Thornton, Esq., of Malling Abbey; he died in October, A.D. 1810, leaving a sole daughter, Frances, born A.D. 1775. Also of Frances, the wife of the said Leonard Bartholomew, who died in October, A.D. 1801.

On her marriage to the Hon. John Wingfield Stratford, in 1797, she carried the ownership of the manor and the patronage of the church to him. She died in 1827.

At the opening of this period Allington was in the possession of Sir Robert Marsham, fifth baronet, who was created Lord Romney by George I., in 1716, and died in 1724; when he was succeeded by his only surviving son, Robert, at the early age of twelve years.

Robert died Nov. 1794, when he was succeeded by his second son Charles, who was created, June 15th, 1801, Earl of Romney and Viscount Marsham.

We find that James Drayton, a well-known botanist of Maidstone, was buried at Allington; he lived from 1681 to 1749. In the register we read : "Mr. James Drayton, of Maidstone, buried September 11th, 1749." His monument is gone.

The memorials in Allington church belonging to this date are mostly in the porch :—

In memory of Sir Edward Austen, of Boxley Abbey, Baronet, who departed this life December the 16th, 1760, aged 55 years. He was descended from Sir Robert Austen, formerly of Hall Place in the parish of Bexley, in this county, who was created a baronet by King Charles the Second, in the twelfth year of his reign. Also Lady Austen, relict of the above Sir Edward Austen, Baronet, died 20 of September, 1772, aged 57 years.

Also :—

Here lyeth the body of Sarah Titheron, who died April ye 19th, 1722, aged 7 months ; also Thomas Titheron, aged three months. Thomas Titheron died July 21, 1732, aged 40 years, left issue one daughter.

Also :—

William Thatcher, Gentleman, another son of John Thatcher, by Ruth his wife, departed this life April xi., 1722, aged 85 years.

Here lyeth the Remains of Robert Thatcher, of London, Distiller, the younger son of Charles Thatcher, of London, Gent, who died 6th July, 1754, aged 50 years.

Here lies the body of Susannah Thatcher, Gent, who departed this life the 26th of June, in the 30th year of her age, 1709.

Also :—

John Russell, 1778, and
In memory of Elizabeth, widow of John Russell, who departed this life 1762, aged 56 years. Likewise on the South side of this stone lieth interred, John Russell, who departed this life . . . . . 178. . , aged 83 years.

Also :—

In memory of John Baldock, who died 1799, 7 Oct., aged 75 years. Here is interred 3 sons of the Body of Edward and Jane Baldock. Two Edwards and Grundy.

Also :—

Here lieth buryed the son of John and Ruth Thatcher, who departed this life 21 Day of May, 1749, aged 3 years.

Sir Thomas Colepeper, was buried on May 24th, in the year 1723, and the estates passed to his sister, Lady Alice ; and thus the family of Colepeper, which had been at Preston Hall for four hundred years, became extinct at her death ; for though she was married four times, she had no children who survived her. She was first married to Sir Thomas Colepeper, as appears by the Aylesford register :—

> 1663. Sir Thomas Colepeper, of Hollingbourne, Knight, and Mrs. Alice
> Colepeper of Aylesford, daughter of Sir William Colepeper, late of
> Aylesford, deceased, were married by virtue of a licence out of the
> prerogative court the 31st day of December.

Her only child mentioned in the register is by this husband; we read :—

> 1665. Frances, the daughter of Sir Thomas Colepeper, Knight, and of Alice
> his wife, was baptised the one-and-twentieth day of February.

She was afterwards married to Herbert Stapely, Esq., M.P. for Seaford; but this marriage is not to be found in the registers. Her other two marriages are in the Ditton register, and are entered thus :—

> 1692. October 6th, Thomas Taylor, of Maidstone, Baronet, and Madame Alicia
> Stapely, of Aylesford, were married.
> 1723. Oct. 16th, Dr. Millner, of Maidstone, and the Lady Taylor, of Aylesford,
> were married. Lady Taylor, the last of the Colepepers, had no
> children.

Lady Alice gave an alms-dish 11¾ inches in diameter to the church; it is inscribed, " The gift of the Lady Taylor to the Parish of Aylesford. T. Tilson, Vic., 1724-5." Her death took place in 1734, and we find it twice entered in the Aylesford register of burials :—

> 1734. The Lady Alice Taylor was buried.

As the family records of the Colepepers in the registers of our parishes are very few besides those we have already mentioned, we subjoin them. The oldest of them is in East Malling :

> 1579. Buried was George Culpeper, gent, the 15th of September.

The rest are from Aylesford :—

> Mr. John Beale, of Maidstone, and Mrs. Ann Colepepy', of Aylesford, were
> married the 8th day of May, in the presence of Sir Richard Colepepyr, of
> Maidstone, Baronet, and of Thomas Crispe, of Dover, gent, by George Duke, one
> of the Justices of the peace for the county, 24 November, 1656. Witness his
> hand : GEORGE DUKE.*
> > 1659. William, the son of Sir Francis Colepeper, Baronet, and Dame Margaret
> > his wife, was buried the eighth day of March.
> > 1660. Sir Richard Colepeper, baronet, was buried the 10th of January.
> > 1661. Helene, ye daughter of Rt. Honourable Sir Richard Colepepyr, Baronet,
> > deceased, and of Dame Margaret, his widow, was buried the vi^th
> > day of December.
> > 1667. John Alchurn, junior, of Boughton Monchelsea, Esquire, and Mrs.
> > Francis Colepepper, daughter of Sir William Colepepper, late of
> > Aylesford, baronet, deceased, by virtue of a licence out of ye court
> > of faculties, were married the five-and-twentieth day of April.

---

* This kind of marriage has been already commented upon.

1667. Mrs. Helen Colepeper, daughter of Sir William Colepeper, late of this parish, baronet, deceased, was buried the xxii[nd] day of October.

1677. Dame Helene Colepeper was buried, not in woollen, Oct. 19.

1691. The Lady Margaret Colepeper, widow of Sir Richard Colepeper, Baronet, was buried September 24th.

1708. Dame Elizabeth, wife of Sir Thomas Colepepper, Baronet, was buried February, 5th day.

1714. Thomas Colepepper Joslyn, of Maidstone, was buried here at Aylesford, Oct. 16.*

1723. Sir Thomas Colepepper was buried May 24.

Dr. John Milner, on his marriage with the Lady Alice, owing to the death of her daughter by her first husband, and of her son by her third husband, inherited the Preston and Aylesford manors. He devised them to his brother Charles, who willed them to the Rev. Joseph Boteler, upon his taking the name of Milner, in 1784. He died childless, and these estates came to Charles Cottam, on his taking the name of Milner, in 1788, and in his family they continued about sixty years.

The priory of Aylesford and other estates came to Heneage Finch, created Earl of Aylesford in 1714. He died in 1719, and left the property to his son Heneage, who died in 1757. His son Heneage, who was M.P. for Maidstone 1739 and 1741, died in 1770. The fourth earl, who was Lord Steward of the Household, died in 1812, and was succeeded by his son Heneage, the fifth earl.

In 1769 Mr. Golding, of Ryarsh, died, and left his property of Tottington to Mrs. Frances Golding.

The manor of Birling descended through this period in the Nevill family. On the death of George, Lord Abergavenny, in 1694, the estates and title reverted to his grandfather's brother's family, in the great grandson of Christopher Nevill, who died in 1649. George, the eleventh earl, succeeded. He died in 1720, and on his decease in 1723 he was succeeded by his brother Edward, who also died without any sons in 1724, and the title went, with the lands, to William, Lord Abergavenny, who died in 1744. He was succeeded by his son George, who was created the first Earl of Abergavenny and Viscount Nevill by George III. in 1784. He died 1785, and was succeeded by his son Henry, Recorder of Harwich.

Sir Philip Boteler died in 1772, and by his will half his property went to Mrs. Elizabeth Bouverie, of Chart Sutton, and the other moiety went to Elizabeth, Dowager Viscountess of Folkestone, and William Bouverie, Earl of Radnor. On the estates being divided the manors of Ditton, Brampton, and Syfleton were allowed to the Dowager Countess of Folkestone, who died in 1782, and was succeeded by her only son, the Hon. Philip Bouverie, who took the name of Pusey. Philip Pusey sold the manor, in 1832, to Charles Milner, Esq., of Preston Hall. Ditton Place was, on the death of

---

* I add this entry because I think Colepepper and Aylesford point to his being an offshoot of the family.

Mr. Thomas Golding, of Ryarsh, in 1769, left by him to his son, John Golding.

The Grange at Leybourne, together with the Castle and Manor, were sold in September 1724 to Francis Whitworth, Esq., by Captain William Saxby,* who was succeeded in this estate by his son Charles, afterwards Sir Charles Whitworth, knight, who again sold the places to Dr. James Hawley,† in whose family they have continued ever since. Dr. James Hawley died December 1777, and was succeeded by his son Henry, who was created a baronet in 1795 and died on January 26th, 1826. He was succeeded by his son Henry, who died in 1834.

The manors of East Malling and Bradbourne descended from Sir Roger to Sir Thomas Twisden, who built over the ancient court in the centre of the house at Bradbourne, and perhaps filled in the moat which appears to have once surrounded the house, and which was fed by the stream which, after turning the paper-mills of East Malling, passes through the pretty grounds of Clare House, and then runs to Bradbourne, from whence it finds its way across the London and Maidstone road near Ditton to the Medway between Aylesford and Snodland. Sir Thomas died in 1728, and was succeeded in the title and estate by his eldest son, Sir Thomas, who, however, died without children. His brother, Sir Roger Twisden, succeeded, who was M.P. for the county of Kent in the fifth and sixth parliaments of George II.; he married, as has been already stated, Elizabeth, daughter and heiress of Edmund Watton, of Addington, Esq., and widow of Leonard Bartholomew, Esq.. Their son, Sir Roger, succeeded his father at his death in 1772, and on his death, seven years after, the title and estates passed to Sir John Papillon Twisden, who, dying in 1810, was succeeded by his son John. The monuments of the Twisden family in East Malling church give the family tree very correctly :—

Sir Thomas Twisden, Bart., died Sept. 12, 1728, aged 58 years. Dame Anne, wife of Thomas Twisden, and daughter of John Musters, of Colwick Hall, in Nottinghamshire, died Oct. 19, 1729, aged 48 years ; they had issue, Thomas, Roger, William and John. Sir Thomas Twisden, Bart., eldest son of Sir Thomas Twisden, Bart.,' died at Grenada, in Spain. July 30th, 1737, aged 31 years. Sir Roger Twisden, Bart., second son of Sir Thomas Twisden, died March 7th, 1772, aged 66 years. Dame Elizabeth, wife of Sir Roger Twisden, died the 4th of October, 1779, aged 41 years. He married Rebecca, daughter of Isaac Wildash, Esq., of Chatham, by whom he hath left one daughter, Rebecca. Dame Rebecca Twisden, relict of Sir Roger Twisden, Bart., died at Jennings, in the parish of Hunton, Feby. 3rd, 1833, aged 74 years, and lies buried in the family vault in this Church, leaving her only daughter, Rebecca, surviving, married to Thomas Law Hodges, Esquire,‡ of Hemsted, in this county.

There is a separate tablet to Sir John Papillon Twisden, which runs :—

Sacred to the memory of Sir John Papillon Twisden, Bart., who departed

---

* See Registers, p. 231.   † See Hawley Registers, p. 226.
‡ He was M.P. for West Kent in 1848.

this life the eighth of February, 1810, in the 68th year of his age. He married Elizabeth, youngest daughter of Sir Francis Geary, by whom he has left issue one son. This monument is erected by his most affectionate and disconsolate widow.

> " O shade revered, O thou lamented dust,
> Say, what in thee shall we regret the most?
> Or thee by what dear title most commend—
> The husband, father, brother, or the friend?
> For in these various characters he shone ;
> None most—for all these virtues were his own.
> Oh, thou on earth his justly fondest care,
> Who with him Hymen's softer bands didst share,—
> Whose torch, when many years their course had run,
> Than at the first with purer lustre shone,
> Nor think that tho' on earth's dissolved the tie
> That so sublime a flame can ever die.   ·
> No ; that to more exalted orbs shall soar,
> And still exist when time shall be no more.
> Then O suppress thy grief, the rising sighs,
> To yon bright heaven lift up thy streaming eyes—
> There view thy happy consort on the plains
> Where love and harmony eternal reigns,
> There with the shade of some loved friend he roves
> Through everlasting sweets and ever-blooming groves."

Also :—

Dame Elizabeth, relict of the above Sir John Papillon Twisden, Bart., departed this life December 19th, 1815, aged 62 years.

Also :—

S.M. Dame Catherine Judith, wife of Sir John Twisden, Bart., of Bradbourne ; · she died in childbed April 13, 1819, aged 29 years.

A neighbouring monument speaks much in favour of this baronet, of whom there is a picture in the family collection at Bradbourne ; the monument is to one of the Papillon family, a well-known East Kent name :—

Near this place are deposited the remains of Philip Papillon, Esq., of this parish, who died April 15, 1762, aged 63 years. This monument was erected in gratefull remembrance of him by his godson, John Papillon Twisden, Esq.

The abbey of West Malling, as we have already stated, was sold to Edward Honeywood, who passed it to his son Isaac. Frazer Honeywood, the only son and heir, and a banker in London, on succeeding, pulled down the old house,—then occupied by a fellmonger named Seager, probably Richard Seager, sen., who is mentioned in the West Malling registers as buried June 2nd, 1737,—and with the materials, Thorpe informs us, built the present house. He died on February 8th, 1764, and was buried in a vault in Town Malling churchyard, where he had previously placed, on August 13th, 1757, the bodies of his wife, Jane, daughter of Abraham Atkins, and the body of his son Isaac, who died in 1756, aged 19 years ; he devised the abbey to Sir John Honeywood, Bart., of Elmsted, who died

1781, and was buried in that church, where is a white marble
monument and bust to his memory :—

Sacred to the memory of Sir John Honeywood, Baronet, of Elvington, in this
parish, who in times when hospitality and simplicity of manners were giving
way to fashion and refinement, maintained them pure and uncorrupted, and
was an eminent example of the virtues of private life. In religion of pure and
unaffected piety ; in morals of strict honesty and integrity ; in social life of
openness and freedom of conversation amongst all ranks of men. His tenants,
neighbours, and those who served him never experienced from him oppression,
but all the good offices of kindness and benevolence. Thus did many partake
of the influence of his example and the affluence of his fortune. Others may
have moved in a higher sphere, but no man ever contributed more to the
advantage, comfort, and happiness of the circle around him. By his first wife,
Annabella, daughter of William Goodenough, of Langford, in the county of
Berks, Esquire, he had issue William, Edward, Annabella, Christiana, Mary,
and Thomasine ; by his second wife, Dorothy, daughter of Sir Edward Filmer,
of East Sutton, in this county, baronet, he had issue Filmer, John, and Mary.
He died June 26th, 1781, aged 71.

Sir John Honeywood, to whom the abbey had descended, died on
April 7th, 1806, and was buried in West Malling churchyard. He
sold it in 1799 to George Talbot Hatley Foote, Esq., but the manor
now no longer went with the abbey. Whether, when the Stewarts
(who acquired Malling Place, as we shall presently show) exchanged
certain lands near the abbey for the tower, they acquired the manor,
or whether Mr. Douce obtained it afterwards, we have not been
able to discover ; but it is certain that Thomas Augustus Douce, Esq.,
possessed both the manor and the advowson of the living.* Mr.
Foote built the picturesque cascade which is so well known to all
visitors to Town Malling, and which bears his initials and the date,
1810. In the year 1822 the abbey came, by devise, from Mr. Foote
to R. Losack, Esq., who died in 1838.

There are few records of the Honeywoods in West Malling : they
are as follows :—

1757. August 13th, the bodies of the late wife and son of Frazier Honeywood,
Esqre, were deposited in his vault.

BURIALS.

1764. Feby. 8, Frazier Honeywood, Esq.
1764. October 25, William Honeywood, Esq.
1803. April 13, Mrs. Elizabeth Honeywood, aged 72.
1806. April 7, Sir John Honeywood, Bart., aged 49 years.

The old mansion of Malling Place was dwelt in by Roger Twisden,
Esq., second son of Sir Roger Twisden, and grandson of Sir Thomas
Twisden, the judge. He appears to have hired the house from
Commodore Stewart, afterwards Admiral of the White ; he died here
in 1728—unfortunately the register of West Malling in that year
is lost. His wife was buried, we learn from the register, in West
Malling church.

1721. Mrs. Jane Twisden, buried in linen.

* Since writing the above we have found out it was in 1809.

Admiral Stewart, it appears, continued the chapel wall of St. Leonard's to the road, and in digging foundations for this and other walls to enclose the premises near St. Leonard's Tower, he came upon bones in the garden, that showed the burying ground of the chapel to have been towards the north.

The Hon. Charles Stewart was buried at West Malling, we learn from the registers :—

> On Feby. 18, 1741. The Hon[ble.] Charles Stewart, Esq., Vice-Admiral of ye White, buried, and 50[s.] paid.

His son appears to have resided here after him, and was buried in Malling, as his register runs :—

> 1779. Feby. 25, Charles Stewart, Esq.

Another person who appears to have lived at Malling Place was Lady Forbes, daughter-in-law of the Earl of Granard ; her son, Admiral Forbes, died here in 1796.

The property appears to have been acquired in this century by Mr. Hubble, who handed it, with the manor, on to the Douces. We have several entries of the Hubble family in the registers :—

> 1713. May 1, Elizabeth, daughter of Benjamin Hubble and Jane his wife.
> 1714. Aug. 13, Baptised Benjamin, son of Benjamin Hubble, junior.
> 1735. Mar. 19, Mr. Benjamin Hubble, sen.
> 1742. Oct. 22, William, son of Benjamin Hubble and Anne his wife, baptised.
> 1744. April 26, Baptised Catharine, daughter of Benjamin Hubble, gent, and Anne his wife.
> 1746. April 29th, Elizabeth, widow of Benjamin Hubble, gent.
> 1746. July 16, Benjamin, son of Benjamin Hubble, and Anne his wife.
> 1748. May 23rd, Anne, daughter of Benjamin Hubble, gent, and Anne his wife, was privately baptised.
> 1749. Jany. 14, William, son of Benjamin Hubble, gent (Burial).
> 1749. Feby. 17, Richard, son of Benjamin Hubble, gent, and Anne his wife.
> 1753. Dec. 30, Savage, son of Mr. Benjamin and Anne Hubble.
> 1755. Jany. 30, Mary, daughter of Benjamin and Anne Hubble.
> 1757. Dec. 30, Hannah, daughter of Benjamin and Anne Hubble.
> 1777. Thomas Augustus Douce, of the Parish of St. Catharine Coleman, London, a Bachelor, and Margaret Hubble, of this parish, spinster, married in the church the thirtieth day of September, in the year one thousand seven hundred and seventy-seven.
> 1780. Nov. 18, Benjamin Hubble, Esq.   ⎫
> 1781. June 30, Mrs. Anne Hubble.   ⎬ Burials.
> 1808. Nov. 28, Mrs. Catharine Hubble, aged 64. ⎭

The Hubbles had two monuments in Malling church. The first, which has now disappeared, ran :—

> Near this place lies interred the body of Benjamin Hubble, of this parish, gent., who died the 15th of March, A.D. 1735, aged 75 years. Hannah, wife of the said Benjamin Hubble, died the 4th of October, A.D. 1734, aged 65 years. Benjamin Hubble, gent, son of the above-named Benjamin Hubble, died September 24th, 1728, aged 47 years. Also Eliza, his daughter, died July the 26th, A.D. 1713, aged 3 months.

Their other monument still exists :—

Sacred to the memory of Benjamin Hubble, Esq[re], of West Malling, who died 11th November, 1780, aged 66 years, and of Ann his wife, daughter of John Savage, Esq[re], of Boughton Monchelsea, in this county, who died 22nd June, 1781, aged 67 years, leaving the following children, who also lie buried in this chancel :—

Elizabeth,	born 1st and	died	20th May, 1741.
Benjamin	„ 16th July, 1746	„	15th May, 1747.
William	„ 4th Octr., 1742	„	7th Jany., 1748.
Anna Jane	„ 11th May	„	7th November, 1748.
Richard	„ 17th Feby., 1750	„	the same day.
Savage	„ 30 Dec[r], 1753	„	1st Jany., 1754.
Mary	„ 30th Jany., 1754 *	„	the same day.
Hannah	„ 2nd Dec[r], 1757	„	11th July, 1758.
Catharine	„ 5th April, 1744	„	21st Nov., 1808.
Margaret	„ 8th December, 1751	„	14th Oct., 1809.

It is here noticeable that Mr. Hubble only left two of these children at his death, one of whom, Margaret, married, as we have seen by the register, Thomas Augustus Douce, Esq., who built the modern mansion of St. Leonards, and formed its pretty artificial water.

There is a long inscription to the Douce family in the church :—

Sacred to the memory of Francis Douce, of Lamb's Conduit Street, and of Nether Wallop, in the county of Hants, who died the 3rd of April, 1799, aged 82 years, and of Ellen his wife, who died the 23rd of November, 1799, aged 83 years. Also Thomas Augustus Douce, who died the 30th August, 1802, aged 55 years. Also of Margaret his wife, daughter of Benjamin Hubble, Esq., who died the 14 of October, 1809, aged 57 years. Here also are deposited the remains of the following children of the above T. A. Douce and Margaret his wife : John Alfred Douce, born 8th Feby., 1794, died June 15th, 1795. Charles Benjamin Douce, shipwrecked in the North Sea, 24th December, 1811, aged 22 years. Jane Douce, wife of F. H. Douce, Esq., died 24th December, 1809, aged 32 years. Emma Douce, infant daughter of the above, died the same day. Francis Hubble Douce, Esq., died 2nd January, 1819, aged 40 years. Also Frances Catherine Douce, died 11th Aug., 1832, aged 46 years.

From the Douces the house at St. Leonards and the manor passed, by purchase, to the late Captain Savage, R.N., whose daughters now hold them.

The manor of Offham continued in the hands of the Tuftons, who, however, lost the earldom of Thanet : to this earldom and the other properties Sackville, nephew of the sixth earl, succeeded in 1729. He died in 1753, and was succeeded by his son Sackville, the eighth earl, who, dying in 1786, was succeeded by Sackville, the ninth earl, his son.

The house on the Green, opposite the Quintain, of which Mr. Tresse died in possession in 1737, takes with it the duty of keeping up that ancient relic, it would appear.

The manor of Goldwell about this time appears to have passed into the hands of the Streatfields of Chiddingstone.

Carews' Court, on the death of Edward Walsingham, passed into

* This at first sight seems curious, but the peculiarity arises from the old commencement of the year on Lady Day. Savage was born 30th December, 1753, and Mary, 30th January, 1755, according to present calendar.

the hands of Sir Edward Austen, of Boxley, baronet, in the year 1760, whose monument in Allington church we have already mentioned. He, dying soon after, devised it to John, son of Nicholas Amherst.

Holloway Court manor, in Snodland, was sold by the Manleys to John Conwy, surgeon, and he devised it to his son Robert, by whom it was sold to Pearce, who alienated it to the Mays, who had already obtained the manor of Veles.

Snodland, Halling, and Trottescliffe still continued in the hands of the see of Rochester, though the palaces of the bishops in this neglectful age were allowed to fall into decay. Trottescliffe was at this time rented by the Whitakers, whose entries in the registers date from early in this century. In 1743 T. Whitaker, Esq., was sheriff, and in 1748 his son also obtained that dignity.

1724, May 11. Then was baptised Thomas, son of Mr. Thomas Whitaker and Martha his wife.
1725, July 30. Then was baptised John, son of Mr. Thomas Whitaker and Martha his wife.
1727, May 4. Then was baptised Edward, son of Mr. Thomas Whitaker and Martha his wife.
1729, Jany. 8. Then was baptised Sarah, daughter of Mr. Thomas Whitaker and Martha his wife.
1729. There was buried in linnen Martha Whitaker, wife of Mr. Thomas Whitaker, affidavit was made by Richard Briggs before Sir Thomas Styles, Jany. 22.
1753, Oct. 27th. Thomas Whitaker, Esq., was buried.
1754. Mr. Thomas Whitaker, of Troterscliffe, bachelor, and Mrs. Anne Walsingham, of Birling, were married by Licence, Jany. 25. [Addington Register.]
1754. Anne, ye wife of Thomas Whitaker, Esq., was buried ye May 25th.
1761, May 1st. Thomas Whitaker, Esq., was buried (sheriff).
1768, April 11th. Mr. Edward Whitaker buried.
1770. John Lock of this parish, and Elizabeth Whitaker, by Licence, 19th February.
1771, April 30. Edward Whitaker, son of Mr. John Lock and Elizabeth his wife, was privately baptised, and admitted into the congregation May ye 21st.
1837, May 28th. Laura Gertrude, daughter of Charles Gustavus and Anne Whitaker.
1876, July 18th. Sophia, daughter of Charles Gustavus and Anne Whitaker.
1877, April 25th. Charles Gustavus, son of Charles Gustavus and Anne Whitaker.
1880, Feby. 7th. Frederick Thomas Lake, son of Charles Gustavus and Anne Whitaker.

There are many matters that must interest us considerably, which we find in the history of the different places in the valley, apart from its owners, at this period.

On Feb. 1st, 1721, William Chapman, one of a family of much importance in Town Malling at one time, who owned the Manor House, claimed the right of a faculty pew in West Malling, mentioning that his family had lived in the same house from time immemorial.

The Addington bell frame is dated 1742. The second and third bells of Birling are dated 1746: on the second is, "Mr. Armigill Whiting, T. Lester made me"; and on the third, "T. Lester made me"; the first bell has on it, "Lester and Pack, of London, fecit 1759." The second bell of Ditton has on it, "Edward Middleton, C. W., 1717." On the first bell of Ryarsh, "Mr. James Thurston, Minister, Jeremiah Heaver, Churchwarden, 1779: Pack and Chapman, of London, Fecerunt." As regards sacramental vessels of this date, the Aylesford flagon is dated 1712, the Ditton cup 1735, the Birling paten 1730, the East Malling flagon and alms-dish 1728, and the Addington paten, on foot, is dated 1728. The flagon given by the rector, Rev. John Boraston, has on it, "Gloriæ Dei Opt. Max., in usum Ecclesiæ Parochialis de Addington, Dat Dicat Dedicatque Johannes Boraston, A.M., Prædictæ Ecclesiæ, Rector, Anno Dom. 1721"—"To the Glory of the Very Good and Very Great God, for the use of the Parish of Addington, Rev. John Boraston A.M., rector of the aforenamed church, gives, grants, and dedicates, in the year of our Lord 1721."

The returns of the plate were made for Rochester deanery in 1741. These entries are very interesting, as showing that at this very sleepy period there was some life in the Church in this part of England. The inventory of the church goods of Ditton in 1759 is very curious, we therefore give it :—

Silver Paten. inscribed, "Tuum est Domine tibi reddo donum, Thomas Tilson, rector, 1735."—[Translated, " Thine is the gift, O Lord, I return to Thee, Thomas Tilson, rector, 1735 "].
Silver Chalice for Communion, the gift of Mary Brewer to the Parish of Ditton, in Kent, for ye use of ye church, Jany. 4th, 1689.
One small Paten of uncertain silver.
N.B.—No proper Carpet to cover ye Communion Table in time of Divine Service.
Pulpit Cushion of Crimson Velvet.
Linen Cloth, with words, "The Communion Table Cloth of Ditton, Will Seagar, Churchwarden, 1721." A napkin with same.
Large Surplice of Holland, no Hood, Cope, or Vestment.
Folio common prayer book, printed 1745.
Folio Bible, last translation, printed by Robert Barker, 1613.
Quarto Common prayer Book for ye Clerk, printed 1746.
Three small Bells. the smallest broke.
The second has date only round it 1656.
The third circumscribed, " Edw. Middleton, Ch. W., 1717."
Deal Chest, for Parish book and Registers.
Font for Baptism.
No Book of Homilies.

Many things must strike us in this inventory : the use of the word uncertain before silver, meaning doubtful ; the carpet to cover the Communion Table ; the careful marking of the linen ; the fact that the surplice is of holland instead of lawn or linen. The fact that the writer seems to be rather surprised to find there is no cope, which he classes with the hood and vestment as perfectly proper, tells us that this part of the dress of the priest was worn so long after the Reformation as the year before George III. came to the throne, and gives proofs of its legality—whether the vestment that he missed were a chasuble or not we cannot say, but the compiler of the inventory evidently thought it the duty of the parishioners to provide all canonicals, and not the surplice only, as seems to be the feeling nowadays. Lastly, we find that he expected that there would be a book of homilies, showing that those treatises had not been yet given up, and that all clergymen, even one hundred and thirty years ago, were not expected to preach.

In the early part of this century a family were living in Malling by the name of Brooke, who gave their name to a house which is still so called. This family had a very handsome property in Holy Rood Street, or Swan Street, which perhaps owed its change of name to the birds swimming on the stream, which here made its way, before the cascade was built, from the abbey grounds into the gardens of this house. The last of the race, Joseph Brooke, who died in 1792, and his widow, in 1796, left this property to John Kenward Shaw, Esq. This family entertained, among other guests, the famous lexicographer Dr. Johnson ; and it was under a yew hedge in the garden, of which a remnant still exists, that he composed the prayers and meditations that are mentioned by Boswell. There are a few records of their names in the registers :—

1738. Nov. 16, Mary wife of Francis Brooke, gent.
1748. Feby. 11, Francis, son of Francis Brooke, gent., and Hannah his wife.
1770. Aug. 25, Mrs. Hannah Brooke, wife of Francis Brooke, gent.
1772. Sept. 6, Mrs. Margaret Brooke.
1782. March 25, Francis Brooke, Esq.

The property of Bavent, in Halling, was sold by the Goldings to Robin Wood, Esq., from whom it passed to Mr. William Baker. In the year 1728 the statue of Hamo de Heth over the palace gate in Halling was blown down ; it was presented by Thorpe the antiquary to Bishop Atterbury, but what became of it is not known. In the year 1756 there was quite a plague in Malling, if we may judge from the extraordinary number of funerals. In the year 1760 we find that there was a remarkable trial and acquittal of Robert Fowler, barber, of Town Malling, before the Hon. Sir Michael Foster, knt., at Rochester Assizes, on Friday, March 21, for poisoning Elizabeth Skinner of the same place.

An Elegy on a Storm which happened in West Kent, on the 19th of August, 1763, by Wm. Perfect, surgeon and apothecary, of Town Malling, was published about this time : he advertised to cure insanity—the first poet, says Gough, that ever pretended to such an art. He also published *The Laurel Wreath*, a collection of poems in two vols., 1766—it has two local poems, upon Barham Court, Teston, and Yotes Court, Mereworth. This Dr. Perfect first established a lunatic asylum here, which has continued ever since : the house in which the author lives was the home of this gentleman.

In 1778 the centre of Malling church fell down, leaving the tower and chancel still standing. In 1779 a Brief was issued for West Malling church for £2200; amongst other places that contributed, we learn, was Cranbrook, that gave 1*s.* 6*d.* We must here observe that the ancient way of collecting money was by these Briefs, which were issued by the Ecclesiastical Court. They were, as we shall show in our chapter on the registers, for various charitable objects ; and our Prayer Book still directs that after the Nicene Creed notice shall be given of Holy Days and Fasting Days in the week, and then of the Holy Communion, and afterwards of all Briefs, Citations, and Banns. The rebuilding of only part of the church gave rise to the adage :—

> " Proud Town Malling, poor people,
> They built a church to their steeple,"

one of the many sayings of Kent.* In 1781 the church was re-opened.

In 1779 William Parry, Admiral of the Blue Squadron, built the pretty house of St. Vincent's. There is the following entry in the register :—

1779. William Parry, Esq., buried May 26.

There is also a monument to him in Addington church that runs thus :—

Near this place are deposited the remains of William Parry, Esqre·, Admiral of ye Blue Squadron of His Majesty's Fleet, obiit 29º April, 1779, Ætatis 74.

Besides the monuments noticed in their proper place, we have belonging to this period, at Addington, this inscription :—

Near this place are interred the bodies of John Petley, Esq., and Jane his wife. He died June the 28th, 1747, aged 75 years. She died September the 6th, 1766, aged 74 years. Also the body of Jane their daughter, who died February the 10th, 1762, aged 44 years. Erected Oct. 1766.

In Aylesford we find :—

Infra lapidem marmoreum, in hoc vestibulo, jacet Thomas Tilson, A.M., olim aulæ Catherinæ apud Cantabrigienses, socius necnon Rector de Ditton, et hujus ecclesiæ Vicarius per annos 47, ob^{t·} 12º Feb., 1749, ætat. 76.

Beneath this marble, in this porch, lies Thomas Tilson, M.A., formerly fellow of Catherine Hall, at Cambridge, moreover Rector of Ditton, and Vicar of this church for 47 years, he died on the 12th of February, 1749, aged 76.

Two Vicars, father and son, from A.D. 1666 to A.D. 1749—eighty-three years.

Rev. Thomas Tilson, died A.D. 1702, aged 61 years ; 36 years Vicar.
Rev. Thomas Tilson, died A.D. 1749, aged 76 years ; 47 years Vicar.

* See notes, p. 249.

Also :—

On the south side of the Church are deposited, in a vault, the remains of Mr. John Spong, who, after an exemplary and well-spent life at the little Hamlet of Mill Hall, in this parish, died Jany. 21st, 1815, aged 64 years, universally respected and esteemed. Also Rosamond his wife, who died on the same day of the same month 1840, aged 92 years. Orta Carolo Rege (*Sprung from King Charles*).

And :—

Elizabeth, wife of William Bowles, Esq., of Fitz-Harris House, daughter of John and Rosamond Spong, late of Mill Hall, in this parish, who departed this life 12th of January, 1814, aged 22.
Also Mary, their daughter, 9 Oct., 1813, aged 7 weeks and 3 days.

The Spongs of Aylesford were said to be the originals of Dickens' "Wardle family" in *Pickwick*. I think there can be very little doubt of this, as Dingley Dell and Muggleton are easily traceable in the vicinity; Dingley being simply an ingenious combination of the two famous Kentish cricketing villages, D[eptl]ing [and Box]ley—the parts in brackets left out, the rest forms the word. That Muggleton was Maidstone we have very little doubt, as several allusions to it are made in *Pickwick*, though the scene frequently shifts to other towns. For instance, the Mayor and Corporation are spoken of, and the "nail and saucepan business" appears to have reference to a worthy ironmonger of that town, who was Mayor at that period.

Another monument is :—

Mr. John Ward, died August 10, 1772, aged 67 years: Mr. Daniel Ward, died October the 4th, 1783, in the 76 year of his age.

Also :—

Near this place lyeth the body of Sir Robert Faunce, Knt., who had three wives, by whom he had 4 sons and 4 daughters. Sir Robert Faunce, Knight, was buried February 16, 1715.

At the time of the repewing of the church, certain of the stones in the south aisle were partly covered, so the vicar (the Rev. W. Tolbutt Staines) of that time informs us. They ran :—

For the wife of Augustine Taylor, Daughter of Edward James Baldock, who departed this life Nov. 19, 1753, aged 41 years. Likewise the same Augustine Taylor, who died the 2nd day of May, 1785, in the 71st year of his age.

On another stone, farther towards the east :—

Here lyeth the body of John Taylor, of this Parish, Yeoman, departed this life May 7, 1733, aged 49 years.
Here lieth the Body of Elizabeth Taylor his wife, died Aug. 5, 1741, aged 57. He left 5 sons, John, Augustine, Thomas, Robert. . . . .
In the year 1794 were buried near this monument Susannah Rebecca and John Evcleigh. Also, in 1795, John; in 1800, James; in 1803, Thomas and William George ; in 1805, Susannah ; and in the year 1815, when the vault was made, Charlotte and Georgiana, children of the Rev. William and Susannah Eveleigh. In the year 1830 the remains of the Rev. William Eveleigh, LL.B., were deposited in the same vault. He departed this life the 29th of October,

in the 74th year of his age, having been 38 years vicar of this parish. Also, in 1834, were deposited the remains of Susannah Eveleigh. She departed this life 19th Feb., in the 69th year of her age.

In the church of Birling we find a monument :—

Sacred to the memory of John and William May, sons of John and Jane May, of this parish. William May, ob. 25th Aug. 1777, æt. 41 ; John May, ob. 2nd Sept. 1803, æt. 71.

A little distance off is a stone which is curious—we read :—

Underneath this stone are deposited the remains of John and Jane May, also their sons, John and William May, to whose memory the monument is erected in this Church.

There is none except this stone.

To the memory of the Rev⁴· Edward Holme, late Vicar of this Parish, and founder of the two free schools at Leybourne and East Malling in this county, who departed this life on the 7th day of January, 1782, aged 71 years.
Nanny Holme, daughter of the above Rev⁴· Edw⁴· Holme ; she died the first of January, 1789, aged 21 years.
Susannah, widow of the said Rev⁴· Edward Holme, and mother of the said N. Holme, died May 17, 1801, aged 65.

On the East Malling School is the inscription :—

Liberos tam Literis bonis quam pietate expolere.
Ἀλλ ὄυν ὁρωσιν οἱ μαθοντες τὰ γραμματα μελετη τὸ λᾶν.

This house and school, and that at Leybourne, were both erected at the sole expense of the Reverend Edward Holme, Vicar of Birling.

On the Leybourne school we read :—

Γνῶθι σεᾶυτον alias Scire tuum nihil est.

The words we suppose mean, "Know Thyself, otherwise knowledge is nought of thine." The expression must be declared obscure, at the least.

In Halling church we have :—

Here lieth the body of Robert Wood, late of this parish, who departed this life 14th of July, 1738, aged 54 years.

Also :—

Here lyeth the Body of George Small, son of Richard and Elizabeth Small, of this Parish, Yeoman, who died January the 21st, 17³⁸⁄₉, aged 17 years.

Also : —

Here also lieth Frances, the daughter of Mr. Robert Wood, late the wife of Thomas Tryland, who died the 3rd of February, 1742, aged 31 years.
In Memory of William, son of William and Anne Baker, of this parish, died April 9, 1775, aged 7 years.

And :—

Near this place lieth the remains of Frances Comfort, who died the 2nd of October, 1824, aged 64 years.

In Leybourne we find :—

In a vault underneath are deposited the remains of James Hawley, Doctor of Physick, who died at the Grange in this Parish on the 22nd day of December, 1777, in the 73rd year of his age. And also of Elizabeth, wife of the said James, who was one of the daughters of Joseph Banks, Esq. She died the 27th day of November, 1766, in the 47th year of her age, and was buried at Isleworth in the county of Middlesex, but was afterwards removed to this vault. And also of Dorothy, the wife of Henry, the son of the said James and Elizabeth ; she was the daughter of John Ashwood, Esq., of Madely, in the county of Salop, and died in childbirth on the 4th day of December, 1783. To the memory of these excellent persons the said Henry hath erected this monument, by whose death he is left to deplore the loss of parents most virtuous and indulgent, and of a wife most faithfull, pious, and affectionate.

Again :—

Sacred to the memory of Anna, eldest daughter of William Humphreys, Esq., of Llewyn, Montgomeryshire, and second wife of the first Sir Henry Hawley, Bart., of Leybourne Grange, Kent. She died November 7th, 1829, aged 72 years. " Blessed are the dead who die in the Lord." This tablet is erected by her sorrowing children.

Also :—

In a vault underneath the gallery are deposited the remains of Sir Henry Hawley, Bart., of the Grange, in this Parish. He died the 26th day of January, 1826, in the 81st year of his age. He was the only son of James Hawley, M.D., by Elizabeth Banks his wife. By his first marriage, with Dorothy Ashwood, he had one son and three daughters, viz. : Henry, the present baronet, who married Catherine Elizabeth Shaw, eldest daughter of Sir John Gregory Shaw, Bart., of Kenward, in this county ; Dorothy Elizabeth, who married Sir Brook William Bridges, of Goodnestone, in this county, Bart., and who died without issue ; Harriet ; and Charlotte, married first to the Rev. Brooke John Brydges, secondly to Thomas Gardner Bramston, Esq., of Skreene, in the county of Essex. Sir Henry married secondly, in September 1785, Anne, the daughter of William Humfreys, Esq., of Llewyn, in the county of Montgomeryshire, by whom he had issue : Frances Anne, married to Captain Shaw, R.N., second son of Sir John Gregory Shaw, Baronet, James, Eliza, and Louisa, married to the Rev. Tatton Brockman, third son of J. D. Brockman, Esq., of Beachborough, in this county. Sir Henry was created a Baronet of Great Britain in May 1795, H.M. P.C. July 17th, 1826.

Also :—

In the vault underneath are deposited the remains of Anne Bond, daughter of Mr. Charles Bond, of Rochester, who departed this life at the Grange in this parish the 30th of October, 1829, in the fortieth year of her age. This is erected as a tribute of respect to her memory by her affectionate children.

In East Malling church we have also several entries of this date, which we now give, besides those to the Twisdens :—

Here lyes ye Body of Mr. Roger Furner, second son of Thos. Furner, of this Parish, Gent. He died on Michaelmas Day 1724, aged 77 years.
Near this place lieth the [body of] Thomas Furner, [of this parish, gent], and Mary his wife ; Thomas, Elizabeth, and Mary [their children]. Under this stone lye the remains of Francis, another of their [children, and] the body of Hester his wife. She departed this life [4 of Dec. 1720] ætat. 61. To whose Me[mory] Roger, their sorrow[ful and only surviving issue], hath affixed this I[nscription.]

8

Two monuments are unfortunately so hidden and so defaced that we can only partly decipher them.

Here lie depo[sited]the remains of James Tomlin, [gent], of this Parish, [who] departed this life April the 29th . . . . aged 74 years. Also Mrs. Eliza[beth] Tomlin, W[idow of] the abovesaid, [who died] 9th of November, aged 76 years. Also [Mr.Thomas Tomlin], son of [Roger Tomlin, of East Malling, who departed] this life September the [thirteenth], 1728, in the twelfth year of his age.

The words in brackets have been supplied by the author. Besides this we have memorials to the same family, as follows :—

In Memory of James Tomlyn, Esq., who died June 30th, 1759, aged 71 years. Also the body of Charlotte Gregory, his granddaughter.
In memory of Mrs. Ann Tomlyn. wife of Mr. James Tomlyn, late of this parish, died 15th March, 1773, aged 82 years.

Outside the church we have a stone, placed in a conspicuous position near the entrance, on which is recorded the death of a real centenarian :—

Near this place lies the body of Mary Baker, of this Parish, widow, who was buried Dec. 14, 1753, at the advanced age of 105 years.

Thorpe also tells of monuments now gone :—

Here lyes the body of Mr. Richard Furner, of this parish, gent. He died on Michaelmas Day 1724, aged 77 years. Here lyeth the body of Mary, the wife of Mr. Roger Tomlyn, who died the 25th of September, 1687. in the 23rd year of her age.

Besides these we find also this one :—

Mrs. Ellen Humphrey, obiit March 13th, 1828, Æt. 83. Stephen William Godmond, her grandson, and youngest son of the Rev^d· Samuel Francis Godmond, and Anne his wife, obiit Oct. 3, 1828, Æt. 17. Samuel Hugh Godmond, their son, died May 7th, 1837, aged 33 years. The Rev. Isaac Singleton Godmond, their son, died June 10, 1839, aged 34 years.

At West Malling this one, though it has no year left, is probably of this period—the first part is erased. It appears to be the last one to the Chapmans :—

. . . .
Anno Ætatis Suæ
Here lyeth enterred ye Body of
MARY CHAPMAN, the wife of
JOHN CHAPMAN,
" She is not dead, but sleepeth."
Obiit December 23rd, Anno domini
Anno Ætatis Suæ 35.

Another one runs :—

H.S.E. Robert Say, gent, who departed this life August ye 17th, 1717. In ye 41st year of his age. This monument was erected by Mrs. Catherine Say, his widow. Near this place also lyeth his daughter, Catharine Say, aged six weeks. Here also lyeth the body of Mrs. Catherine Say (relict of ye abovesaid Robert Say), who departed this life the 27th of February, 1730, aged 41.

Another inscription, under a coat-of-arms of three tigers' heads, is:—

Underneath is interred Mary, only child of George and Mary Smith, late of Wanstead, in the county of Essex, gent, and granddaughter to John Weckley, gent, and Ann his wife, of this parish. She died December the 21st, 1718, aged 12 years. Her loving mother erected this monument. Also the body of Dr. Handisyed, who died September the 29th, 1748, aged 69 years.

On another tomb we find the record :—

H.S.E. John Weckley, of the Inner Temple, London, Esq., who died Nov. 21, 1738. To whose endearing memory his mother, Mrs. Matthew Weekley, dedicates this gravestone.

Another monument to this family, now destroyed, was :—

Here lieth the body of Mary Weckley, who died August the 9th, 1740, aged 25 years. Also the body of Margaret Weekley, who died August 11th, 1750, in the 37th year of her age.

Near the vestry door, almost worn away by treading, we also read :—

Rev^{d.} [CHARLES] BOWLES, Rector of Ditton
[June 3, 1786.]
Revd. CHARLES BOWLES
Jany. 21, 1809
Aged 79 years.

So far are we able to restore this monument. The younger Mr. Bowles married a Miss Weekley, hence we have introduced this monument here : they must have resided at the house near the church. The Weekley entries in the register are very few :—

1710. August 26, Ann, ye daughter of Mr. John Weckley, gent, and Margaret his wife, baptised.
1713. May 28, Margaret, the daughter of John Weekley, sen., and Margaret his wife.
1718. Dec. 27, Mrs. Mary Smith (buried).
1719. Nov. 22, John Weekley, senior, a very antient gent, he died the 15th and was buried the 22nd.
1721. Jany. 6th, Catherine, the daughter of Mr. John Weckley and Mrs. Margaret his wife (baptised).
1731. Ann, the widow of John Weckley, gent, Oct. 14, buried.
1737. Jany. 2, John Weekley (gent), buried.
1738. Jany. 3, Captn. Thomas Weekley, buried.
1738. Nov. 26, John Weekley, gent (buried).
1740. Aug. 15, Mrs. Mary Weckley (buried).
1743. May 8, Mattheia, widow of John Weckley, gent (buried).
1744. Sept. 2nd, Margaret, widow of John Weekley, gent (buried).
1748. Oct. 3rd, Mary, wife of Dr. Handisyde (buried).
1750. August 18th, Margaret Weckley, a single gentlewoman (buried).
1778. Jany. 4th, George Weekley, Esq. (buried).
Charles Bowles, of this Parish, and Catherine Weekley, of this parish, spinster, married in the Church by licence, this eleventh day of February, in the year one thousand seven hundred and sixty, by me, Daines, minister.

Another inscription is :—

Here lies the body of Peter Elliston, of this parish, gent, who departed this life December 10, 1729, in the 41st year of his age.
Here also lies Kender Elliston, his son, by Katherine his wife, who died an infant three days old, January 14, 17⅘.

To whose dear memory his widow, Catherine Elliston, erects this monument, Dec. 22, 1720.

His only surviving son is Peter Elliston, born four months and twelve days after his father's decease, which said Peter Elliston, gent, married Mary, only daughter of Philip Hedman, of Kingston-upon-Thames, Esq., and departed this life 10th of October, 1746, aged 26 years, leaving one son, Peter Hedman Elliston.

Likewise Mrs. Mary Elliston, grandmother (aged 94 years) ; John Elliston, gent, and Elizabeth his wife, father and mother ; and John, Francis, and Hannah, brothers and sister of the aforesaid Peter Elliston, deceased, lye here interred.

Also we read :—

Underneath lies the Body of Mrs. Ann Herne, who was the wife Mr. George Herne, of New Inn, London, gent, and one of the daughters of William and Mary Mason, late of this parish, who departed this life the 14th day of August, in the year of our Lord 1742, in the 38th year of her age.

Besides we have :—

Here lyeth ye body of John Ecle, once a writing-master of this place, who died Feby. 17th, 1766, aged 59 years. Also Mary, his beloved wife, who died December 30th, 1746, aged 46 years. Likewise Mary, his mother, who died 10th July, 1756, aged 73 years. Also Ann, his youngest daughter, who died 13th March, 1767, aged 26 years.

Still more we have the following, which is interesting as showing the connection between the Bartholomews of Oxonhoath and Addington :—

Near this place are deposited the remains of Humphrey Bartholomew, M.D., youngest son of Leonard Bartholomew, Esq., and Eliz[h.] his wife. late of Oxonhoath, in this county, who departed this life the 15th day of December, 1764. To whose memory his nephew, Leonard Bartholomew, of Addington, Esq., erected this monument.

Another person commemorated is Rev. Dale Lovett : his tablet says :—

Sacred to the memory of the Rev. Dale Lovett, of Pankhull, in Staffordshire, many years curate of this Parish, who departed this life 1st August, 1797, aged 75 years.

Mr. Dale Lovett seems to have thus died here a curate, never having won the bishop's smile, which would have given him the position that a life of hard work as a curate deserved, but did not obtain ; who thus, in common with others, proved indeed that of him the world was not worthy.

Besides those already mentioned, we have a tablet :—

To the memory of Robert Thomas Cromp, Esq., late of Frinsted, in this county. He died on the 12th of February, 1808, in the forty-second year of his age.

Also one to the last vicar but one :—

Sacred to the memory of the Rev. Richard Husband, A.M., who was Vicar of this Parish 44 years. He died the 25th of March, 1814, aged 78 years. Also of Sarah, his wife, who died the 11th of February, 1814, aged 83 years.

Mr. Husband, of whom we shall say more hereafter, cannot be too much praised for the trouble that he took with the registers.

One of the entries of this period in the parish register is very curious; it is :—

Jany. 22, 1784. A strange woman, name unknown, found dead in the cage.

The lock-up, or cage, formerly stood on the piece of ground that lies on the side of the hill at the entrance of the town, opposite Ryarsh lane—it has long been pulled down. It no doubt succeeded the old abbey prison, and has been succeeded by the police station. Two things strike us about the entry : one is, we are not told how the woman came to be in the cage, and the other is, that there is no mention of a coroner's inquest having been held to examine into the cause of her death.

The monuments of this period in Offham church are not very numerous; there are only two. The first is to a former incumbent :—

Here lies interred the body of the Rev. Mr. Wm. Miles, Rector of this Parish, and of Westgate Holy Cross, and of St. Peter's United, of Canterbury. Vita sua Optima Laus.* Departed this life Oct. 16, 1746, aged 39. Here also lieth ye body of Rose, wife of ye said Wm. Miles, who departed this life ye 9ᵗʰ of Decemʳ· 1762, aged 66 years.

The second is interesting, as it gives us a lasting memorial of coaching days, though by a sad accident; it is :—

Sacred to the memory of Elizabeth, wife of Robert Spearman. of Newington; in this county, Esqʳᵉ· She was the second daughter of Mr. John Smith, late of this parish, and met with an instant and untimely death on the 18th December, 1800, by the overturning of a stage coach near Woolwich ; age 19 years, leaving issue two children, John and Elizabeth.

We should not give a complete record of this period if we did not mention some of the uncouth verses on tombstones, of which Offham supplies us with one or two, viz. :—

> My sledge and hammer lie reclined,
> My bellows, too, have lost their wind ;
> My fires are burnt, my forge decayed,
> And in the dust my vice is laid.
> My coal is spent, my iron's gone;
> My nails are drove, my work is done.
> WILLIAM SMITH, of this parish, blacksmith.

Another, to Mr. John and Mrs. Mary Brooker, 1810 :—

> An honest pair beneath this spot of earth .
> Contented rest, unmoved by pain or mirth.
> No useless subtleties perplexed their brains,
> Nor fraudful habits to increase their gains.
> They lived in Peace, and died devoid of fear,
> And left a name which all their friends revere

Another :—

> Here rests from all the cares of life
> A husband faithful to his wife,
> A father to his children dear,
> A neighbour honest and sincere.

---

* His Life is his Best Praise.

Another of this style, though much later, is to Philip Morphew :—

> He had his share of the toil and strife,
> Of the fears and cares that compass our life.
> We may miss the clasp of his gentle hand,
> Which now strikes his harp in a better land;
> But we feel, in the midst of our deepest pain,
> That our own great loss is our loved one's gain,
> And perchance those graves are but landmarks given
> Directing our hearts and our hopes to heaven.

In addition we may mention two in Addington churchyard. One on Sarah Still, 1802 :—

> Tho' young she was,
> Her youth could not withstand
> Nor pardon her
> From death's impartial hand.

The other on George Still, of Ryarsh, who died 1809 :—

> To praise the Lord the good man went
> From home with pious zeal;
> But ne'er returned, his lamp was burned,
> Death followed at his heel.

We find the following inscriptions in Ryarsh church that belong to this period :—

Here also lyeth interred the body of Felix Walsingham, wife of Edward Walsingham, who departed this life the 17th day of September, 1717, aged 60 years.

It is worthy of notice that Mrs. Walsingham's * name is always recorded as Felix, and not Felicia.

Besides, to this family there were erected the following inscriptions :—

Here lyeth the body of Edward Walsingham, late of Callis Court, in this Parish, gent, who departed this life the 9th day of May, A.D. 1718, ætatis suæ 30 years.

Here lies interred the body of Mr. John Walsingham, late of Birling, who departed this life January the 10, 1764, aged 72 years, and left surviving two daughters, viz.: Jane and Mary.

Here lyeth the body of Rebekah, wife of Mr. John Walsingham, of this parish, and daughter of Abraham and Rebekah Standen, of the parish of [illegible], in the county of Sussex. She departed this life the 13th of July, 1726, aged . . . . years, and left issue four daughters . . . . Rebekah, Jane, and . . . .

Most of the Walsinghams' tombs, through being made of soft stone and placed on the ground, have been rendered illegible.

Besides them there is an inscription to Rebekah Standen :—

Here lyeth the body of Rebekah Standen, who died the 7th of September, 1737, aged 67 years.

And one or two to a family called Penury :—

Here lieth interred the body of Rebecca, wife of Solomon Penury, of this

---

* See notes, p. 234.

parish. She departed this life Jany. yc 9th, 1761, aged 39 years. She left issue one son and one daughter, viz., Solomon and Mary.

Here lieth interred Solomon Penury, of this parish, who departed this life the 10th of March, 1767, aged 53 years. He left issue one son and one daughter, Solomon and Mary.

And one to John Miller :—

Here lieth interred the remains of John Miller, gent, of this parish, who departed this life September the 13th, 1786, aged 43 years.

In Trottescliffe church we read :—

In this chancel lies interred John Morgan, B.D., rector of Medburn, Commissary of Richmond, and Precentor of St. David's, he died 20th September, 1773, aged 73 years. Mary Philips, widow, sister of the above. John Morgan is likewise interred here, who died 6th November, 1744, aged 70 years. In the same vault rest the remains of Francis Lloyd, M.A., late rector of this parish.

The old church of Ditton contains quite a history of the Brewer and Golding * families, nearly every part of the church being taken up with their memorials or those of their connections. The Brewers had become extinct before this era, but of the Goldings we have the following memorials :—

> Under th[is stone lie]
> The remains [of William]
> Son of John [Golding, Esquire,]
> And Ann, [his wife,]
> Of Ditton [Place],
> Who dep[arted this life]
> The 27th of [February,]
> Aged 14 years,
> 1786.

This inscription, partly buried under the reading desk, we have supplied as best we could from the burial register. The words which are buried we give in brackets as we conclude them to have existed.

Also :—

In a vault beneath this monument are deposited the remains of John Golding, Esqr·, late of Ditton Place, in this Parish, who died the 12th of November, 1807, aged 80 years. Also of Ann, his wife, who died the 1st of August, 1807, aged 61 years. They had issue the following children : Oliver, who died the 8th·of February, 1777, aged 4 months ; William died the 27th of February, 1786, aged 14 years ; Mary Ann, died the 24th of November, 1805, aged 23 years; Frances, died the 17th of December, 1809, aged 34 years. All of whom are interred in the vault below. Elizabeth, the wife of William Alexander Dunning, died the 22nd of June, 1814, aged 35 years, and is buried at Boxley in this county. Thomas, who died the 3rd of January, 1818, and is buried in the vault beneath. John, who died the 17th of February, 1856, at Ditton Place, aged 85 years.

To two of those on this monument there is this second tablet :—

Sacred to the memory of Mary Ann Golding, youngest daughter of John

---

* See notes, p. 238.

Golding, Esqre., of Ditton Place, Ditton, in the county of Kent, who died November the 24th, 1805, aged 24 years. Also to the memory of Frances Golding, eldest daughter of the said John Golding, who died December 17th, 1809, aged 34 years.

The first of these ladies is made one year older on her second monument.

There is a third slab just inside the belfry door—this we think is a unique and exceptional case—it runs :—

Underneath this stone are deposited the remains of Frances Golding, of Derby Street, Westminster, daughter of John Golding, Esqre, and Anne his wife, of Ditton Place, in this parish, who departed this life December the 17th, 1809, aged 34 years.

Also a third to Mary Ann Golding :—

Underneath this stone lie deposited the remains of Mary Ann Golding, daughter of John Golding, Esq., and Ann his wife, of Ditton Place, in this parish, who departed this life November the 24th, 1805, aged 23 years.

There is also a second to Thomas Golding :—

Sacred to the Memory of Thomas Golding, late of Ditton Place, son of John Golding, Esqre., and Ann his wife. He died on the third of February, 1818, aged 46 years, and lies interred in the vault below. This monument is erected, as well as the adjoining one, by his affectionate brother, John Golding, Esqre. of Ditton Place.

The monument mentioned as the adjoining one is the tablet of the family, of which we have given the inscription above.

There are also one or two to the family of Luck belonging to this period :—

In memory of Rebekah, wife of William Luck, of this parish, Yeo. She died June 25th, 1756, aged 72 years. Also to the memory of the abovesaid William Luck, who departed this life Sept. the 5, 1763, aged 68 years.

The other monument to the Luck family is, strange to say, also repeated, like those to the Goldings. The first reads :—

In Memory of George Luck, of this parish, yeoman, who died October the 4th, 1771, aged 44 years. He left issue, by Sarah his wife, four sons and one daughter, viz. : George, William, Thomas, John, and Sarah. Also Nathanael John Luck, son of the above, who died December the 26th, 1831, aged 62 years.

On the wall of the church we read virtually the same :—

In memory of Mr. George Luck, of this parish, who died October 4th, 1771, aged 44 years. Left issue, by Sarah his wife, four sons and one daughter, George, William, Thomas, Nathanael John, and Sarah. Also Mr. Nathanael John Luck, son of the above, who died December 26th, 1831, aged 62 years.

And this one to the family of Cox :—

Here's interred the body of Mrs. Mary Cox, wife of William Cox, gent, of the parish of Snodland, and daughter of Mr. Edward and Ann Sedgwick, of this parish, who departed this life September 24th, 1724, in ye 22nd year of her age. Here lieth the body of William Cox, gent, who died June ye 11th, 1737, aged 35 years.

In Snodland there is a memorial belonging to this period to a former rector; it runs :—

Here lyeth the body of Mr. John Walwyn, Rector of this Parish 31 years, who departed this life ye 8th day of [January, 1713].

The last part of the inscription is buried under the altar step, but is supplied by calculation from the dates in the Rochester registers. He is not entered in the Snodland register.

CHAPTER X.

THE records of this era in our valley are not so numerous as we might expect, still there are some things worthy of recording in these reigns. During this period the cricket ground of Malling, which had been rising in importance, became one of the favourite grounds of the county cricketers, and continued so for a long period, the natural beauty of the place, and the good cricket of many of the inhabitants, rendering it long a popular place of resort for lovers of the most manly of sports. We cannot do better than follow the history of the different places in alphabetical order.

By the marriage of Miss Bartholomew with the Hon. John Wingfield-Stratford, second son of Richard, Viscount Powerscourt, Addington Manor passed into his family.

His monument in the church commemorates him in this manner :—

Sacred to the memory of the Hon. John Wingfield-Stratford, 2nd son of Richard, Viscount Powerscourt, and formerly a lieutenant-colonel in the Cold-stream Guards. He married Frances, sole daughter and heiress of Leonard Bartholomew, Esq., of this place, by whom he had three children, Frances Amelia, Isabella Harriet, John. He married 2ndly, Harriette, the daughter of Henry Grant, Esq., of the Knoll, Glamorganshire. In 1803, by royal sign manual, and in obedience to the will of his uncle, Edward Augustus, 2nd Earl of Aldborough, he was authorised to assume the name and arms of Stratford in addition to his own. He died 3rd August, 1850, aged 78 years, leaving his wife Harriette surviving. Also to the memory of the said Frances, his wife, who died 13 July, 1827, aged 51 years. Also in memory of the said Harriette, his second wife, who restored this church A.D. 1856, and who died April 6th, 1863, aged 80 years.

This monument is interesting as giving us the date of the restoration of the pretty church of Addington, and also as pointing out the reason for the double name of Wingfield-Stratford. The son of the Hon. John Wingfield-Stratford succeeded his father, and is commemorated by the beautiful reredos in the church, as is shown by the inscription :—

To the glory of God, and in loving memory of John Wingfield-Stratford, Esqre·, of Addington Park, born Dec. 10, 1810, died May 8, 1881. This reredos was erected by his widow, Christmas 1881.

In the year 1887 the Manor of Addington, which, with the advowson of the church, had been so long in the one family—for from the time of Robert Charles, in the fourteenth century, it had descended by heirs male or female—was alienated to C. J. S. Whitburn, Esq.

Besides the monuments we have already mentioned, there are in the church at Addington the following :—

Anna Maria Pickering, widow, of Rev^d. E. Hayes Pickering, born 28 Jan., 1813, died 27 Nov., 1872. " Then are they glad because they are at rest."

To the glory of God, and in loving memory of Ellen Elizabeth Boys, who died in this parish Dec. 29th, 1886, aged 83 years.

To the glory of God, and in memory of William Wells, who acted as Clerk to the Parish of Addington for 23 years. He died at the age of 93, respected and beloved by all who knew him, and by none more than her who raises this tablet to his memory. " May he rest in peace." Also in memory of Kezia Wells, who died July 30th, 1888, aged 72 years.

To the glory of God, and in memory of George Baldock, died May 7th, 1891, aged 34. " Well done, thou good and faithful servant, enter thou into the joy of thy Lord."

Besides this there is an inscription under one of the handsome stained glass windows in the church, which is commemorative of the loyalty of the inhabitants of Addington at the time of the late jubilee :—

To the glory of God, as a thank-offering to Him in this 50th year of Queen Victoria's reign, this window was erected by the parishioners of Addington, 20th of July, 1887.

In the parish church of Aylesford at this period we have a number of monuments to the Milners, who, it will be remembered, were the descendants of Charles Cottam, Esq., upon his taking the name of Milner :—

Beneath this marble are interred the remains of the Reverend Joseph Milner, D.D., of Preston Hall, in this Parish, who departed this life July the 26th, 1784, aged 54 years. He was the first who inherited this estate upon his taking the name, Milner. Beneath also are interred the remains of Sarah, his wife : she died September the 27th, 1803, aged 72 years.

Then, in point of time, follows :—

Sacred to the memory of Charles Milner, Esq., of Preston Hall, in this parish, who died Jan. 7th, 1836, aged 72 years. Also of Harriett, his wife, the daughter of Sir John Dixon Dyke, Bart., of Lullingstone Castle, in this County, who died August 1808. They had issue Harriet Philadelphia, born December 6, 1792, died June 27, 1793 ; Charles Joseph, born August 10, 1796, died September 6, 1796 ; Mary Anne, born March 13, 1799, died June 5th, 1836 ; Harriett Sarah, born September 19, 1800, died May 31, 1802 ; Charles, born September 16, 1801, died September 19, 1844 ; John, born February 4, 1804, died October 6, 1846 ; Caroline Elizabeth, born July 2, 1802, died January 28, 1843. Henry Robert, a Lieutenant-Colonel in Her Majesty's 94 Regiment, the last survivor of his family, has erected this monument as a tribute of affection to their memory.

And :—

This tablet is erected to the memory of Henry Robert Milner, Major-General in the army, and formerly Lieutenant-Colonel of Her Majesty's ninety-fourth Regiment. He was the youngest and last surviving son of the late Charles Milner, Esq., of Preston Hall, in this Parish, by Harriett, his wife, the youngest daughter of the late Sir John Dixon Dyke, Bart., of Lullingstone Castle, in this county. He was born at Preston Hall on the 29th day of January, 1805, and

died at Plymouth on the fourteenth day of January, 1855, in the 50th year of
his age. Major-General Milner entered the army on the 7th of February, 1822.
He joined the 94th regiment in 1828, and from that period served uninter-
ruptedly with it in the Mediterranean, Ceylon, and the East Indies, until the
return of the regiment from Madras in June 1854. He commanded the regiment for fourteen years, and was employed on the staff of the Madras army
for five years as Brigadier, commanding at Aden, Cannanore and Bangalore.
His brother officers testify to the many estimable qualities of their late commanding officer and friend, by uniting with his relatives in dedicating this
tablet to his memory.

Charles Milner, Esq., who had inherited the manors of Preston
and Aylesford from his father, acquired in 1832 the manor of
Ditton, which he transmitted to his son Charles, who, as the above
monuments show us, died in 1844. The manor then went to his
brothers John and Henry Robert successively. These estates passed
to Edward Ladd Betts, Esq., by purchase in 1848, who in 1849 built
the present mansion, and on his failure they came afterwards into
the hands of T. Brassey, Esq., in 1865, who gave it to his son, H.
A. Brassey, Esq., in 1870. Mr. Brassey died May 13th, 1891, when
his son, H. L. C. Brassey, Esq., came into possession.

Mr. Betts is not buried in the church, but in the churchyard,
where a handsome monument was raised to him by the parishioners
of Aylesford for his great generosity to them in his time of prosperity,
and this is a remarkable instance that worth is sometimes still
respected in the days of adversity, and that even when persons are
absent.*

Besides, we have two monuments to two vicars of Aylesford :—

In memory of Rev. William Tolbutt Staines, M.A., formerly fellow of Queen's
College, Cambridge, who for upwards of eight years was Vicar of this parish.
He died Sept. 24, 1840, in the 55th year of his age. His clear judgment, well-
stored mind, and pious and devoted spirit, rendered him a bright example as
a Christian pastor ; and he has left behind him a name which will ever be
cherished with affection and respect, not only by his relatives and friends, but
by all who knew him and could appreciate the real excellence of his character.

And :—

In memory of Edward Garret Marsh, M.A., formerly fellow and tutor of Oriel
College, Oxford, Canon of Southwell, and for twenty-one years Vicar of this
parish. Born at Salisbury, Feby. 8, 1785, died at Aylesford, September 20,
1862. "I know in Whom I have believed."
Also of Lydia Marsh, his wife, born at Portsea, January 17, 1788, died at
Aylesford, Dec. 1, 1859. "Of a meek and quiet spirit."
In memory of Thomas Henry Marsh, of Lincoln's Inn, barrister-at-law,
their third son. Born at Hampstead, May 31, 1824, died at Aylesford, March 10,
1852. "Blessed are the pure in heart, for they shall see God."
Also of Joseph Samuel Marsh, their youngest son, born at Hampstead,
January 22, 1826, died at Aylesford, January 27, 1847. "Surely I come quickly.
Even so, come Lord Jesus." In hope of eternal life, which God, who cannot lie,
promised.

---

* On Oct. 28th, 1892, a window subscribed for by his neighbours was unveiled
in Aylesford Church, and a new wing to the Trinity Hospital, given by his
widow and son, was opened in memory of Mr. Brassey.

The first three bells of Aylesford were cast by Gillett & Co., of Croydon, and used on Saturday, 9th December, 1869, and were given by H. A. Brassey, Esq., J.P.   The priest's bell is blank.

Two small alms-plates are inscribed, "A gift to the parishioners of Aylesford, from Thomas Franklyn, Esq., of Cobtree, 1859.   Rev. E. G. Marsh, vicar."   In 1838 Aylesford church was reseated.   In 1852 was the cholera year, when we learn that twenty-five people were buried in Aylesford in September only.   The window at the east end of the church contains sentences from the *Te Deum*, all in Latin.   On the south side of the chancel is a window with "Fides" and "Spes" in the different lights.   Then comes another window, with "Charity," and then, as we walk round the church, we find on the south and north sides, two in each window: St. Peter, St. Paul, St. Matthew, St. Mark, St. Luke, St. John, St. Thomas, St. Bartholomew, St. Simon, St. Andrew.

The second Earl of Romney, in whose hands the castle, manor, and patronage of the living of Allington still continued, succeeded his father in 1811, and, dying in 1845, was succeeded by his son Charles, the third earl, who was M.P. for West Kent 1841-45; at his death, in 1874, Charles, his son, succeeded to the title and estates, which he still holds.

The Rev. E. B. Heawood, the present Rector of Allington, kindly forwarded the following inscription, and the melancholy tale it records :—

To the memory of Robert Chapman, late scholar of C. C. College, Cambridge, who was accidentally drowned while crossing the Medway near this place on the 14th August, 1833, aged 22 years.   This tablet was erected by friends who sincerely deplore his loss, as a testimony of their high regard for his unassuming manners, promising abilities, and amiable disposition.   "Therefore, be ye also ready."—Matth. chap. 24, v. 44.

The history of the tablet is as follows : "There was a large evening party at Allington Castle, given by the Packs to their citizen friends at Maidstone.   In those days there was a ferry at the Gibraltar Inn, over against the castle—there was an ancient punt for its service.   Eleven ladies entered the punt with Mr. Chapman, under-master of the Grammar School, Maidstone—the punt was safe for five or six—and when they reached the middle of the river the punt sank.   The eleven ladies floated by reason of the buoyancy of their dresses, and they were safely landed.   Chapman was forgotten, and he sank : his body was found sooner or later afterwards.   Mrs. Pack held a lantern from the castle walls, and saw the punt disappear."

Birling at the beginning of this ·period was, with Ryarsh, in the hands of the second Earl of Abergavenny, Henry ; he was succeeded by his son John, third earl, in 1843, and he by his brother William, the fourth earl, for many years rector of Birling.   He succeeded in 1845, and died 1868 ; his son William succeeded him as earl, and was created a marquis in 1886.

On the cover of the font in Birling church we read that it was carved by the Ladies Caroline Emily, Henrietta Augusta, and Isabel Mary Frances, Nevill, the daughters of the Earl and Countess of Abergavenny, A.D. 1853. It was first placed upon the font, and the font moved into the centre of the church, at the christening of the eldest son of the Hon. Thos. E. M. L. Mostyn, M.P., and Lady Augusta Mostyn.

The monuments in Birling church, besides those to the Nevills, are as follows :—

Beneath are deposited the remains of Maria, second daughter of John and Emily Selby, of this parish, who was born 10th of January, 1827, and died 15th January, 1830.

And two other memorials of former incumbents, one to the vicar himself, the other to the wife of another. Both are windows :—

To the glory of God, and in memory of Rev. H. D. Phelps, M.A., Vicar of Birling, who died October 28th, 1864. This window was erected by his parishioners, as a tribute of affection and of gratitude for his devoted attention to their interests, both temporal and eternal.

"Blessed are the dead which die in the Lord."

To the glory of God, and in affectionate memory of Annette H. Madden, this window is erected by parishioners and friends. A.D. 1889.

The memorials to the Nevills, as many of them are only shields, we give in detail below. Other remarks upon the family will be found in the foregoing history and in the notes. The entrance to the vault of the Nevills is of cast iron, with coronet, shield, and roses, supported by bulls, with the motto : *Ne Vile Velis.* We give the inscriptions in chronological order :—

Sir Geo. Nevill, Milit. Dom. Burgavenny, obᵗ anno 1492, born at Raby Castle in co. Durham, son of Edwᵈ Lord Burgavenny who was sixth son of Ralph, 1ˢᵗ Earl of Westmoreland, Lord of Staindropa, Branxspeth, Sherriff Hotton, Middleham, Warkworth and Coverham, Earl Marshall of England, Knight of the Garter, Constable of the Tower of London, and Warden of the Forests North of ye Trent, by Lady Beaufort, dau. of John of Gaunt, Duke of Lancaster, 4ᵗʰ son of King Edward ye 3ʳᵈ. Edward Nevill married Elizabeth, daughter and heiress of Richard Beauchamp, Lord Burgavenny, and Earl of Worcester, by Isabel, daughter of Thomas, Baron Le Despencer and Earl of Gloucester. Sir Geo., Baron Burgavenny, married Margaret, daughter and co-heir of Sir Hugh Ffenne, Under Treasurer of England, and was the father of George, Lord Burgavenny, buried here in 1536.

This lord directed by his will, made at Birling : "His body to be buried in the monastery of St. Pancras, called the Priory of Lewes, Sussex, on the south side of the altar, where he had made a tomb for his body. Disce mori mundo vivere disce Deo (*Learn to die to the world, learn to live to God*). Ou je tiens ferme (*Where I hold firmly*)." Sir Geo. Nevill, to whom this monument is raised, was knighted at the battle of Tewkesbury.

George Nevill, Dom. Burgavenny, Knight of the Garter, Constable of Dover Castle and Warden of the Cinque Ports, married 1ˢᵗ Joan, daughter of Thomas

Fitzallan, Earl of Arundel; 2ⁿᵈ, Mary, daughter of the Duke of Buckingham, obt. 1536.

This Nevill preserved Kent from joining the Cornish rebels, and was present at the Field of the Cloth of Gold.

Henricus Nevill, Dominus Burgavenny, married Ffrances, daughter of the Earl of Rutland and Baron Roos, obiit 9th February, 1586.

This baron opposed the proceedings of Wyatt, and met Sir Thomas Isley at Blacksole Field, in Wrotham, and by this probably saved Mary her throne. He afterwards sat in judgment on Mary Queen of Scots. He died at his seat, called Comfort, in Birling, seised of the manors of Birling, Ryarsh, Yalding, Luddesdown, West Peckham, Mereworth and Old Hæie, alias Holehaie, and the advowsons of Birling, of the church of Maplescomb in West Peckham, and of Mereworth.

We may remark here that the church of Maplescomb, now in ruins, which is joined to the living of Kingsdown, being described as part of the parish of West Peckham, is extremely interesting, as giving us an example of how detached chapelries became associated with certain parishes in days gone by—we know not why. It was not for the convenience of the people of the places, neither could it have been very convenient to the clergyman, as, for instance, Maplescomb was at least six miles from West Peckham, as was Comp, at the shortest, two miles from Leybourne.

Here lieth Lady Frances Burgavenny, obt. 1576, wife of Henry, Lord Burgavenny, and daughter of Thomas Manners, Earl of Rutland and Baron Roos : also Margaret Nevill, daughter of Edward Nevill, obt. 15 Oct., 1616 : also Lady Elizabeth, first wife to Sir Henry Nevill, Lord Abergavenny, and after wife of Sir Willᵐ· Sedley, Knt. and Baron, ob. 15th Aug., Anno 1617. Also Lady Katherine, 2ⁿᵈ wife of Sir Henry Nevill, Lord of Abergavenny, and daughter of George, Baron Vaux, of Harrowden, ob. 10 July, 1641. Expectamus adventum Domini Jesu Christi.

Sir Edward Nevill, Dominus Burgavenny, ob. 1st December, 1622.

On the death of Lord Henry, as already stated, his daughter disputed the title of Lord Edward, when, after much litigation, it was settled she could hold the barony of Le Despencer in her own right and satisfaction, but Lord Burgavenny goes in tail male with the office of Lord Marshal.

Sir Thomas Nevill, Knight of the Bath, married Frances, daughter of Henry, Baron Mordaunt, ob. 1628.
Henry Nevill, Lord Abergavenny, married Mary, daughter of the Earl of Dorset, Lord Treasurer of England, obt. 24th Dec., 1641.
Sir Christopher Nevill, Knight of the Bath, ob. June, 1649.
John Nevill, Lord Abergavenny, ob. 12th Deceʳ·, 1662.
George Nevill, Lord Abergavenny, ob. 14th June, 1666.
George Nevill, Lord Abergavenny, ob. March 1720, ætat 63.
George Nevill, Lord Abergavenny, obt. anno 1723.
Edward Nevill, Lord Abergavenny, obt. October 1724.
In memory of Henry Nevill, aged 3 years, and Augustus Nevill, aged 14 months, who both died 28th of March, 1828. This window was erected by their

parents, the Earl and Countess of Abergavenny. "Suffer little children to come unto Me, and forbid them not, for of such is the kingdom of God." In the vault beneath are interred the bodies of Henry Nevill, aged 3 years, and Augustus Nevill, aged 14 months, the children of the Honble. and Rev. William Nevill, Vicar of Birling, youngest son of Henry, Earl of Abergavenny, by Caroline his wife, who both departed this life 28th of March, 1828.

Besides these memorials, there are handsome windows to the late Earl of Abergavenny (who was many years Vicar of Birling) and his wife, with these inscriptions :—

To the glory of God, and in memory of William Nevill, IV. Earl and XLVI. Baron of Abergavenny, who died August XVII., MDCCCLXVIII. This window was painted and erected as a humble tribute of affection to the memory of a kind father, by his three daughters : Caroline, Augusta, Isabel. "The memory of the just is blessed."

Also :—

To the glory of God, and in memory of Caroline, wife of the IVth Earl of Abergavenny, who died 19th of May, 1873. This window was erected in grateful and affectionate remembrance of a beloved mother, by William, Caroline, Augusta, Isabel and Ralph. "Those also which sleep in Jesus will God bring with Him."—1 Thess. iv. 14.

And :—

To the glory of God, and in grateful memory of William, IVth Earl of Abergavenny, and Caroline his wife. Erected by the parishioners, 1874. "Lovely and pleasant in their lives, and in death they were not divided."

A tankard five and a quarter inches high, and four and a quarter inches across, of the date 1697, once the property of the Dampiers, was presented by Mrs. Jane Phelps to the church of Birling, on Christmas Day, 1854. Mrs. Phelps was a Miss Lupton. Mr. Lupton married Elizabeth Dampier, who was cousin to Thomas Dampier, Bishop of Rochester, 1802, translated to Ely, 1808, died 1812. He was brother to Judge Dampier; they were both of the old family of Dampier of East Hall, Blackford, Somerset. On the handle are the initials $\frac{D}{WM}$, probably referring to the Dampiers. It will be remembered that at this time the Rev. H. D. Phelps was vicar of Birling.

In Ditton church the memorials of this age are numerous :—

Beneath this monument, in the family vault, are deposited the remains of Mary Ann, wife of John Golding, Esqre·, of Ditton Place, in this Parish, who departed this life on the 22nd day of May, 1837, in the 50th year of her age, a true Christian, an affectionate wife, and a sincere friend ; also Marianne, their daughter, wife of the Revd· John Barrow, who died the 25th of October, 1842, aged 26. Also Clementina, their daughter, some time wife of Alfred Luck, Esqre·, who died the 16th November, 1842, aged 28. "Upon whose soul may God have mercy." Also Caroline, their fifth daughter, wife of Robert Tassell, Esqre·, of the Inner Temple, Barrister, who died the 7th of May, 1851, aged 31 ; John, died 1807, Ellen, 1822, infants ; John Henry, died October 20th, 1820, aged 12 ; John Golding, Esquire, of Ditton Place, died 17th February, 1856, aged 85 years ; Henrietta Golding, died 3rd August, 1866, aged 40 years.

Another :—

In Memory of Robert Tassell, Esqʳᵉ·, of the Inner Temple, Barrister, only son of Robert and Mary Tassell, of Blackland, in the parish of East Malling: He died on the 2nd day of September, 1852, aged 34 years.
Also of Caroline, his beloved wife, fifth daughter of John Golding, Esqʳᵉ·, of Ditton Place, she died on the 7th day of May, 1851, aged 30 years.

Another, to one of the former incumbents :—

Sacred to the memory of the Revᵈ· William Hamilton Burroughs, B.A., second son of the Venᵇˡᵉ· Newburgh Burroughs, Archdeacon of Derby, and 16 years Rector of this Parish, born 1795, died Oct. 20, 1856.

And another :—

In Memory of Robert Tassell, Esq., J.P., of Cob Down, in the Parish of Ditton, late of Blacklands in the Parish of East Malling, who died on the 13th day of January, 1874, aged 89 years.

Just inside the church door is a stone on which we can yet read "Margaret Godden," but the rest of the inscription is illegible.
In the wall of the church, on the south, outside, is the following :—

Hic jacent corpora Annæ, Jacobi, Gulielmi, Marthæ et Henrici Boghurst, singuli pietate et honestate conspicui.
Here lie the bodies of Anna, James, William, Martha and Henry Boghurst, each conspicuous for piety and honesty.

The church of Ditton is almost filled with monuments, and these are chiefly to the Brewers and Goldings, as above shown. On the village green the pound may still be seen. The priest's bell at Ditton is dated, " Borodino, 1825."
In Halling church we have this inscription only :—

Peggy Towers, died 18th March, 1834, aged 70 years.
Thomas Towers, died 31st August, 1837, aged 76 years.

This parish, owing to the cement works of Formby & Co., and Lee & Son, has during this period developed a large industry, and with the neighbouring parishes of Cuxton, Strood, Snodland, Birling, Aylesford, Burham and Wouldham, is becoming a large manufacturing district, which, like the great centres of business in the North, is peopled principally by factory hands. The centre of the cement trade may be said to be Halling, Snodland, Burham and Wouldham, which, with the Ham Mill portion of Birling, go to form what may be called Cementopolis, containing about 12,000 people : this population is ten times what it was in the days of William IV. The great thing to be regretted over these works is that Messrs. Anderson, the owners of the works behind the quaint old church of Halling, have, in order to make the way for their wharfage along the river, swept away nearly every remnant of the ancient palace of the Bishops of Rochester. Here is a Working Men's Institute, which is in a very flourishing condition, and possesses a commodious building for a club-house in the village.

9

In Leybourne church the monuments of this period are to the family of the Hawleys and their connections; the oldest of these is :—

In a small vault beneath this church are deposited the remains of Sir Henry Hawley, Bart., of the Grange, in this parish, who departed this life the 29th day of March, 1831, aged 54 years. Sir Henry succeeded his father as second Baronet the 20th day of January, 1826. Sir Henry married, the 29th day of November, 1806, Catherine, eldest daughter of Sir John Gregory Shaw, Bart., of Kenward, in this county, by whom he had issue three sons and eight daughters. Also of Catherine Elizabeth Hawley, wife of the abovenamed Sir Henry Hawley, Bart., who departed this life March 16. 1862, aged 75 years.

When opening the vault for the reception of Sir Henry Hawley's body, we are informed that the tower fell down, which necessitated its being rebuilt. Sir Henry's sister, who married Mr. Brockman, has also a memorial in this church ; it reads :—

To the memory of Louisa, the wife of the Reverend Tatton Brockman, M.A., of Beachborough, and Rector of Otham in this county, youngest daughter of Sir Henry Hawley, of Leybourne Grange, born 1793, died 1837. "If we believe that Jesus died and rose again, even so them also who sleep in Jesus will God bring with Him."

Also there is an inscription :—

Mabel Diana Hawley, died April 12th, 1852, in her twelfth year. " The flower fadeth."

There is a memorial window :—

In memory of Sir Joseph Hawley, Baronet, who restored this Church, 1874. Born Oct. 27, 1813, died April 20, 1875.

Sir Joseph was one of the best known racing men of the period, and established the famous racing stud with which he won four Derbys, namely, with Musjid, Teddington, Bedesman, and Blue Gown. After winning with Bedesman Sir Joseph put up the well-known clock in Leybourne Grange, which can be heard a long distance off. He repaired the church, as stated on his inscription, and still more the castle, both of which works cause him to deserve well of posterity. He was succeeded in the manors and title by his brother Henry James.

The reredos in the church is :—

In memory of the Rev^d. Henry Charles Hawley, by his loving wife, 28 years Rector of this Parish. He died Feby. 16th, 1877.

To the Cusack-Smiths, the relations of this Mr. Hawley's wife, there are these memorials :—

Sacred to the memory of Michael Cusack-Smith, Esqre., of the 14th King's Own Light Dragoons, who departed this life at Meerut, in the East Indies, on the 14th day of March, 1851, aged 24 years and 6 months. Sincerely and Deeply regretted by His brother officers, who have erected this tablet as a token of their esteem and attachment for their late comrade.

The side window nearest the east end is, we learn :—

To the memory of Sir Michael Cusack-Smith, ·Bart., by his daughter, Mary Hawley, January 1877.

The next is :—

In memory of Hester Augusta, daughter of Sir M. Cusack-Smith, Bart., and wife of Fred. W. Craven, Ord. Major of the Royal Artillery, by whom this window is erected, died 15th Oct., 1863.

There is also, in the vestry, a monument to Mr. Hawley's predecessor, with this inscription :—

To the memory of the Rev^d. Charles Cage, Fifty years Rector of Leybourne. " I go to prepare a place for you, that where I am there ye may be also." Firmly relying on this promise, our deceased friend slept in peace. A few friends with whom he walked in happy companionship on earth, anxious to record the rare simplicity of character, and the numerous endearing qualities which bound their hearts to his, have placed this window over his grave : their love has ended not with life ; it survives to cheer with soothing remembrance the hours which yet remain till they also are called from their labours, to be reunited, as they humbly hope, in Jesus, to him whom they loved on earth, in a new existence where there shall never be separation.

The first bell of Leybourne bears the date 1826—it has no other inscription ; it was probably a gift of the Hawleys. When Sir Joseph had Leybourne church restored, by some accident several ancient monuments and glass windows got buried or destroyed ; but fortunately the old building, which shows the church to be Norman at least, was not interfered with sufficiently to destroy the marks of its antiquity.

Sir John Twisden succeeded his father, Sir John Papillon Twisden, in the baronetcy and in the manors of Bradbourne and East Malling. Unfortunately there is no tablet to him : he died in 1841, and is the last who held the baronetcy.   On his death, Captain John Twisden, R.N., second son of Sir Roger Twisden and Mary his wife (of whom there is a picture at Bradbourne, in a green dress, with roses), succeeded to the estates.   With him we are introduced into another scene in English history, for he served with Kempenfeldt in the *Victory*, and with Hood in the *Queen Charlotte*, and commanded the *Fearless* in Sir Sydney Smith's gallant defence of Acre against the French, under Bonaparte, 1819.   It is curious that Sir Sydney was educated at Tonbridge School, and thus perhaps we may surmise that he and the captain were known to each other early in life.   There is this monument in East Malling church to him and his branch of the family :—

Sacred to the memory of William Twisden, second son of Sir Roger Twisden, Baronet, who departed this life 30th of December, 1771, in the 30th year of his age ; and of Mary, wife of William Twisden, who departed this life July 2nd, 1771, aged 27 years ; and of five of their children, who died in infancy.   Also to the memory of Anne, wife of Captain John Twisden, R.N., only surviving son of the above William and Mary Twisden, who departed this life June 13th,

1843, in the 71st year of her age. Also in memory of the above Captain John Twisden, R.N., of Bradbourne, who died on the 22nd of June, 1853, in the 86th year of his age.

There is also a tablet to the last Lady Twisden :—

S.M. Dame Catherine Judith, wife of Sir John Twisden, Bart., of Bradbourne. She died in childbed, April 13, 1819, aged 29 years.

On the death of Captain Twisden the estates passed into the hands of the Misses Twisden, the last of whom still holds them and the manor.

The pictures at Bradbourne are a very fine historical collection, and not only give us the family of Twisden, but also some of the well-known characters with whom their family was connected, and other well-known persons in English history. Amongst the persons of the family (or connected with it) to be seen here, who are historical, are :—

Judge Sir Thomas Twisden.	Sir Roger Twysden, the Antiquary and
Colonel Tomlinson.	Royalist.
Sir Thomas Wyatt, the elder.	Sir Harry Vane, the elder.
Sir Thomas Wyatt, the younger.	Sir Harry Vane, the younger.
Sir Henry Wyatt.	

Besides, there are these historical ones :—

Charles I. (Vandyck).	The Duke of Marlborough.
Henrietta Maria (Lely).	Countess of Carlisle.
James II. (Godfrey Kneller).	Nell Gwynne.
The Duchess of Portsmouth.	Catherine de Medici.
The Duchess of Richmond.	John, Duke of Saxony.
The Duchess of Monmouth.	The Earl of Essex.
The Duchess of Cleveland.	Archbishop Tillotson.
The Duke of Monmouth.	Pope Julius II.
The Chevalier de St. George.	Cosmo de Medici.
Sir Beville Greenville, born 1601,	Martin Luther.
killed at Landsdown, whose son was	Dante.
the bearer of the letter to Charles II.	William Noy, a celebrated Lawyer of
concerning his restoration, and to	the 17th century.
whom he gave a valuable jewel ; and	Mrs. Price, Maid of Honour to
who with only one other was allowed	Catherine of Braganza.
to remain in the room while Charles	Algernon Perey, Earl of Northumber-
II. received the Sacrament, after the	land (Vandyck), 1602-68.
manner of the Papists.	

Besides, there are studies from Van Eyck, Murillo, Cuyp, Paul Potter, Sir Godfrey Kneller, Turner, W. Vandervelde, Domenichino, Rembrandt and Snyders, and many others of less note. The pictures, the fine oak staircase, and the situation of Bradbourne, together with the great historical connection of the old Kentish family that live there, render it one of the most interesting spots in England.

Besides the monuments to the Twisdens, there is this monument to the Wigans in East Malling church :—

In a vault beneath the porch of the church are deposited the mortal remains of John Alfred Wigan, of Clare House, in this parish, born Sept. 11, 1787, died Nov. 16, 1869, and of Elizabeth Pratt, his wife, born March 11, 1793, died St. James' day, 1864. Also of the following of their children : Eliza Lewis, born July 16, 1816, died May 3rd, 1826 ; Henrietta, born June 1, 1830, died April 18, 1844 ; Julia, born July 17, 1826, died March 10, 1846 ; Emily, born St. John Evangelist's day, 1822, died Dec. 7, 1846 ; Harriet, born Jany. 21, 1824, died Dec. 31, 1850 ; Georgiana, wife of Nathaniel Dimock, clerk, born June 4, 1825, died July 14, 1853 ; Amelia, born St. Luke's day, 1828, died St. Thomas' day, 1868.

The windows of the south aisle are filled with handsome stained glass to the Wigans.   The first is to the last incumbent : we read :—

"They immediately left the ship and followed Him."
" Are ye able to drink of the cup that I shall drink of ? We are able."
To the glory of God, and in memory of William Lewis Wigan, 28 years vicar of this parish, who died Jany. 8, 1876, aged 58.

Then we read, on the next :—

To the glory of God, and in loving memory of Amelia Wigan, by her sisters J., L., W., A.D. 1870.
" There shall be joy in heaven over one sinner that repenteth."
"I have found my sheep which was lost."

The next is to

Bernard Wigan, born and died Aug. 7, 1868.
" He shall gather the lambs with His arm, and carry them in His bosom."

The west window is put in to the present vicar's wife :—

" Whosoever shall receive one of such children in My name receiveth Me."
To the glory of God, and in memory of Sarah Wigan, who died July 28th, 1890.

There is also another window :—

To the glory of God, and in memory of James Thornhill, formerly of this parish, and late of Camberwell, Surrey, who died Sept. 30, 1875, aged 63 years. Erected by his son, James Alfred Thornhill, Xmas 1876.

Over the chancel arch there is a mural painting of the crucifixion, ascension, and resurrection, and the words, " I am He that liveth, and was dead. Behold I am alive for evermore" ; and round the arch runs, " Blessed are they that do His commandments, that they may have a right to the tree of life." There is also painted on the walls, " The Gentiles shall come to Thy light."   " He saith to them, Be of good cheer ; it is I ; be not afraid."   " Ye are no more strangers and foreigners, but fellow-citizens with the saints, and with the household of God ; and are built upon the foundation of the apostles and prophets, Jesus Christ Himself being the head corner stone."   " In Whom all the building, fitly joined together, grows into an holy temple : in Whom ye also are builded together for an habitation of God through the Spirit."   " This beginning of miracles did Jesus in Cana of Galilee."   " Alleluia."   " Unto us a Child is born."
The first bell of East Malling is marked " Mears, of London, fecit, 1831." The old sanctus bell was done away with at this date.
The parish of East Malling has prospered much during this period

owing to the paper trade. At East Malling, we are informed, millboard was first made. The hamlet of Larkfield contains many ancient houses; and the old turnpike gate-house, which has still its wooden porch, reminds us of this annoyance to travellers, done away with eighteen years ago. Between this place and the hamlet of Newhythe-on-the-Medway, once one of the most miserable and neglected corners of Kent, now much improved, there stands the church of Newhythe, which was consecrated Oct. 4th, 1854. Revs. N. Dimock, S. Wigan, F. H. D. Ness and W. F. Woods have been curates, and O. C. Legge-Wilkinson is the present curate.

Mrs. Losack lived at Malling Abbey till 1844, when Aretas Akers Esq., J.P., rented the abbey, which he afterwards bought. He died in the year 1855, and was buried at West Malling. His widow lived here till her death in 1891, when the property came into the possession of Aretas Akers-Douglas, Esq., M.P., at one time M.P. for Mid Kent, but since the new distribution member for St. Augustine's Division of Kent and Conservative Whip, and Political Secretary to the Treasury in both Lord Salisbury's first and second ministry. It has been disposed of this year to Miss Boyd, who intends to keep it for a home in connection with her Orphanage of the Infant Saviour, at Kilburn.*

During the time that the Akers have been at Malling Abbey, discoveries have been made of lids of coffins, with circles on the top, and a line running down the centre, crossed in two or three places with foliage : in all likelihood the resting-place of the Abbesses. The gateway chapel was renovated in the year 1858 by the Akers. It had passed through various vicissitudes of fortune : it would appear that it was at one time a tannery ; then, in 1773, it was a meeting-house ; then a carpenter's shop. Since All Saints' Day, 1858, when the Akers family and their friends, together with the present worthy Vicar of West Malling, saw their work of renovation completed by its being opened for public worship, continual prayer has been said here. It is about twenty-two feet long and thirteen broad, and there is an ante-chapel behind the screen about fifteen feet long. The walls are handsomely decorated. In the window are scrolls, with the initials of the various contributors and the day of their decease.

As we here take leave of the abbey, we may mention that the ancient seal of the abbey was a figure of the Blessed Virgin, crowned, under a Gothic canopy, with Jesus on her right hand and a sceptre in her left. In the niche a figure praying. Underneath, " Sigillum Commune Monasterii, Beatæ Mariæ de West Malling," *i.e.*, The Common Seal of the Monastery of the Blessed Mary of West Malling.

There are still many ancient houses in Malling, notably at the back of some of the shops between the High Street and the abbey,

---

* Since writing this, Miss Boyd has informed the author that she has given the Abbey to the Benedictine sisterhood of the Church of England, the trustees of which are the Cowley Fathers.

especially those of Mr. Carman and Mr. Jarvis, where the remains
are decidedly Norman. Besides these we have many dating from the
seventeenth century; one especially noticeable on the Offham Road
bears date 1675.

In the church of West Malling are one or two monuments
belonging to this period. One to the last vicar, his father and
mother, and his own wife :—

In a vault near this place are deposited the remains of Benjamin Bates, Esq.,
of Town Malling, and formerly of Brunswick Square and Thames Street,
London, who departed this life the 29th of October, 1821, aged 88 years : also
of Elizabeth Theresa, his wife, on the 11th of February, 1837, aged 87 years.
Also of the Rev^d· George Fern Bates, son of the above, many years Vicar of
the Parish, and of South Mims, Middlesex, who departed this life the 18th
November, 1841, aged 66 years : also of Lydia Amy, his wife, on the 19th of
May, 1833, aged 53 years. "These all died in faith."—Heb. 11, 31. "Examine
yourselves whether ye be in the faith."—2 Cor. 13, 5.

On a stained glass window in the chancel we read, "Sarah
Charlotte Savage, et Ellen Duppa, pinxit Anno Dom. 1849." On
two others, in the chancel, "Charlotte Savage, 1852," and "Sarah
Charlotte Savage, 1858." The east window has no inscription.

We have also these tablets :—

Sacred to the memory of Richard Kennard, M.D., M.R.C.S., who departed this
life in London, June XVI., MDCCCL. With a mind enlarged by study and
travel, he united a warm heart, strict integrity, and practical benevolence. To
the Parochial School of Town Malling he bequeathed the sum of two ·hundred
pounds. His end was peace. His remains are interred in the family grave at
East Farleigh.

Besides we have one stone :—

Sacred to the memory of Thomas Luck, Esq^ro·, of Went House, in this Parish,
who died Nov. 13th, 1857, aged 92 years. He had issue seven sons, of whom
Edward, Thomas, Charles, and Alfred alone survive. This tablet was erected
by his Executors, Aug. 13, 1858.

The organ was placed in the church by Mr. Luck's widow and
son, in memory of him.

There is one more tablet, which is :—

In loving memory of Henry Montague Randall Pope, born May 21, 1849,
died Nov. 18, 1880, buried at Sea.

The church of West Malling was reseated in the year 1852 by
the contributions of certain subscribers, amongst whom were the
Rev. J. H. Timins, Mrs. Akers, Mr. R. B. Stedman, Rev. H. F.
Foster, Mr. Dutt, Mr. Allchin, Mr. Hodges, and the Trustees of
Mr. Lawson.

The Union was erected in the year 1836, and in the year 1872
the present beautiful chapel was built, in which are three handsome stained-glass windows in the west end ; the memorial stone
was laid by Lady Caroline Nevill, May 1st, 1872. Rev. John Manus
was chaplain 1873, when the baptismal register was commenced.
He was succeeded by the Rev. John Stuart Robson, 1875-80.

Rev. Henry Frederick Rivers was chaplain 1880-89, at the end
of which last named year he was appointed vicar of St. Faith's,
Maidstone, and the Rev. Cecil Henry Fielding, the author of this
work, was chosen his successor at the Union.

Bells one, two, and eight of West Malling are all inscribed, " Mears
& Stainbank, founders, London 1869." The magistrates sit here
still on Mondays, but the market has been done away with ; so
that the meetings of the Bench, and the new Police Station, built
in 1866, alone declare Malling to have a right to the dignity of a
town. In 1865 a Masonic Lodge, called the Abbey Lodge, was consecrated here. The Kent Nursing Institution was started in the
year 1875 by the Rev. J. H. Timins and the late lamented Lady
Caroline Nevill. There is also here an Athenæum, which has for
some years afforded amusement and instruction in the winter even-
ings, by means of concerts, lectures, and discussions. The cricket
ground of Malling has been already alluded to. The last county
match played there was between Kent and Sussex, in 1890, when
the former county was victorious. A Horticultural Society has
been some time established, which holds several flower shows in
the course of the year. Malling had once a good many tan-yards ;
these, however, have been done away with of late years. The great
staple of industry now in the town is brewing, and the Paper
Mills of East Malling, already mentioned, employ a good many
hands.

The little village of Offham still keeps the quintain on the green,
which, we learn from Hasted, the dwelling-house opposite was bound
to keep up—of which house Mr. Tresse died possessed in 1737. The
Manor House, by the church, where the Court Leet is held, has a
fine Elizabethan room.

In the church there is one monument of this date :—

Sacred to the memory of John Smith Addison, Gent. of this parish, who
departed this life the 25th of October, 1834, aged 41 years.

There are no monuments left in Paddlesworth, nor in Dode ; the
one having become a barn, the other a ruin. Hussey tells us that
the church of Paddlesworth was desecrated in 1852, but perfect, except that the north porch was removed. There is a door in the south
wall—it is Early English. The rest of Hussey's remarks, like these,
are not very correct : it never was a chapelry of Birling, and an
incumbent was presented to it, with Dode, in the person of Edward
Aldey, as proved by the register of Bishop Buckeridge in 1623.
The last record of the church is its mention as being valued at
nothing, and belonging to Sir J. Watton, in 1659, in the register of
Bishop Warner, of Rochester. Of the church of Dode we have
given all that is known. The Manor House of Dode is still called
Buckland : in it there is some fine old wood-carving, a relic of the
times of its ancient lords.

There are two monuments of this date in Ryarsh. One :—

In memory of Thomas White, Esq., of Congelow, Yalding, and of Calais Court in this parish, also of 53, Portland Place, London, who died Sept. 9th, 1883. This tablet was erected by Faithful friends.

The other is to the last rector but one. It is a brass, on which is inscribed :—

In memory of the Rev. Lambert Blackwell Larking, M.A., 37 years the beloved Vicar of this Parish, who died August 2nd, 1868, aged 71 years. He was son of John Larking, of Clare House, Esq., High Sheriff of this county, and Dorothy, daughter of Sir Charles Style, Bart., and married Frances, daughter of Sir William Jervise Twysden, of Roydon Hall, Bart., who, together with his brother, John Wingfield Larking, Esq., dedicates this tablet. Si Deus vobiscum quis contra nos (*If God is with us who is against us*).
To the glory of God, and in memory of the Rev. L. B. Larking, the interior of this Church was restored.

Rev. Lambert Blackwell Larking was born at Clare House, East Malling, Feb. 2nd, 1797. He graduated at Oxford, taking a second in Literis Humanioribus in 1820; he went on the grand tour with Mr. Lowthe, was licensed as deacon on his return in 1823 to East Peckham, and was presented by Col. H. J. Wingfield-Stratford to Ryarsh in 1830, and by Charles Milner, Esq., to Burham in 1837, after which he held the two livings together. He married, as stated above, the eldest daughter of Sir W. J. Twysden, Bart., of Roydon Hall. The Surrenden MSS. were presented to him by Sir Edward Dering, and the use he made of this present can be found in the Maidstone Museum. He founded the Kent Archæological Society in 1857, at Mereworth Castle, and it is to him that Kent owes her prominent rank among the counties for her accurate researches into the past. He indexed the *Pedes Finium* and the *Inquisito post mortem* records in the Archæological Society's volumes, and contributed several valuable papers, especially one on the Heart Shrine in Leybourne church. Many curious trees, such as the red cedar, the tulip, and others, were planted by him in the garden of Ryarsh, and still flourish there ; he was also a great observer of birds. He died, as above stated, on August 2nd, 1868.
There is also a window in Ryarsh Church :—

Dedicated to the glory of God, and in memory of James and Mary Phillips. James Phillips, b. 12 Aug., 1803, died 2nd Dec., 1886, Mary Phillips, born 6th March 1807, died 9th June, 1884. Dedicated by their children.

The third bell of Ryarsh has on it "Mears & Stainbank, founders, London 1879 "; but the donor's name is not inscribed.
There are several monuments of this date in Snodland church, and one of these shows that the men of this part are determined that they will not sink into insignificance, if we can quote this man as one of its heroes.
His monument is inscribed as follows :—

In memory of Thomas Fletcher Waghorn, Lieutenant R.N., who by extraordinary abilities and self-devotedness became one of the greatest benefactors of his country, by exploring and bringing to perfection the short overland route to the East Indies: he entered the royal navy at the age of 12, and served as Midshipman in the *Bahama* under Captain Wilson. He was afterwards engaged in the Bengal Pilot Service ; this first suggested the utility of the Overland Route, and enabled him to acquire the knowledge necessary for bringing this great undertaking to perfection, which he did under circumstances of peculiar difficulty and discouragement in the year 1841, and nothing could exceed the perseverance, energy, and untiring vigilance of the high-spirited officer when on duty, or his warmth of heart and most kindly feelings in private life. He died Jany. 7, 1850, aged 49 years.
This tablet was erected by the disconsolate widow, who felt much supported under her bereavement by the assurance that, "Whom the Lord loveth He chasteneth. and scourgeth every son whom He receiveth."
Also Harriet, widow of the above, died Jany. 19, 1856, aged 54 years.

Lieutenant Waghorn was the discoverer of the so-called Overland Route, viz., down the Mediterranean, across the Isthmus of Suez, then down the Red Sea. His genius pointed out to the world the necessity, in consequence, of a Suez Canal.
The Eastern window has this inscription under it :—

In commemoration of four stedfast witnesses to the truth of the Protestant faith, this window, showing the dates of their respective martyrdoms. was given by Henry Dampier Phelps, M.A., Rector of this Parish, died July 30, 1863, aged 88. Anne Ayscough, 6 July, 1546. Nicholas Ridley, 16 October, 1555. Hugh Latimer, 16 October, 1555. Thomas Cranmer, March 21, 1556.

The handsome stone cross which once stood nearly opposite Veles, was moved by Mr. Phelps to its present position in the churchyard ; when, on account of his people interpreting this as a step toward Rome, to pacify them the rector put in this window, to assure the people of the firmness of his views. The cross is one of the few market crosses remaining in this part of England.
Another of the church windows has this dedication :—

This window was placed by his bereaved wife to the endeared memory of Captain W. H. Roberts, Royal Engineers, a most affectionate and beloved husband, A.D. 1840.

There are also these inscriptions :—

Sacred to the beloved memory of Samuel Lee, of Holborough, who departed this life, August 15th, 1852, aged 26 years. "Blessed are they that mourn, for they shall be comforted."
And J. M. William Lee, of Holboro' Court, J.P. and D.L. for the county, for 17 years M.P. for Maidstone, died Sept. 29th, 1881, aged 80.
Also Christiana his wife, died Dec. 14th, 1871, aged 70.
This brass is placed in loving remembrance by their only surviving child, Sarah Smith, the Gleanings, Rochester, and her eldest son, Samuel Lee Smith, Larkfield.

Another window is inscribed :—

In memory of J. G. Le Marchant Carey, M.A., for eight years Rector of the Parish, obiit March 17, '85. This window was given to the Church by Miss

Mary Ann Poynder, Miss Isabella Rebecca Poynder, Miss Sarah Matilda Poynder, Miss Frances Ann Poynder. This family used to live at Holboro'.

William Henry Poynder, Esq., is the lord of the manor of Holborough at this present time.

Another window has :—

This window is placed to the memory of Ann Roberts, who died Aug. 11, 1881.

We may mention that the window to the eastward of the northern aisle has two figures—one a bishop, the other with a cockle—while above them is St. Martin. The two lower figures are no doubt the two St. Jameses.

On another window we read :—

Pater non est Filius

est est

Deus

est

Spiritus Sanctus

The church of Snodland has been lately repaired, but there are still many things to be done to this ancient edifice to render it thoroughly renovated. A former restoration, made about forty years ago, was, we must say, a most unkind handling of the ancient masonry of the church, and is quite out of keeping with this venerable edifice. The trade of Snodland in cement has largely increased the place in the last thirty years: this and the paper-making and paper-bag-making employ a large number of hands. About three years ago a battalion of the Cinque Ports Artillery was formed here. We cannot leave Snodland without giving one or two stories of the place, which was very wild and desolate in the early part of this century; and this will aid to explain the first story, which is taken from *Gleanings in Natural History*, by Edward Jesse, published in 1842, who says that he took it from the *Sportsman's Cabinet*. " Mr. Henry Hawkes, a farmer residing at Halling, in Kent, was late one evening at Maidstone market. On returning at night, with his dog, who was usually at his heels, he again stopped at Aylesford, and as is too frequently the case upon such occasions, he drank immoderately, and left the place in a state of intoxication. Having passed the village of Newheed (hythe) in safety, he took his way over Snodland Brook—in the best season of the year a very dangerous road for a drunken man. The whole face of the country was covered with deep snow, and the frost intense. He had, however, proceeded in safety till he came to the Willow Walk, within half a mile of the church, when by a sudden stagger he quitted the path, and passed over a ditch on his right hand. Not apprehensive he was going astray, he took towards the river; but having a high bank to mount, and being nearly exhausted with wandering and the effect of the liquor, he was most fortunately prevented from rising the mound, or he certainly must have precipitated himself (as it was near high water) into the Medway. At this moment, completely overcome, he fell among the snow in one of the coldest nights ever known, turning upon his back. He was soon overpowered with either sleep or cold, when his faithful defendant, who had closely attended to every step, scratched away the snow so as to throw up a sort of protecting wall around his helpless master; then mounting upon the exposed body, rolled himself round and lay upon his master's bosom, for which his shaggy coat proved a most seasonable covering and eventual protection during the dreadful severity of the night, the snow falling all the time. The following morning a person who was out with his gun, in expectation of falling in with some sort of wild fowl, perceiving an appearance rather uncommon, ventured to approach the spot; upon his coming up the dog got off the body, and after repeatedly shaking himself to get disentangled from the accumulated snow, encouraged the sportsman (a Mr. Finch), by actions of the most significant nature, to come near the side of his master. Upon wiping away the icy incrustation from the face, the countenance was immediately recollected; but, the frame appearing lifeless, assistance was procured to convey it to the first house upon the skirts of the village, when, a pulsation being observed, every possible means was instantly adopted to promote his recovery. In the course of a short time the farmer was sufficiently restored to relate his own story, as already recited, and in gratitude for his miraculous escape ordered a silver collar to be made for his friendly protector, as a perpetual remembrance of the transaction. A gentleman of the faculty in the neighbourhood, hearing of the circumstance, and finding it so well authenticated, immediately made him an offer of ten guineas for the dog, which the grateful farmer refused, exultingly adding that so long as he had a bone to his meat, or a crust to his bread, he would divide it with the faithful friend who had preserved his life : and this he did in a perfect conviction that the warmth of the dog, in covering the most vital part, had continued the circulation and prevented a total stagnation of the blood by the frigidity of the elements."

The next tale is one that happened in the time of Mr. Phelps. The rector was very friendly with a former Rector of Wouldham, with whom he used frequently to dine, and after that Mr. Phelps would walk home after dinner by the river side. He had been making himself very unpopular amongst a low class who inhabited Snodland at that time, by aiding the police, then newly established, to issue search warrants into the houses in the districts. Now it so happened the Rector of Snodland was a little man, and had a

sidelong gait, which was also the case with a tailor belonging to the
parish. One evening the rector, instead of walking home, after
crossing Wouldham ferry, by the river side, went into Rochester and
came over Rochester bridge, and returned through Cuxton; the
tailor, on the other hand, came back to Snodland by Wouldham
ferry, when it is supposed, being taken for the rector in the dark,
he was thrown into the river. His lifeless body was recovered
two or three days after, and was buried by Mr. Phelps, who looked
upon him as his scapegoat.

In the year 1873 another dreadful occurrence happened. A man
who had been frequently threatened by a policeman, Israel May,
that if he found him again in a state of intoxication he would
receive some punishment at his hands, was found, as it would appear,
where there was then a hop-garden (now a row of cottages), near
where Birling stream crosses the Malling road, by the policeman.
Both were very powerful men. The intoxicated man stated that he
saw May standing over him when he awoke, and threatening him
with his truncheon, and that he struck him : be that as it may, in
the course of a fierce struggle the policeman lost his truncheon, and
the following morning was discovered near the road in the hop-garden,
dead. The other man escaped into the Malling woods, but two or
three days afterwards gave himself up, and was tried and committed
to penal servitude for fifteen years for manslaughter.

On the first bell in Snodland we read, " Mears & Stainbank,
founders, London, presented Rev. G. Le M. Carey, Easter 1878."
On the second bell is inscribed " Mears & Stainbank, founders,
London, W. Lee, Esq., Holborough." The fourth has the same inscription as the second. The church has been lately partly renovated
by the exertions of the worthy rector, Rev. J. G. Bingley. Near
Snodland, but in the parish of Birling, is rapidly rising the new
church of Christchurch, Birling, on St. Catharine's Bank. The
church has been built by subscription. The foundation stone was
laid by the Hon. Mrs. Ralph Nevill, on April 30th in this year (1892).
In Trottescliffe church we notice the curious old pulpit, with a
sounding board, surmounted by a palm tree; and we learn by a
memorial in the church that " This church was beautified in
October, in the year of our Lord 1824 ; and the pulpit was the gift
of the Dean and Chapter of Westminster Abbey, and presented to
this parish by James Seager, Esq. S. Shrubsole, churchwarden, W.
Smith, overseer."

There are inscriptions under the windows :—

To the memory of our dear parents Edward John Shepherd, and Catharine
his wife, 1875.

And :—

To the memory of Charlotte Dalton, and her grandson, Francis Henry
Hayman Shepherd, 1873.

In the year 1866 the church of Allington was entirely rebuilt, and only one of the ancient monuments in the church left intact. The stones in the porch are almost effaced by the treading of feet. The tower, the only mediæval work allowed to remain, it is proposed now to sacrifice.

The modern monuments in the church are windows. The memorials are :—

Ad honorem Dei et in dilectissimæ uxoris Ameliæ Elizabethæ Holmes, quæ obiit xviii. die mensis Novembris, A.D. 1863. Memoriam hunc fenestram dolens sed fidens posuit maritus suus.

To the honour and glory of God, and to the memory of the Rev$^{d.}$ F. J. Marsham, Rector of this Parish, who died Jan. 29, 1852, aged 45, this window was erected.

To the honour and glory of God, and to the memory of Elizabeth Marcia Marsham, wife of the late Rev$^{d.}$ George F. G. Marsham, deceased April 20, 1849, aged 38, this window was erected.

Another is inscribed :—

By the children of Henry Godden, in affectionate remembrance.

In the churchyard all the ancient monuments are gone but this:—

Hic Jacet Richardus Thomas, Magister in Artibus utriusque academiæ, nuper Pastor hujus ecclesiæ, qui obiit Feb. 8th, 1656.

Another monument we notice is to the Rev. John Earle, Vicar of Aughton. Drayton's monument is gone, but we trust the Maidstone people will erect another to their distinguished fellow-townsman.

In Birling churchyard, near the entrance, we find this memorial:—

In memory of John Black, Esq.,* who was born at Dunse, in Berwickshire, on the 7th day of November, 1783, and died at Birling on the 25th day of June, 1855, aged 71 years. He was Editor of the Morning Chronicle for more than twenty-three years, and was highly esteemed by some of the most eminent men of his day. Also to the memory of Anne Croonek, who died at Birling on the 29th of August, 1852, aged 79 years.

The two noble yews, about 18 feet in circumference, must attract the notice of all who visit this churchyard.

There are few remnants of the ancient episcopal palace of Trottescliffe to be found, and the materials have been scattered : in a builder's garden in Malling may be seen a fine piece of carving belonging to it.

The palace of Halling is being quickly destroyed for the convenience of the cement works on its site.

We have spoken of what remains of St. Blaise and Newhythe chapels. St. Laurence, in Halling, has become a cottage, a solitary window alone telling its old use. The Aylesford chapels we have shown to have also disappeared, and we have described the state of Longsole chapel.

* He was editor of the *Morning Chronicle* when the late Charles Dickens, Esq., was appointed sub-editor.

The ancient parish church of Dode still remains, and the farmers pay tithe for Dode to Mr. Curtis, of Paddlesworth, as 'lord of the manor. It is decidedly a Norman church, but has remains of where there was a chancel arch, which my friend Mr. Livett pronounces an uncommon thing in the Norman churches. The late Vicar of Luddesdown erected a new church at Leywood, not far from here : it is a pity that he did not, instead, restore this venerable ruin. There is enough of the church still remaining to form the basis of another, and we do hope that in time it will be rebuilt : there is an aumbrey, and the ruin of a niche where was once probably the piscina.

Paddlesworth church (the twin church of Dode, the parish having been so long administered as Paddlesworth-cum-capella-Dodecirce) is another desecrated sanctuary, which has long been used as a barn. The parish is, for civil matters, stitched on to Snodland, but ecclesiastically it has nothing whatever to do with Snodland ; nor is it, nor ever was it, as some have maintained, a chapelry of Birling, but it is quite a distinct parish. We notice in it the Norman manner of laying stones, and the tufa quoins ; but the tufa ceases six feet from the ground, and we find it finished with Caen stone. Here is an aumbrey : there are traces of a rood screen : the arch was evidently restored and inserted not long before the church fell into disuse. It is to be hoped it will be re-consecrated and re-opened shortly.

In Aylesford church we have no Norman work, though there has been a church here from Saxon times. It is a handsome building of the fourteenth century, and has been recently restored. Three stone coffin lids are in the churchyard under the east window.

Allington church, as we have already stated, has been reconstructed *ab imis fundis.*

Addington church shows us many traces of old Norman work, but there are distinct signs of an addition to the westward when they built the tower. The porch has a very handsome piece of mediæval wood-carving over the entrance.

Ditton church, I am told, is as fine a specimen of Norman work as can be found in the neighbourhood.

East Malling church is a very handsome building, and will repay the visit of the archæologist ; there appear a few traces of the originals in it, Norman and Saxon.

Halling is an old mediæval church, which has, however, been so pulled about that there are no very early relics in it.

Leybourne still shows signs, especially in its northern wall, of its Norman origin.

Offham gives many proofs of an early Norman date, though the tower was probably a later re-erection. The different periods when this church was erected are well shown in the interior. Ryarsh shows by its ground-plan, its tufa quoins, and its masonry, that it is a Norman church to which a mediæval tower has been added. Those who restored the church some years ago sacredly kept its Norman landmarks; some modern innovations that have been made since injure the venerable structure.

Birling has a few traces of Norman work, notably in a window over the great east window.

Trottescliffe is a very early Norman erection; perhaps, as some antiquaries have thought, part may be Saxon. The sarson stone in the foundation of the south wall is worthy of notice.

Snodland does not appear to contain any relics of its ancient Norman church, and the repairs, which were done at a period when men did not leave old work untouched, but put their own work in anyhow, teach us how necessary it is to have a building repaired in keeping with its own history.

The chancel and tower of West Malling are Norman, and perhaps the work of Gundolph, or his school. The Georgian body of the church must be considered most ugly.

Those who wish to survey this valley, so rich in the relics of the past, and also in natural history—as we shall show in the other part of our book—can travel by the North Kent Railway from London to Maidstone, to the stations of Halling, Snodland, or Aylesford, all of which are in our district—this line was opened in 1855. The London Chatham and Dover, opened in 1875, also passes through the district, traversing the parishes of Addington, Offham, East and West Malling, Ditton, Aylesford and Allington, and having stations at Wrotham, Malling, and Barming on its route from Sevenoaks to Maidstone. Some of the stations about here are misnomers, and in our district Aylesford on the North Kent line is really in Ditton parish, and Barming station is in Allington or Aylesford parish. In concluding this sketch of the history of this valley down to our own time, we feel that all must agree with us in what a wonderful way every passage of the history of our country is illustrated here—by varied relics, the churches, and the monuments that we find around us.

CHAPTER XI.

"YOU should never look a gift horse in the mouth, nor for a lady's age in the parish register," says the old English proverb. We can assure our lady friends that we do not intend to make any disclosures of so mean a character, our sole purpose being to show how valuable these parochial records might become if properly utilised. It is a pity that some registers have been allowed to moulder away uncared for in the damp corner of a church, or thrown about in an old parish chest; I have much pleasure in stating that in this neighbourhood the majority of these treasures of the past are not thus treated by the various incumbents. The different Acts to regulate the registers are as follows: First, Cranmer's Act, in the days of Henry VIII., 1536, which instituted the parish registers; and, as already observed, Offham registers begin two years, and Trottescliffe four years afterwards. The registers were next legislated upon by Cromwell in 1653, of which we have already spoken, when he ordered registrars to be appointed in every parish; and further, that marriages should be solemnised before a Justice of the Peace after that year, the banns to be published in the market place, or in a place of worship. We have already referred to this order in the other parts of the book, and given examples of the entries. This interference with the marriage laws appears to have worked anything but well, as people after this went far and wide to be married; and it was not for a long time after these laws had been forgotten that anything like the old custom of " being married in your own parish," now fortunately once more sanctioned, though not nearly enough insisted upon, became the rule instead of the exception. We could mention several cases of the ways in which the registers were affected by these laws, but prefer to leave them to be mentioned under the heads of their respective parishes. After this no interference with the registration of the baptisms, burials, and marriages appears to have taken place till that most elaborately formal register for marriages was promulgated in 1751, of which nearly every church contains a copy, certainly all the parishes of this district of ours do. The following is a specimen of an entry under the Act, taken from the church of West Malling :—

Charles Bowles of this Parish, Clerk, and Catharine Weekly of this Parish Spinster, married in the Church by Licence this eleventh day of February, in the year one thousand and seven hundred and sixty, by me, J. Dennis, minister. This marriage was solemnised between us, Charles Bowles, Catherine Bowles, in the presence of B. Hubble, Anne Hubble.

In 1783 the Government fixed a duty of 3*d.* on all baptisms and burials, which is especially accounted for in some registers, notably that of Snodland. In the Higham register there is a curious entry, declaring the father of the child to be a rebel, for refusing to pay this duty. The custom of inquiring what was to pay when a clergyman christened a child, which the author remembers so well in the early days of his ministry, may have possibly been a remnant of the memory of this duty; because we never remember the most greedy ecclesiastic ever positively desirous of a fee at baptism. The Church of England charges *nothing* for her Sacraments, this being as proud a boast of our Church as there is no slave beneath the Union Jack is the boast of our nation. This iniquitous tax was repealed in 1791.

Soon after this came the orderly baptismal register, which is complete and useful, in 1812 ; but the burial register, which has been unfortunately carried down to this day, is woefully deficient, the more especially as now persons not belonging to the Church of England have to be recorded as buried by some particular person. The error that exists in these registers, which might have been corrected by attention to the registers of the last century, was the omission of a column stating the employment of the deceased. In one of our parishes we shall show the reader that this omission, had it been the form of our registers in the sixteenth century, would have caused almost inextricable confusion to persons of one name. To identify the deceased, though of course the Government registers may be very useful, still our old parish registers in the country should be kept up, and in them stated the station, as well as age and race, of the deceased ; and, moreover, whether committed in the body to the ground by a clergyman of the Church of England, or by any other minister, or by some person who agrees with his notions as to the future upon questions concerning which many persons would here wish me to enlarge. I fight shy, not because I am afraid, but because as this is not a controversial work, I do not wish to introduce into it any arguments upon such questions : the only thing I would remark is, that the late alteration of the Burial Law, 1880, has not been followed by a new burial register, which is sadly wanted.

The marriage register in 1814 followed soon after the baptismal and burial register, which, being defective, was improved by the one of 1837, which contains all that is necessary for the marriage register. Since that date there has been no alteration made in the form of any of our registers.

We shall now proceed to deal with the registers of the different parishes, prefacing our remarks with a description of the church, giving the names of the clergy, and peculiar entries—whether of subjects foreign to the register, rigmarole, incompleteness, or briefness —while at the end we shall mention the principal families of the parishes, the names of other gentry recorded in the registers, and the professions and trades entered therein. We shall deal with the parishes according to the age of their registers. The oldest registers in this district, as already stated, are those of

OFFHAM.

The church of Offham is of very ancient date, and contains relics of early Norman, if not of Saxon work. It was intended to build a north aisle, of which the arches can be distinctly seen, but this was abandoned, and an Early English porch shows that it was some time after the original erection. The hagioscopes or squints are another feature of this church, by which it differs from any just round here. There are two shields and a figure in old stained glass. It is dedicated to St. Michael. The various incumbents of Offham are as follows :—

In the time of Edward II.

> Master Bartholomew.
> Richard de St. Quentin.

In the time of Edward III. :—

> Robert Randolph.
> 1336. Robert Joy succeeded Randolph.
> „ Robert de la Chambre.
> „ Edmund de Harwedone, changed with
> 1372. Peter de Burton Leonards (Vicar of Beatrichsden).
> „ Nicholas Balsham.
> 1400. John Miller changes with Balsham ; he had been rector of Keston in this county.
> „ John Carter.
> 1424. Thomas Westhorp ; he was vicar of Throwley.
> 1427. William Estryngton, vicar of East Farleigh, by change with Westhorp.
> „ Clement Willis.
> 1436. Henry Esthalbe.
> 1442. John Haselor succeeded Esthalbe.
> 1444. John Newbot succeeded Haselor.
> 1454. Thomas Brown.
> 1458. William Bele.
> 1493. William Spayne.
> 1498. Richard Wantone, on the death of Spayne.
> 1512. William Whiteacres, on the death of Wantone.
> 1534. Thomas Dixon.

On the death of Whiteacres (in his days the register of Offham commences)—

> 1545. Thomas Burrell.
> 1554. Richard Kidde.

1567. Henry Barnes.
1569. John Moore.
1572. John Baxter.

The cup and paten date from this rector's time : he is recorded as buried here.—" John Baxter, pson of Offham, was buried the xviiith day of November, 1587."

1586. Robert Holder.
1590. John Cooper ; presented by the king.

The first bell of Offham dates from this rector's first year ; he is not buried at Offham, but his son, John Cooper, is.

1632. Robert Brownell : his wife Anne and he are both buried at Offham.

The registers run :—

Anne, the wife of Robert Brownell, was buried the three-and-twentieth day of May, 1641. Robert Brownell was buried the 20th of May, 1647.

In his days the second bell of Offham was hung.

1647. Edward Masters.

The wife of this incumbent is buried here, but not himself, it would appear from the register :—

1649. Mary, the wife of Edward Masters, was buried the three-and-twentieth day of May.

Perhaps he was one of the ejected ministers.

1660. William Polhill.

This rector was also Rector of Addington from 1673. The paten of Offham is dated 1675. In the Offham register we read that Mr. Polhill was married at Addington by his predecessor there :—

1666. William Polhill and Margaret Deane, married at Addington, by Mr. Peter Davys, Rector, *ibidem* on Monday, 16 Jan.

Mr. Polhill was no doubt one of the Polhills of Sundridge. He is also buried here, when the clerk corrupts his name to Polly :—

1675. William Polly, Rector of the Parish Church of Offham, and of the Parish Church of Addington, was buried the 15th day of October.
1675. Henry Miller.
1708. Samuel Bickley.

It would appear from the registers that Mr. Bickley was twice married, as both his wives, himself, and his infant son are buried here :—

1716. Jany. 4, Buried Mrs. Bickley.
1721. August 7th, Buried Thomas Bickley, infant.
1736. Sept. 20th, Buried Mrs. Bickley.
1740. Feby. 23, Obiit Saml. Bickley. M.A., Hujus parochiæ rector et 27°. sepultus est. He was Curate of Snodland, as well.
1740. William Miles.

He and his wife are both buried here, the clerk showing an amount of carelessness over the register, as we read :—

Omitted in October last, the Rev. Mr. Miles, Rector of the Parish, died Oct. 15, 1746, was buried Oct. ye 21st.

1762. Dec. 19th, Mrs. Miles, relict of ye Rev. Mr. Miles, Rector of this parish, was buried.

He was previously curate of East Malling.

1746. Bessworth Liptrott.
1777. John Liptrott.
1830. John Cecil Hall.
1832. Frederick Money.
1869. William Pellowe Philp, buried in the churchyard, as appears by the register, Nov. 26.
1873. William Frederick Chambers Sugden Fraser, previously chaplain of the prison at Maidstone. His youngest daughter was christened here.

Besides its own clergy, the parish register of Offham has the following entries. Two to Mr. Rabbett, formerly vicar of Birling :—

1639. Michael Rabbett, clerke, and Katherine Hunt, widow, were marryed Jany. 2.
1670. Mr. Michael Rabbett, dyed at Birling, but was buryed in Offham, on Monday, 6th Feby. ; he was layed in the Chancell, close to the groundwork of the Chancell, on the north side, near the belfry.

Besides these, in 1838 there is the marriage of Rev. Henry William Steele, and Augusta Graham Hutchinson.

The registrations of Offham have not been regularly kept. There is no entry in 1543, and again none are found in the years 1603 to 1608 in the baptismal register. Again there are no burials in 1555, 1556, 1566, 1568, 1573, 1574, 1578, 1583, 1585, 1586 and 1589; likewise from 1676 to 1683. There are no marriages entered also in 1539, 1540, 1542, 1544, 1545, 1546, 1547, 1548, 1549, 1557, 1558, 1562, 1567, 1568, 1570, 1571, 1572, 1573, 1574, 1585, 1586, 1587, 1588, 1589, 1590; they are also wanting in 1601, 1605, 1609, 1619, 1621, 1634, 1635, 1637, 1638, 1646, 1670; and during the whole period from 1675 to 1690, and again in 1700, 1701, 1702, 1706, 1707, 1708, 1713, 1714, 1727, and from 1732 to 1736. This will show, together with the register above mentioned, that the parish clerk did not keep the registers in first-rate style, but probably left them not filled in, and then wrote down from memory or from scraps of paper.

The preface of the books of this parish simply states : " The register book of the parish of Offham, beginning 1538."

1666. Thomas, ye sonne of Thomas and Mary Clarke, strangers : came a hopping, baptised 4 September.

This has already been referred to as giving us a very early mention of hopping.

1660. Charles Osmer, the son of Lawrence Osmer, baptised in the parish of Offham, 24 June. To be referred to its proper place.

Here we have the parish clerk again forgetting to put his entries in properly, as this entry is placed after those of 1660.

> 1731. March 19th, Bapt. Elizabeth, daughter of Clement Jarvey, of the parish of Maltham, in Stagg hundred, in Norfolk, about 12 miles beyond Norwich, as you go from London.

Here the parish clerk treats us to a rigmarole description, as he does, too, in the next following :—

> 1732. July 23rd, Bapt. James, son of John Rogers, and Mary his wife, of All Saints, as he says, in ye town of Evesham, in county Worcester.

The following entries show that very little trouble was sometimes taken to ascertain anything :—

> 1543. The third day of January, were two children of a stranger buryed.
> 1585. A little gyrle, of the psh. of Wrotham, was buryed ye first day of May.
> 1586. Grant, a nurse child, was buried the xviith day of January.
> Anne, the daughter of John Marshaller Stow, an infant, buryed in the Chancell with my leave, 27 June, 1667, being Thursday, and by me, William Polhill.

The next two are interesting as giving the names of Holy Days :—

> 1660. Ash Wednesday, 27 February, Mrs. Frances Deeritz, buried in the Chancell.
> 1661. Annunciation of the Blessed Virgin, commonly called Lady Day. Munday 25 March, Martha, the wife of William Martin, was buried.
> 1669. Richard, the son of Richard Evans, dyed at Town Malling, and was buried at East Malling, on Monday, 30 December.

Why this is inserted in the Offham register we are not informed.

> 1674. Mrs. Clarke, buryed in her husband's grave, under the great Tombstone in the Churchyard, on Friday, being the 3rd of April.
> 1698. William Coleman, Churchwarden, buried Dec. 5th.

To describe the parish clerk as such is by no means uncommon, but this description of the churchwarden we have not met elsewhere. We have many examples in this register of persons being styled "stranger," "widow," "single person," "single man," "single woman," "poor traveller," "poor man," "poor woman," "poor travelling man," "poor travelling woman," and of "foundlings" with or without their names; but Offham only shares these eccentricities of the parish clerk with other parishes, and these entries are so common and would take up so much room that we cannot mention them again. But to proceed :—

> 1732. November 2, Buried Michael Smith, Vid. (Latin *Viduus*, this is the short for, and means, widower; this term in English we find sometimes used.)

1741. In this year Affidavits are first mentioned ; in other parishes, there are numberless entries of them long before ; they require explanation. By an act of Charles II., it was ordered that all persons, in order to encourage English trade, should be buried in woollen, and affidavit of the fact be given to a Minister, or a Justice of the Peace, or a fine was to be paid, which we shall give some instances of being levied; this absurd law was not repealed till George III.'s reign, A.D. 1814. Pope makes the vain beauty of the day say :

> "Odious in woollen, 'twould a saint provoke,
> Were the last words that poor Narcissa spoke.
> No, let a charming chintz and Brussels lace,
> Wrap my cold limbs and shade my lifeless face :
> One would not, sure, be frightful when one's dead ;
> And, Betty, give this cheek a little red."

> 1669. Samuel McDood, and Catherine Pywell of Wrotham, were married 1 June, by Licence, from Gilbert Archbishop of Canterbury, dated anno translationis sexto, 1669, W^m· Polhill.

This entry is misplaced :—

> 1661. John Ashton, and Catherine Gardener of Seale, were married by licence, out of ye spiritual Court of Rochester, 13 June, by me W^m. Polhill.

These marriages show us that though this parish was in Rochester diocese, it mattered not whether you went to the bishop or archbishop for a licence.

The next is very rigmarole :—

> 1669. Lawrence Sales, and Elizabeth Byrshott, were marryed by licence, granted by Gilbert Archbishop of Canterbury, dated nono die Aprilis, Anno Domini, which said marriage was celebrated in the parish church of Offham, on Thursday the 15th of April, 1669.

The next is a curious entry as regards the licence :—

Nicholas Hamon, jun. of Hadlow, and Anne Gilbert of Tonbridge, were married by licence, signed Guill. Trumbell, May 15, 1659.

We have a good number of marriages, as appears to have been commonly the case with churches in the country districts, solemnised in this church, of which the bride and bridegroom were from different places ; thus we have them from Aylesford, Maidstone, Yalding, East Malling, Hadlow, West Malling, Trottescliffe, and Tonbridge.

The descriptions of the bride at this period are very curious. She is described as a "spinster," a "single woman," and a "widow-woman"; the bridegroom is put down as a "single man," if his status is given at all.

Besides in the old book finding baptisms, marriages, and burials, we have a few extraneous matters. First, there is the appointment of a registrar in time of Cromwell. It is as follows :—

These are to certify that I do approve of the choice of Andrew Dunning to be register for the parish of Offham, he having taken his corporal oath before me this 27th day of June, 1657.

The ancient way of collecting money in church was by briefs, which were documents issued by the Government and read by the clergyman together with banns and citations in the Communion

Service, after giving out the Holy Days and notice of the Lord's Supper. We shall have occasion to mention several of these briefs in different parishes. In the parish of Offham we learn that they were read for fires and for the rebuilding or restoring of the churches and other purposes; amongst other churches for which aid was sought here, were Gravesend, Gillingham, and Cliffe in this county, and assistance for the Protestants in Lithuania. Besides these we have a memorandum of Mr. Polhill going the bounds with the *ancient* men of the parish, Charles Easdown, 73, William Robinson and Hugh Tresse. Moreover, there is an agreement between the said Mr. Polhill and John Austen for rating a hop-ground—a case perhaps of extraordinary tithe in the seventeenth century.

> 1719. October ye 17th, married Anthony Chambers, a lodger in this parish, and Elizabeth Wells of the same, single woman.
> 1720. October 2, married John Row of the Parish of Ryarsh, and Eliza Wells of the parish of Offham, spinster, with banns.
> 1743. Dec. 26th, Baptised John Clarke, aged 24 years.
> 1747. Dec. 18, James, son of John and Mary Girum, a soldier in Captain Noble's company of Scotch fusiliers, was baptised.
> 1753. The new style which had been used in ye greatest part of Christendom from the year 1583, was introduced into England by Authority of Parliament, ordaining that all Entries, Deeds and Records, should be dated accordingly.

This is interesting as giving the date of the change of the commencement of year in this country, from March 25th to January 1st. Moreover, the eleven days by which our calendar was wrong were ordered to be omitted. This caused great disturbances, and since the Pope, Gregory XII., had issued a brief to enforce this new calendar in Papist countries, it was foolishly considered a movement towards Rome.

The next entry's meaning has been already commented upon :—

> Oct. 21, 1783. New Act of Parliament took place with respect to Burials 3d on each, and also on Baptisms.

This was taken off in 1787.

> Jany. 1st, 1792. Thos. Blake, titheman, buried.
> Sept. 18th, 1792. Thomas Hodges, aged 85.

This is the first burial mentioning the age of the person. The following burial entries are largely descriptive :—

> 1800. Elizabeth, widow of the late John Knell, died ye 28 of ffeb., about four in the afternoon, buried March 8.
> 1800. Died within a few hours of each other, William Broad, yeoman, and Rachel his wife, buried May 11th.
> 1800. Elizabeth Spearman, wife of Robert Spearman, overturned in the Chatham stage going to London, and killed on the spot in a moment ; buried in Offham church, Dec. 26, aged 49. For burying in the church and putting up a memorial monument, I charged, and Mr. Spearman paid me, four guineas or £4 4s. 0d. John Liptrott, rector of Offham.

This is one of the many little histories of accidents we find in the registers, as also is the following :—

Catherine Ray, who, being brought in by the Coroner Mills and his inquest a lunatic, disordered in her senses, drowned herself, buried April the 18th.

The next is interesting as a saint-day memorial :—

1801. Addison Frances, wife of Friend Addison, the elder, buried March 17th, St. Patrick, aged 62, buried at Ryarsh.
1809. Frances Douse, wife of William, buried Jany. 30th.

The next is an early mention of Reception into Church :—

1769. Eliza, daughter of John and Eliza Longford, privately baptised Sept. 27, and admitted into the publick congregation Oct. 9th.

And the following another early hopping record :—

1780. Uriah, son of Henry and Hannah Styles, baptised, who came here hopping from Mayfield, in the county of Sussex.
1863. March 2, Elizabeth Broad, aged 103, buried.

The next oldest register to Offham is

TROTTESCLIFFE.

This commences with the words, " The Register-Book of Trottescliff, of all the Christenings, Weddings, and Burials in the year of our Lord, 1599. Anno Elizabethae Quadraginta secundo et Decimo Septimo mensis, February." This register, however, begins in 1540, or 59 years before this date.

The church of Trottescliffe, dedicated to St. Peter, appears to have in it some remains of the Saxon period. In the foundations on the south side, we find one of the stones which form the neighbouring monuments of Coldrum and Addington. There is a wonderful absence of memorials. They consist of one to a Rev. John Morgan, and family, and Mr. Lloyd, a rector of the Parish, a brass to William Cotton, and windows to the present rector's family. Thorpe has omitted the brass from his antiquities of the diocese of Rochester. There is a small ancient glass window light representing the Father as an old man, the Son Crucified, and the Holy Ghost hovering as a Dove between the Father's mouth and the Cross. The pulpit we learn from an inscription came from Westminster Abbey; the sounding board is supported by a carved oak palm-tree : the inscription runs :—

This church was beautified in October, in the year of our Lord 1824, and the pulpit was the gift by the Dean and Chapter of Westminster Abbey, and presented to the parish by James Seager, Esq. S. Shrubsole, Churchwarden, W. Smith, Overseer.

The following is a list of the rectors of Trottescliffe :—

Robert 1176.
John 1185.
James 1235.   }   These four rectors appear in deeds.
Nicholas de Rokclunde 1256.

Richard Poynz, 1327.
Edmund Rober, pro Richard Poynz, 1327.
Richard de London.
1337. John de Dennyngton. He was also rector of Snodland.
1337. William de Middleton. He also followed Dennyngton at Snodland.
1341. John de Everyng.
1347. John Gilbert, changed with the last. He was previously vicar of Tenterden.
1348. John de Bradewey.
1349. John de Cranbourne, exchanged with Bradewey, he was rector of Ichene (Itchen, Hants).
—— Robert de Vaghne.
1355. Stephen Randulf, on resignation of last, afterwards rector of Cowden, one of the executors of John de Sheppey.
—— John Wolfetche.
1361. William atte Dene. previously of Stopham, in the diocese of Chichester.
1361. Robert Fynchecock, on Dene's resignation.
1361. John de Hanneye, prebend of Wenlakesbarn.
1369. John de Whytecherche, previously rector of Lower Hardres, by exchange.
1400. John Cheyne.
1413. John Puttencye (by exchange with Cheyne), previously vicar of Milton, Canterbury.
—— Thomas Wale.
1425. John Mankyn (by exchange with Wale), previously rector of Fetcham, Surrey.
1425. Henry Adesham (by exchange with Mankyn), previously vicar of Wadhurst, Sussex.
—— Andrew Malton.
1434. Roger Haynes, (exchange with last), previously vicar of Woking, Surrey. He desired in his will to be buried here, but no monument remains.
1439. Marmaduke Skelton.
1471. John Bolun.
—— Richard Bonde. At this time, William Cotton, whose brass is in the church, left a silver-gilt chalice and two silver cructs to the church.
1497. Thomas Cartewrighte, on resignation of Richard Bonde.
1500. Richard Carpenter.
1500. Alexander Bukley, on resignation of Carpenter.
1513. Marmaduke Waldeby, afterwards, 1520, vicar of Brenchley, Kent.
1514. Thomas Shawe, on resignation of Waldeby : his burial is entered in the register. " 1543. Thomas Shawe, parson of Trottescliffe, was buried 5th day of April."
1543. Thomas Bull, Prebend of Rochester, on death of Shawe. .
1546. Bartholomew Bowsfell, was deprived on accession of Queen Mary.

As we find a gentleman in this parish connected with the Wyatt rebellion, we may not be wrong in concluding, that he was deprived for his share in this rebellion. He was restored on the accession of Queen Elizabeth, and Robert Salisbury, rector of Addington, who had filled his place in his absence (1554-1560), was deprived of Trottescliffe in his turn. Nicholas Heath, Bishop of Rochester, conveyed the patronage to Henry Bowsfell, John Sibell, Esq., Thomas Ffurnes, and Thomas Bowsfell. Sibell presented Salisbury, but Henry Bowsfell having left the patronage to Bartholomew, he presented himself the second time.

The old cup belonging to Trottescliffe bears the date of this period; about this period the bishops of Rochester appear to have left Trottescliffe palace.

1578. Thomas Bowsfielde, presented by Edward Webb.
1598. Thomas Rither, presented by the Lord Chancellor.
1608. Thomas Busfield. The Bishop of Rochester once more patron.
1610. In March, Thomas Alchin, minister, is mentioned.
1621. Edmund Jackson, previously rector of Norton, Kent ; he was chaplain to the bishop of Rochester, and prebend of Rochester cathedral.

In the register we read :—

1626. Edmund, the sonne of Edmund Jackson, Dr. in Divinitie, the parson of this parish, was baptised the 10th of November.

It appears that owing to his official duties elsewhere he kept curates at Trottescliffe, as we read :—

Mr. Godden, Curate of the Parish of Trottescliffe, was buried the fifth of September, 1635.

Then :—

Susanna, filia Magistri Johannis Mann, Clerici, sepulta fuit 24to, die Augusti supradicto anno. (Susan, daughter of Mr. John Mann, Clerk, was buried on 24th day of August, in the above named year, 1635.) Jana, filia Magistri Johannis Mann, Clerici, sepulta fuit primo die Feb. 1638. (Jane, daughter of Mr. John Mann, Clerk, was buried the first day of February, 1638.) Constantia filia Nathanieli Starke, Clerici renata fuit septimo die Junii, Anno D$^{ni\cdot}$ 1640. (Constance, daughter of Nathaniel Stark, was born anew on the 7th day of June, in the year of our Lord 1640.)

And again :—

1651. John Clarke, Cleric, was buried July 3rd.
Dr. Jackson died in 1652.

Trottescliffe Bell dates from this period.

John Head appears in 1652 as Minister ; in the registers under this name he is mentioned several times.

1653. Richard, the son of John Head, minister, was baptised the twenty-ninth day of June.
1654. Sarah, the daughter of John Head, minister, was baptised the twentieth day of August.
1657. Ann, the daughter of John Head, minister, was baptised the twentieth day of March.
1658. Francis, the son of John Head, minister, was baptised the 18th day of July.
1658. Francis, the son of John Head, minister, was buried the 21st day of November, in the year of our Lord 1658.

Margaret, the wife of John Head, minister, being also the eldest daughter of W$^{m\cdot}$ Bysshe, of Hengetake, in Worth parish, in Sussex, Esq$^{re\cdot}$, and about 40 years of age, dyed the 25th day of May, and was buried in the Chancell of Trottescliffe church, near ye Communion table, the 28th day of the said month, in the same year of our Lord.

Mr. Head, not being mentioned as instituted by the Bishop of

Rochester, was probably a minister under the Commonwealth, as was also his successor William Woodward.

The Rochester registers speak of Edward Archbold as instituted in 1652, on the death of Edmund Jackson. In the library of the Society of Antiquaries MS., it is stated that he was inducted to Trottescliffe in 1662, on the deprivation of Woodward, who no doubt was a Puritan minister. Edward Archbold was also rector of Kingsdown with Maplescomb, and chaplain to the bishop. This gentleman had several curates, with whom he appears from the register not to have got on, or else they got preferred.

1666. Francis Norton, Clerk, and Curate of this Parish, buried Dec. 19.

In 1667, we find "Samuel Attwood, Curate." On the opposite page of the register we read, "John Freeman was Curate 1671-2"; who adds, "there was four between us" (himself and Attwood).

1690. John Cooper was instituted on the death of Archbold.
1691. Edward Roman was instituted on the resignation of Cooper. Roman was also vicar of All Saints, Maidstone.
1692. Thomas Brett was instituted on the resignation of Roman. He was a native of Betteshanger, in this county, and was born 1667, and educated at Wye, and Queen's, Cambridge, and Corpus; he was rector of Betteshanger, 1703, afterwards vicar of Chislet, and held the rectory of Ruckinge, 1705. He resigned his preferments on the accession of George I., and was a nonjuror; he was also an author.
1695. John Warren was instituted on the resignation of the last.

In his time a statement of the patronage and value of the livings was entered in the registers of the Rochester diocese; and we learn that Trottescliffe was in the patronage of the Bishop of Rochester, 1695, and was worth £10 2s. 11d. Mr. Warren was vicar of St. John's, Margate, 1703—1705, and in 1709 was prebend of Exeter. The silver paten dates from this period (1699), it was perhaps given by Mr. and Mrs. Baristow, the initials P.A., Paul and Ann Baristow being engraved thereon. Mr. Baristow was curate of Trottescliffe, and afterwards vicar of Graine. His wife and he are both buried here.

April 20, 1705. There was buried in woollen Mrs. Ann Baristow, of ye Precincts of ye Cathedral Church of Rochester; affidavit was made before D. Hill, Vicar of St. Margaret's.
Feby. 23, 1716. There was buried in woollen the remains of Paul Baristow; affidavit was made before Mr. Babb, Vicar of West Malling.

In the church of Trottescliffe, we read :—

"*Benefaction.* The Rev. Paul Baristow, by his will dated 30th day of March, 1711, gave £100 for the purpose of purchasing an estate, the rent of which was to be applied in procuring the instruction of the poor children of Trottescliffe in reading and the church catechism, and his executrix, Mary Goodwin, likewise contributed £50 for the same purpose, which with the said £100, and the improvement thereon amounted to £180, and which, after having established a school for the said purposes, she converted into an annuity of £9 for ever, for the support charged on the following lands in the said parish : viz., the parcels of land called Upper Crocklands and Lower Crocklands, containing ten acres more or less with their appurtenances, and also two parcels of land containing three acres, and also a field or close called Street-end, containing two acres, and which said lands are more particularly described in an indenture dated the 28th of January 1719, and remaining among the records of the Dean and Chapter of Rochester."

Hasted mentions a certain Worlidge as incumbent of Trottescliffe after Warren; but he has reversed the order of the next two incumbents Cockman and Lamb, and therefore he is not much to be trusted. However, Robert Worlidge, who in 1701 held the parishes of Ryarsh and Addington, had three children christened here as the registers state :—

> 1676. Hannah, daughter of Mr. Robert Worlidge, clericus, and Elizabeth his
> wife, was baptised Jan. 2.
> 1680. Elizabeth, daughter of Mr. Robert Worlidge, clericus, and Elizabeth his
> his wife, was baptised May 23.
> 1683. Mary, daughter of Mr. Robert Worlidge, clericus, and Elizabeth his
> wife, was baptised December 8.

I think, therefore, it is most probable that Worlidge was curate of Trottescliffe. Warren, it is certain, had curates, as Malling register speaks of Mr. Thomas Harper as curate here 1706.

> 1709. Charles Lamb was instituted to Trottescliffe on the resignation of the
> the last.
> 1723. Bartholomew Hughes, afterwards vicar of Barnston and Laver Parva,
> Essex.
> 1724. Thomas Cockman; he had been presented to East Malling in 1718, and
> obtained a dispensation to hold Trottescliffe with that vicarage.

In the year 1738, the West Malling register speaks of a certain Mr. Morrison as curate here. Perhaps Mr. Hugh Pugh, who was vicar of Birling, was only curate also, though Hasted speaks of him as rector in 1722.

> 1744. John Elton succeeded on the death of Cockman, in 1747 he was
> instituted to the rectory of Speldhurst.
> 1747. James Webb, who had been curate of West Malling and was presented
> to that living in 1748, succeeded on the death of Elton.
> 1759. Francis Lloyd was instituted at this date. He was buried here Oct. 2nd,
> 1778, as we learn from register.
> " 1778. The Rev. Francis Lloyd, A.M., rector of this parish, was buried
> October 2nd."

And also we read :—

> 1773, September 20th. John Morgan, B.D., rector of Medburn, commissary of
> Richmond, and precentor of St. David's, aged 73.

As this latter parson is declared by their memorial tablet to have been buried in the same vault with Francis Lloyd, and Mary Phillips sister to Mr. Morgan, is declared as also buried here, we have a tolerable proof that Mr. Lloyd was connected with one of the chief clerical families of his day.

In 1778 we read in the register :—" Edward Evans, son of the Rev.

Mr. Leach, curate of this parish, was baptised," showing that Trottescliffe had its curates even at this time.

In 1779 Francis Taynton was instituted. He was formerly vicar of Frindsbury and afterwards vicar of West Farleigh, which he held with this till 1794.
1794. William Crawford, he was examining chaplain to Bishop Horsley. He was archdeacon of Carmarthen from 1793 ; he held Milton with Trottescliffe from 1797.

The silver alms dish is inscribed : " This plate was presented to the parish of Trottescliffe by the Rev. Wm. Crawford, D.D., rector, Sept. 11th, 1821." In 1824, as above stated, the church was repaired, and the pulpit from Westminster Abbey given by John Lys Seagar, Esq., who was buried here 28th October, 1872, aged 88 years.

1827. Edward John Shepherd was presented by the Lord Chancellor ; he was rector of Luddesdown 1840—1856. In 1844 the church was repaired. On Advent Sunday 1866, the present altar table was given by the rector.

In 1873 the following inventory of church goods was made and placed in the register :—

Plate—Flagon, cup, plate and paten.
Four old register books, 1540—1813.
Two marriage register books, }
One of baptisms,   } in use.
One of burials,   }
One folio Bible.
One folio Prayer Book.
One quarto Communion Service.
Book of the Offices.
One vestry book.
One church book.
Deed of Assignment to Trustees of an annuity for perpetuating Trottescliffe
    Charity School.
Two surplices. Oak communion table and one chair.
One chair in vestry and one table.
Eight open seats for the people and six forms for the children.
One white communion tablecloth ; organ with five barrels; iron chest for
    register books ; three cushions for communion rails.

In the year 1868, the patronage was acquired by C. W. Shepherd, Esq., who is now the rector. His father Mr. Shepherd was author of *The History of the Church of Rome to the end of the Episcopate of Damasus.* He died in 1874, and was buried in the churchyard. As the register shows, " Edward John Shepherd, buried Dec. 4th. 1874, aged 73." Mrs. Shepherd is also buried here and one of their sons; and three of their sons, including the present rector, were baptised here.

In 1875, Charles William Shepherd, the present rector, was instituted. The east window was inserted in memory of the late rector in 1875. The south window, next the tower, originally in Luddesdown church, was inserted in memory of Francis Henry

Heyman Shepherd. Ten years later the west window was inserted. In 1887, the western window of the north wall was inserted in memory of the Queen's jubilee by the rector ; and, in 1855, the west wall of the church was entirely rebuilt by him. Besides the inventory given above, the present rector has added :—

1875. One red communion tablecloth.
Sixth barrel added to the organ, and first used on Advent Sunday.
1882. Four wooden six-light coronæ lights, with chains to suspend them from beams.
Two brass side-lights for pulpit.
One glass decanter for communion.
One carved oak Elizabethan chair, given to the church 1877, and now placed at north of communion table.
1885. Two parchment register books, one for baptisms and one for burials.

The present rector cannot be too much praised for the care he has taken in copying out a new edition of the registers, and for the great trouble he takes with all the affairs of his church and parish and their history. Besides the clergy already mentioned, we have the entry :—

1618. John Cramp, vicar of Thornham, and Ellen Gibbon, were married ye fifteenth of October, anno predicto.

In the old registers of Trottescliffe, we have mentioned two or three of the Latin entries, which are rather numerous in these books :—

1677. Jana filia William Champion, baptis 16th November. (Jane, daughter of William Champion, baptised.)
1677. Johannes Bell, filius Johannis Bell, sepult September 23. (John Bell, son of John Bell, buried September 23.)
1678. Filia Rogeri Tomlyn Gener(osi), sepult September 24, 1678, in laneis sec(undum) formale statuti in eundem finem facti et provisi quam quidem legitime sepulta Maria Uxor Henrici Martin de Addington, suo warranto testata est serenessimo, W. Twysden, Milit et Baronett, 27° die Mensis Presentibus, {Thomas Sharbrook,}qui sub suis sigillis, {et Johanne Luxford,}idem attestate sunt. (The daughter of Roger Tomlyn, gentleman, was buried September 24, 1678, in woollen, according to a form of statute made and provided for the same end. Mary, the wife of Henry Martin, of Addington, on her own oath, testified her indeed legally buried to the most worshipful W. Twysden, Knt., and Baronet, the 27th day of the month, in the presence of Thomas Sharbrook and John Luxford, who witnessed the same under their own seals.)

One of the Latin entries mentioned before is interesting, viz. :—

1576. Thomas Sanctilis filius Anthonii de Sanctilis nobilis Majoricensis capitanci Brabantiorum mense, July xvi., baptisatus fuit. (Thomas Sanctilis, son of Anthony de Sanctilis, a noble of Majorca, captain of the men of Brabant, was baptised the 16th of July.)

Anthony probably was one of the prisoners brought over to England from the struggle in Flanders, and hence we have the fact that, while he belonged to Majorca he captained the men of Brabant, explained.

Besides Latin entries, which Trottescliffe perhaps possesses more than her share of, we have in addition to the common quaint entries the following :—

(Before mentioning other matter in this register, we should notice the number of Goddens to be found in the books. In order to distinguish them at an early period they had to be described; and we find sometimes their place of residence, sometimes their calling mentioned.) We have :—

1606. Thomas Godden, of the Court Lodge.
1608. Thomas Godden, at the Nut Tree.
1612. Thomas Godden, the tannor.
1617. James Godden, the warrener.
1654. Thomas Godden, the baker.
1656. James Godden,* Mingo.
1599. James Godden, the yeoman.
1635. Mr. Godden, curate.
1640. James Godden of Rouses.
1640. James Godden, ye butcher.
1647. James Godden of the banke.
1680. John Godden, gent.
1600. The 6 daye of July was baptised, Elizabeth Godden, the daughter of George Godden.
1603. Baptised, John Baker, who died.

Sometimes only name and date are given, as :—   .

1605. John Goldsmith, 27th of December.

The next are peculiarly rigmarole :—

1608. James Latter was baptised 27 daye of November, and was buried the 30 daye.
1618. John the son of Thomas Skudder, was baptised the xviith of September : this John was christened the 17th of January.
1618. John had his sonne, James Monck, baptised the 18th of May, ann. predict.
1640. William ye son of Robert Hilles, clerke of ye parish, was baptised 30 day of August, Ann. Dni. 1640.

This is one of the earliest mentions of a parish clerk in this district ; and though they are often recorded in the burial register, we rarely find them mentioned elsewhere.

The next is curious :—

1647. Robert, son of John Pye. a traveller or harvester at Birling, baptised Aug. 22.

The following shows the loyalty of Mr. Jackson, or Mr. Clarke :—

Mary, ye daughter of James and Mildred Attwood, was baptised Jan. 30, 1649 ; the same day that King Charles I. was beheaded.
1654. Mary, the daughter of Thomas Wellard, under the Hill, was baptised on Thursday, being the 20th day of July, in the same year of our Lord.

---

* This one is distinguished by his nickname : besides these there are several Goddens n ot distinguished. This seems to have been the home of the name.

1659. Henry, the son of Henry Henfield, of the pish of Meopham, was baptised here for the better convenience of the parents and people, they being remote from their parish church, the 4th day of April, in the year above said.

1663. Elizabeth Attwood, the daughter of William Attwood, and Sarah Attwood, his *deare beloved* wife, was baptised the 31st day of July.

1540. John Hills, and Anne Fullor, was wedded the 20th of June.

This is the earliest way of entering the marriages. Afterwards the man's name alone is mentioned after this style for several years :—

1592. Richard Baggace was married the xxv. of August.

In 1601 we have this peculiarly spelt entry :—

Moycese Helbe and Marye Stone was married 31st day of August.

Then we have from 1645-1647 this simple statement :—

William Warren and Jane Taylor.

After this we have notice of Cromwell's Act in the registers.

According to an act touching marriage, and ye registering thereof, also touching births and burials bearing date Aug. 24, 1653. Robert Hills being chosen registrar by the parishioners of Trottescliffe, was sworn before me Justice of yo peace, to be register of ye said parish, October ye 5th, 1654. W. James.

A purpose of marriage between Richard Daniell, of ye parish of Luddesdown, and Joan Miller, of ye parish of Trottescliffe, had been published three Lord's days, in ye parish church of Trottescliffe, and nothing was objected to hinder, and forbid the same ; for testimony whereof I hereunto subscribe my hand, Oct. ye fifth, 1654. Uppon ye aforesaid certificate, also of another from ye Registrar of the parish of Ludsdown, of ye like purpot, the marriage between ye said Richard Daniel and Joan Miller, was solemnised before me uppon ye 5th of October, 1654, in ye presence of Robert Hilles, and John Granger. W. J. James.

There are no marriages in 1657, 1659 and 1660. In 1663 we read :—

Richard Ellyet was weded the ninth day of June 1663, and Eleanor Codd, (written in a later hand) : this explains the next.

1664. John Webb was wedded the 21 day of June.

Either at this date they did not enter the name of the bride, or the clerk not knowing the bride, wrote down the name of the bridegroom, expecting to be able to fill in the register afterwards.

1683. 22 of April. Then was married Edward Stephens and Susanna Ben, both of Marden, with a licence, in the parish church of Trottescliffe.

April ye 13th, 1686. Then was married with a provincial licence, Thomas Thompson and Anne Pattenden, both of East Peckham, by Lisons, 1700.

Nov. 29, 1750. James Norman Patch and Lydia Pollard, spinster, both of West Malling, were married by licence, from the Bishop of Rochester.

1594. Johan (Joan) Attwood, uxor (wife of) Nicholas Attwood, was buried the first of May.

1599. James Godden, the yeoman, was buried the 23rd day of August, whose soul Jesus have mercy upon.

11

This addition is very uncommon in registers, though frequent on the tombstones.

> 1600. Samuel Attwode the *good youth*, was buried the 13th day of May.
> 1629. Alice Totting, servant to Mr. Godden, of the Courtlodge, was buried 27 November, Anno Dni.

The word servant is frequently used in registers of this date.

> 1640. Nicholas Neave, who came to his immature death, by falling from a haymow, was buried ye 5th day of September.

An instance of recorded accidents.

> 1641. A child of James Godden ye butcher, not baptised, was buried ye 12th day of February.

We do not find persons at this time often mentioned as unbaptised.

> 1647. Anne Pope, a poor lame widow, from Wm. Cod's house, was buried March 30.
> June 9, 1647. A son of James Godden, on the Banke, not baptised, was buried.
> Anno D^{nl.} 1647. John Medley, the grey hunter, was buried August the first.

The old name for the badger was the grey, as we shall have occasion to show; but this is the only record we have of a man's special calling being to hunt this harmless creature.

> 1649. Old Thomas Stone, and ye wife of Thomas Hues, at ye Bell, were buried Oct. 3.
> 1647. Dorythy, ye little daughter of John Godden, of Rowses, was buried Nov. 27.
> 1651. Robert Cofton, Londoner, was buried here.
> 1651. Francis, a nurse child buried here, June 23.
> 1654. John Warner, of Meopham, being hurt by a wagon in the lane leading from the street to the hill, and dying of his wounds, was buried in Trottescliff churchyard, the 4th day of May.

Here we have the record of another accident, by the village chronicler.

> 1669. Edmund Attwood, the honest gent., was buried September ye 13th.
> 1671. John Gilburne, a wayfaring man, buried 21st of December.
> 1682. Dec. 9. There was buried in woollen, a vagrant man.
> 1700. Nov. 6. Mary Davis, of Trosley, made an oath that, James Maddox, was buried in nothing but what was made of sheep's wool only.
> 1730. Nov. 27. John Robinson, an ancient inhabitant.

This is not an uncommon entry at this period; indeed we often find the adjective ancient used where we should now use old.

> 1752. Jany. 21st. John Hills, householder.

This is a very usual description of persons at this date.

> 1758. Dec. 31. Roots Thomas, orphan and pauper, was buried.

In 1804, we have in the burial register, no less than six entries of infants, beginning with two on Oct. 14th.

> 1762. Then was admitted into ye congregation, Ann, daughter of Robert and Mary Hills, which was privately baptised February 24th.
> 1765. July 21st. Thomas Sands, an infant, was privately baptised.
> 1778. Aug. ye 10th. Mary, daughter of one Smith, beggar, baptised.

The oldest registers after Trottescliffe, are those of

ALL SAINTS', BIRLING.

The south aisle has a piscina in it, showing that there was once a side altar. There are few traces of Norman architecture; though a church existed here, both in Saxon and Norman times, as is proved by the mention of Birling, both in Domesday, and the Textus Roffensis. We will now proceed to mention the incumbents of Birling, so far as we have ascertained them :—

> 1329. John Knots.
> ——. John Combe.
> 1338. John Moleton, succeeds John Combe.
> 1349. John Odford.
> ——. Richard atte Brigge.
> 1395. William Chappel, pro Richard atte Brigge.
> 1397. William Tany, instead of William Chappel, who resigns.
> ——. Henry Spencer.

This clergyman having for some reason refused to pray for his bishop, had to make a public confession of his fault, when he was pardoned. He soon after, however, resigned, when :—

> 1458. John Brompton was appointed.
> 1487. Richard Mann.

This vicar is mentioned by Thorpe, but is not in the Rochester registers.

> 1496. William Rednys, was appointed, by the Abbot and Chapter of Bermondsey.
> 1499. William Watson, in place of William Rednys, resigned.
> 1507. George Brinley, succeeds Watson.

The registers commence 1558, but we cannot find the name of the then vicar.

> 1567. John Ellis.

He is mentioned as buried at Birling 1570.

> John Ellis, the Vicar of Birling, was buried the viith day of April.

When he died most probably Joseph Moore, who is mentioned in the Rochester register, succeeded.

> 1574. John Savill succeeds Joseph More.
> 1585. Matthew Heton.
> 1586. Thomas Lloyd, appointed by Henry Lord Burgavenny.

This vicar has several entries, in the registers :—

1595. Mary, the daughter of Thomas Lloyd, Vicar of Birling, was baptised the xxiii. day of February.
1597. Hester, daughter of Thomas Lloyd, Vicar of Birling, was baptised, ye xiiith day of December.
1606. Hester Lloyd, the daughter of Thomas Lloyd, buried the 30th of October.
1638. Mary, the wife of Thomas Lloyd, clerk, was buried the 24th day of March.
1642. Thomas Lloyd, Vicar of Birling, was buried the 16th of January.

It will thus be seen, Mr. Lloyd held Birling for the long period of fifty-six years, which he undoubtedly possessed for a more extended period than any other beneficed clergyman kept his living, in the parishes whose chronicles we are penning. In his days the third, fourth and fifth bells were given.

1642. Philip Satterthwaite, succeeded Thomas Lloyd.

In 1659, Birling was valued at £6 9s. 4½d., and was in the gift of Lord Abergavenny. Philip Satterthwaite's name does not appear in the registers ; he was ousted in 1652, and Thomas Gunn held Birling, who appears to have been a Puritan minister, not recognised by the bishop of Rochester, as in

1660. Michael Rabbett succeeded, vice Satterthwaite resigned.

This vicar's monument in the church reads :—

Here lyeth interred the body of Michael Rabbett, Vicar of this Parish the space of thirty-two years, who departed this life the twenty-fifth day of March 1692, aetatis suae 84.
1692. Theophilus Beck.

Mentioned in the register as buried here :—

Mr. Theophilus Beck, Vicar of Birling, buried Oct. 19th. 1715.

The Birling flagon is of his date, 1697. The clock and dial were also given partly by Mr. Beck. He appears to have been curate of Bexley and vicar of Barming.

1715. Thomas Winterbottom. (Not mentioned in the registers of the parish.)
1722. Hugh Pugh, M.A.

We do not know on what authority Hasted mentions him also as rector of Trottescliffe ; we find the entry of his burial in the Birling burial book :

Dec. 23rd, 1743, Rev. Mr. Pugh, Vicar of Birling.

In his days the paten was given to Birling.

1743. Gregory Sharpe, L.B., succeeded Mr. Pugh.

He is only mentioned in the parish registers at the end. The second and third bells of Birling were hung in his incumbency.

1756. Edward Holme, M.A., was inducted.

He built the parish schools of East Malling and Leybourne and endowed them. He was buried here; the entry runs:

1782. Jany. 12, Edward Holme, Vicar.
1782. William Humphrey. (Was also Vicar of Kemsing and Seal in 1766.)

His name is not in the registers.

1817. The Honourable William Nevill was rector till 1844.

We find the entries often signed by his name between these two dates. He afterwards became Earl of Abergavenny in the year 1845. Besides being rector of Birling he was also vicar of Frant, Sussex. He was buried here August 25th, 1868, as was also his widow in 1872, and two of their infant children in 1828. There are several of his descendants mentioned in the register of the names of Bligh, Mostyn, and Nevill.

William Corfield was vicar from 1844 to 1850, when he died, and being buried here, he was succeeded by Henry Dampier Phelps.

1850. Henry Dampier Phelps was appointed vicar, and also rector of Snodland.

We have already referred to the cup given by his family to the church of Birling.

1865. On the death of Mr. Phelps the Hon. E. V. Bligh, second son of the fifth Earl of Darnley, and son-in-law of the previous rector, William, Earl of Abergavenny, was instituted to the living. His son was christened here. He resigned the living in 1876.

In 1876 W. Madden was appointed vicar. He first set on foot the movement for the New Church at Hammill. In the year 1890 he resigned the living; his wife had previously died here.

1890-92. Stuart Churchill. In August 1892, Mr. Churchill was appointed to Christchurch, Kilburn, and Charles Forbes Septimus Money,* formerly rural dean of Deptford, vicar of St. John's, Deptford, and of St. Luke's, Cheltenham, and of Christchurch, Kilburn, and honorary canon of Rochester, was appointed.

The Birling registers are very incomplete, though beginning at such an early period as the first words testify, "1558. The Church Book containing Christenings and Burials."

1618. Margaret, the daughter of one John Fricht, a poor wandering man, was baptised the xxiii^rd daie of February.
1638. Joan, the daughter of George Crauford, a poor travelling man, was baptised the third day of June.
George and Joan Ray were married at Paddlesworth, 1688.
1695. Anne, the daughter of Archibald Yorke, a soldier, and Susanna his wife, was baptised July 20.

The burial register commences: "Such as were Burials of the said parish of Birling."

1613. William, a strange lad, vagrant, about the age of sixteen years, was buried the xvii. of October.
1625. A poor wandering woman that died at Walter Gregorie's was buried the iii. day of September.
1636. Richard Clo, servant to Mr. Littleboy, drowned in the Brewhouse pond, was buried the 23rd day of September.

There is a later book of registers which is prefaced :—

This book was bought by Mr. Henry Knowles and Mr. William Newman,

---

* Died at Birling January 18th, 1893.

Churchwardens for the parish of Birling, pretium 15*s.* at Maidstone in the county of Kent, Anno Dom. 1723.   Began to register in it out of a paper book after the old register was complete in ye year 1707 for Christenings, for Marriages 1711, for Burials 1712.

Theophilus Beck, Vicar, M.A., Rector of Birling.
Succeeded by Hugh Pugh, Vicar. M.A., Anno 1722.
Succeeded by George Sharpe, L.B., Anno, 1748.
Succeeded by Edward Holme, M.A., Anno 1756.

The clock and dial of the church were given at the charge of these benefactors.

T. Beck, Vicar.	A. Currall.	H. Knowles.
G. Wray.	W. Castret.	W. Hoath.
Nich. Lyffe.	R. Knowles.	R. Tomlyn.
P. Castret.	H. Castret.	T. Newnham.

1728. David, whose surname is Seamark, a grown person, aged 41, was baptised January the 21st, by me. Hugh Pugh.

1733. Isabel Coleman, an old maid, was buried on the 9th day of January, in woollen only ; affidavit was made by Goodwife Coleman before Mr. Bickley, Curate of Snodland. Hugh Pugh, Vicar of Birling.

1713. March 15. A stranger that died in Wibley's barn, March 15, 1713.

1716. A certain traveller, ignotus, was buried March 13, and an affidavit was brought me from Mr. Samuel Spateman, Rector of Leybourne.

Old Mary Jeffrey was buried, in woollen only, on the 20th of March, 1728 ; affidavit was made by Margaret Coleman, before Mr. Babb, Vicar of Malling.

1733. Elizabeth Dimsee, a hopper, was buried here on the 29th of August, 1733. Hugh Pugh, Vicar.

1741. A hopper, unknown, was buried here, September the seventh day.

These entries give us hopping as forming part of the farming of Birling one hundred and fifty years ago; though this is not nearly so old as the hopping mentioned in the Offham registers.

1758. Feb. 10. Son of a stroller.
1758. Sept. 13. Stranger-boys two.

These entries are peculiarly curt and incomplete, neither giving name nor anything beyond the bare fact of a burial of persons not belonging to the parish.   In this year we have several foundling hospital entries, giving the name and number of the child, *e.g.,*— "John St. Mark, Foundling Hospital, 6473."   There are several of these in the neighbouring parishes.

The burials from 1719-1733 have disappeared, if ever recorded.

Oliff, the daughter of Elisha Lee and Phillis his wife, of Kent Street, as they said, was baptised on the 2nd day of September, 1735. Hugh Pugh, Vicar.

1734. Ann Simmonds, a child of John Simmonds, of Hammill, was buried on 25 January ; according to the Act of Parliament affidavit was made by Widow Sanders, before Mr. Bickley. Hugh Pugh, Vicar of Birling.

1755. William, son of a vagrant.

1760. Dec. 25. John Murphy, being bred up in the faith and profession of the Church of Rome, after renouncing the errors of the said Church, publicly, in the Parish Church of Birling, was admitted a member of the Church of England, by Edward Holme, Vicar. Witness, Nicholas Newman, Robert Austen.

Besides we have these entries :—

Charles Robert, Son of Charles Ryves and Alice Graham, born in the parish of St. Athan, in the county of Glamorganshire, on the 7th day of July, in the year of our Lord 1801, was baptised on the 20th of March, 1807, by me, Charles Graham, Curate of Birling ; his godfathers, Richard Robert Graham and John Dudlow ; godmother, Maria Ursula Graham.

After this we have the entries of three others of the same family, namely :—

Mary Ann, born Jany. 31, 1804, baptised March 10, 1807.
Emma Georgiana, born March 30, 1805, baptised March 20, 1807.
Maria Louisa, born March 1807, baptised December 31, 1808.

The first entry of this family is the longest rigmarole we have come across.

Besides this we have a notice of the gift of the font cover.

The cover to the font was carved by the Ladies Caroline Emily, Henrietta Augusta, and Isabel Mary Francis Nevill, the daughters of the Earl and Countess of Abergavenny, A.D. 1853. It was first placed upon the font, and the font moved into the centre of the church, at the christening of the eldest son of the Hon. Thos. E. M. L. Mostyn, M.P., and Lady Augusta Mostyn, July 20th, 1856.

After Birling the next parish register to commence is that of

RYARSH,

which begins in the year 1559. The church, which is dedicated to St. Martin, is a fine specimen of Norman work. The Norman piscina is engraved in Parker's " Glossary of Architecture." It was, however, originally without the tower, which was added in the Early English period.

The incumbents of Ryarsh are as follows :—

1237. Peter de Sausinton.
1242. Andrew de Wyntone.
1314. Walter, mentioned in Kentish Fines as " de Ryershe Clerk."
1328. Wilfred de Denton.
—— John Roger.
1391. John Humphrey, in place of John Roger.
—— John Gote, alias Birton.
1400. William Godard changed with John Gote.
1411. John Thobyn.
1414. Philip Home, rector of Elmstead, changes with Thobyn.
1418. John Fynch exchanges with Philip Home. He was vicar of Sheldwich.
1421. Thomas Shene, vicar of Fullmere, Bucks, changes with Fynch.
1425. Richard Tarton.
1426. Robert Finch.
1450. William Snary.
1453-1479. John Sutton.
            John Kokk.
1501. Henry Watkyn, on the death of John Kokk. (The Convent of Merton are patrons.)
1515. William Walker, on the death of Henry Watkyn.
1524. Richard Whyte, on the resignation of William Walker.
1534. Robert Covert.
1538. Henry Singleton, on the resignation of Robert Covert.
1550. Thomas Bolton.

Another authority gives him as Bote. In his days the registers were commenced.

1565. John Alwyn.
1568. John Ellis.
1570. Robert Salisbury, prebend of Rochester ; rector of Trottescliffe 1554—1560 ; he was also rector of Addington 1559—1583. He left xii*d.* to every poor householder in Addington and Ryarsh. He was presented by George Watton, Esqre, of Addington.
1583. John Blackburne.

He was buried here as we find by the registers : " John Blackburne, vicar of Ryarsh, was buried on Sunday the xxxist day of August, anno domini, 1589."

1589. John Parker.

This vicar tells us in the beginning of the Ryarsh registers of the pains he took with them, which was most unusual for that age ; the introduction runs :—

" The register book of the parish of Ryarsh within ye countie of Kent and diocese of Rochester, contains all those names that have been christened, married, and buried within ye said parish : ye said register beginning the xxth day of November, in the yeare of our Lorde God 1559, copied out of ye original by John Parker, Vicar of Ryarsh."

He is mentioned in the parish register in this way :—

" Anno regni Elizabethæ tricesimo primo Johannes Parker, Vic. de Ryarsh."

And again as buried :—

" 1603. John Parker, which was the late vicar of Ryarsh, was buried the seventh day of March, in the year aforesaid."
1603. George Shawe.

He tells us in the commencement :—

" That upon ye four and twentieth day of April, a thousand six hundred and three, and in ye first yeare of ye high and most zealous sovreign Lord Kynge James by the grace of God the first of that name king of England, France and Ireland, defender of the faith. George Shawe, Clerke, Vicar of Riarsh, in the county of Kent, being inducted to the said Riarsh on the xxiii. daie of the said moneth did read the articles of religion accordinge to the lawes of this realm, the sed xxiii. daie of April, being Easter daie, as aforesaid, in the psence of us whose names are here underwritten, as mentioned," (etc.).

Mr. Shawe is also buried here.

" 1617. George Shawe, Vicar of Ryarsh, was buried the eighth day of October."

Mr. Shaw was presented by Thomas Watton, Esq., of Addington.

1617. Henry Livett succeeded George Shawe.

In his day the second bell was presented to the church ; he also gave himself the pewter flagon to the parish. He obtained the living from Thomas Watton, Esq.

1632. Abiezer Herbert.

We learn from a memorandum in the parish book :—

"That in the register from the year 1632 and forward, during ye time of Mr. Herbert's ministry here in this parish, ye yeare ends as it does in our Almanacks at ye Nativity of Christ, or last day of December, and begins at ye first day of January commonly called ye Circumcision of Christ, or New Year's day. . . . Abiezer Herbert. That the same account of time was observed by " (rest illegible).

This clergyman was buried here ; the entry in the register is :—

1641. Abiezer Harbord, minister, was buried August 17th.
1646. Herbert Trott was incumbent of Ryarsh in this year,

but whether he was instituted after Mr. Herbert's decease or not, seems rather obscure.

1647. David Sibbald, or Kibbald.

He with Trott are not mentioned in the parish registers.

1653. John Emerson was instituted.

The entry of his wife's burial exists :—

"Margery, the wife of John Emerson, the minister of Ryarsh, was buried May the second, 1656."

And his daughter's :—

"Anna, the daughter of the said John Emerson, was buried July 30, 1656."

In his days the returns of the Rochester diocese were made, when Ryarsh was returned as worth £8 10s., and the patron W. Watton, Esq.

1660. Robert Godden was instituted.
1661. William Deane.
1676. Robert Worledge : also rector of Addington 1701 : presented by William Watton, Esq.
1702. John Dacie.
1730. Henry Burville, previously vicar of West Peckham.
1742. Thomas Buttanshaw ; he was curate of Trottescliffe 1732, then minor canon of Canterbury, rector of St. Peter's, vicar of Westgate, and vicar of St. Stephens, or Hackington : in 1741 he was appointed rector of Addington.
1768. James Thurston : the date of his induction is mentioned in the register ; he died in 1802.

We find these entries :—

"James, son of James and Sarah Thurston, Sept. 3rd, 1775.
1802. Thurston, James, Vicar of this parish, aged 61, buried March 15th."

† An awfully short space intervened between Mr. Thurston's perfect health and his dissolution. In his days the first bell of Ryarsh was hung.

1802. John Liptrott. He says, " I became Vicar of Ryarsh in 1802, planted the churchyard with elms and limes, the glebe with oaks, did the same at my Rectory of Offham 1819. John Liptrott, Vicar and Rector of Ryarsh and Offham. The parishioners planted a yew tree about the same year. Mark well ; in three hundred years it may be a venerable yew."

Mr. Liptrott was appointed rector of Offham in 1777.

1830. Lambert Blackwell Larking, was presented to Ryarsh by Colonel H. J. Wingfield Stratford. He was appointed vicar of Burham in 1837.

We have given a full account of this vicar in our last chapter. Mr. Larking was buried here, as was also his wife. His monument is in the church. March 1832, Rev. L. B. Larking adds in the register :—

"I planted a few Turkey oaks in the skirts of the field next the house, and in the triangular plantation at the south end of my glebe, the limes and Æsculus Pavia in the centre of the same field, and many ornamental shrubs near the house. The cedar of Lebanon near the entrance gate I found there, apparently about fifteen or twenty years old, completely choked up with other wood. The walnuts Mr. Liptrott sowed about fourteen or fifteen years ago."

The burial register runs :—

1868. Lambert Blackwell Larking, Ryarsh Vicarage. Aug. 11, aged 71. R. Garland.
1873. Frances Larking. 6, Brondesbury Terrace, Kilburn, March 29, aged 87. R. Hay Hill, Curate of St. Martin's in the Fields, Westminster.
1868. Henry Welsford Snell. After being nine years at Ryarsh he changed with the present incumbent for the Vicarage of Mendlesham, Suffolk, in 1877.

Mr. Snell says, in the register :—

"Following the custom of my predecessors, I record that I became Vicar of Ryarsh in October 1868 : in 1869 the school was built by voluntary contributions ; in 1872 the interior of the parish church was reseated and restored (see brass in the chancel) ; the work was paid for by voluntary contributions, aided by a grant from the Incorporated Church Building Society, and the Diocesan Society, February, 1873. H. W. Snell, Emman. Coll. Camb."

Two of Mr. Snell's children were christened here.

1877. Edward Henry Roger Manwaring White, was Vicar of Mendlesham 1861—1877, and is patron of that living : he came to Ryarsh in 1877.

His wife and several of his family are buried here; one of his daughters was married here; and one of his grandsons was baptised here.

In 1879 a third bell was hung in the church.

Besides mention of the parochial clergy, the following minister is mentioned :—

1608. John Bridger, the parson of the parish of Mereworth, and Marie Walsingham ye daughter of Edmond Walsingham, were married here the nineteenth day of September.

Of the commencement of the registers of Ryarsh, we have already spoken. The baptismal register commences at once with :—

First and foremost ye xviimo day of January, anno domini 1560, was christened Joane Oliffe ye daughter of George Oliffe.
Monday ye xiiith day of March in that aforenamed year, was christened John Sanrock anno 1560 from Christ's Incarnation.

There are two in Latin :—

Elizabeth Yonge, filia Edward Younge baptizata fuit secundo die Martii anno supradicto 1579. (Elizabeth, daughter of Edward Younge, was baptised the second day of March in the above-mentioned year 1579.)

Katerina filia cujusdam peregrini nata in porta templi baptizata fuit 1mo Dec. 1617. (Catharine, daughter of some stranger born in the gate of the Church, was baptised the first of December, 1617.)

1751. Turpin, son of a gipscy.

Ryarsh has plenty of entries describing the baptised as "strangers," "travellers," and "travelling parents' children," in common with other parishes.

The next two give examples of calling children after places :—

1769. William Fartherwell, an unknown traveller's child, baptiz July 2.

1780. Mary Rash, a traveller's child, bap. Jan. 2. (Rash is the common name of Ryarsh.)

1774. Elizabeth, daughter of John Aaron Broad, bapt. at Addington May 20.

It is by no means unusual at this time to find what is done in another church, entered in the registers of the parish of which one of the parties is a native.

1788. Francis, son of Pawley, bapt. Nov. 3. The christian name inserted from verification of them, when 21 years of age, in the Prince of Wales' regiment.

Marriages in the parish church of Ryarsh from the vi day of November in ye yeare of our Lord God, 1559.

First and foremost, on Monday the xx. day of November, in the abovesaid year, Anno 1559, Thomas Godden was married to Elizabeth Littell.

The next is interesting as mentioning the saint's day :—

Gyles Symons, of Ryarshe, and Marie Crispage, of Stone, widow, was married on St. George's day, 1599, by a licence granted out of the office at Rochester.

As in other marriage registers, so at Ryarsh, we have frequently marriages that took place in other churches mentioned.

The following is a certificate of the period of Cromwell used as the marriage register.

"I, John Birch, do testify that whereas there is a contract of matrimony between Thomas Martin and Elizabeth Johnson, Ellen, the mother of the said Elizabeth, and Joane, the mother of the said Thomas, have given their consent and approbation to the same contract, and therefore I know noe let unto the publication of their purpose of marriage, February the fourth, 1645."

The names of all such as be buried in the parish of Ryarsh from the vith day of Aprill in the yeare of our Lord God, 1560.

The sixth day of Aprill, in the year of our Lorde God, 1560, was buried Jane Boorman, the wife of John Boorman.

The third day of April, 1562, was buried William Byshop, clark of the said parish of Ryarsh, and the said third day of April was Maundy Thursday.

This is perhaps as old a record as we shall find of a parish clerk.

The third day of Marche, 1563, was buried one Thomas Filpot, a poor man, a stranger which came late out of Yorks., that died in Richard Boorman's house.

The xxiind day of Aprill, in the year of our Lord 1573, there was buried Mother Wyborne.

Henry, a poore man, borne at East Malling, was buried ye xxvith day of June, anno 1602.

This entry shows how unsatisfactorily incomplete the registers were kept, as the very important point, the man's sirename, is the only fact left out.

> 1677. There was buried goodwife Curd, in ye parish of Ryarsh.
> 1748. Mr. John Miller, buried in the church by leave from the vicar, Sept. 11.
> 1761. Rebecca, wife of the above [Solomon Penury], buried in the church ; but they did not ask leave of the vicar, which I think they ought to have done.

These last two are interesting as showing how much importance was then attached to the permission of the incumbent on ecclesiastical questions.

1781—1783. There are several entries of persons buried simply as hoppers, without any mention being made of their name or whence they came.

The following will give an example of the form used for twenty-two years in this church for burial :—

> 1678, September 7, there was buried in woollen, Margaret Marsden of ye parish of Ryersh, spinster. Affidavit being made by Elizabeth Turner, of ye same parish, before Sir Willm. Twisden, Bart., one of His Maj^{sty'e} Justices of ye peace, in ye presence of Edward Walsingham and Will Caysior.

Amongst these burials we have one stating that the person was buried in Trottescliffe parish.

> 1813. A man unknown who died upon ye road by ye visitation of Providence, May 17th, in years.   John Liptrott, Vicar.
> 1815. Thomas Hatton, a soldier's son, Nov. 23, 2 years.
> 1816. James Chapman, a stranger, his abode not known, May 5, about 60.
> 1817. Sarah Welsh, hopper sojourner, Oct. 8th, 26.   John Liptrott.
> 1821. Elizabeth James, hopper from Barming, Oct. 11, 47.
> 1826. A woman called Allen, surname unknown, stranger, Sept. 11, about 17 years.   Robert Cobb.
> 1849. O. Hearn died of Asiatic cholera, a stranger come to hopping from Malling Union, Sept. 17, age unknown.   Lambert Larking, Vicar.

The next oldest register to Ryarsh is that of

SNODLAND.

The church here has been dedicated in the name of "All Saints." The mediæval work of this church dates much later than when there was first a building here, as the records of Snodland go back far beyond the Conquest ; and in Domesday and Textus Roffensis the church of Esnoiland, as we have seen, is continually mentioned. There are a quantity of Roman tiles run into the church at different places, which were no doubt borrowed from the Roman villa (just re-discovered) hard by.   The restorations of this church in the last

and present centuries have obliterated what was probably the oldest
building of the church. The present rector and his hard-working
curate are trying all that they can do to repair the church, and
especially to restore those parts which are the more ancient portions
of the edifice. There is a handsome old market cross standing in the
churchyard, which, however, we have already shown, was moved
thither by the last rector but one. The rectors of Snodland that we
have discovered were as follows :—

1274. Sir John de Eastwud, rector of Snodland.
1295. Wynard de Dryland.

It seems, that this rector, was the one who poisoned a Justice of
King Edward I., who had been dismissed from the bench for
bribery.

1330. John Hirlasativer.
1337. Edmund de Dygge.
1337. John de Dennyngton, he was also rector of Trottcscliffe.
1346. William de Middleton, also rector of Trottescliffe.
1349. Robert Carllimon.
——. Stephen Randolph.
1360. John Alcham.
1375. Peter de Lacy, prebend of Swerds.
1388. Bartholomew Waryn.

He was Secretary to Bishop Thomas de Brinton; and, in 1401,
he changed away the living of Snodland for Hadstocke.

1401. Roger atte Cherche.
1427. Richard Mountain.
1447. John Aston.
1453. Thomas Dalby succeeded John Aston : he is buried at Snodland ; and
  his monument tells us that he died 1472.
1464. John Perot succeeded Thomas Dalby.
1499. William Barker succeeded John Perot.
1526. John Addison, pro William Barker.
1533. Robert Truslove, succeeded John Addison in 1531 ; he was chaplain of
  St. Lawrence in Halling.

1571. William Hall died this year as rector of Snodland, he is
entered in the parish registers in these terms :—

" Sir William Hall, pson of this parish, was buried 22 June."

The entry " Sir William Aspley was buried 14 April, 1574," leads
me to believe that Aspley was his successor ; but neither of these
is mentioned in the Rochester registers. These entries give the
clergyman the title he once possessed in common with a knight
and which, though fallen into desuetude, is his by right still.

1576. John Swone appears in both Registers.

His daughter's baptism is thus entered :—

" 1585. Sarah filia Johannis Swonei Rectoris hujus ecclesiæ de Snodland
  baptizata fuit " and her burial " Obiit eadem die 12 die Februarii in
  anno prædicto, sepulta, jacet in coemiterio de Addington quoniam

infra limites parochiae cnutricbatur viz., in ædibus Thomæ Whiting.
(Sarah, the daughter of John Swone, Rector of this Church of Snodland, was baptised. She died the same day the 12th of February
in the aforesaid year, and being buried, lies in the churchyard of
Addington, since she was nourished within the boundaries of that
parish, namely in the house of Thomas Whiting.)

The first bell of Snodland, appears to have been given, at this
date. Rembron Griffin (1600) describes himself as " verbi minister "
in the register.

1608. Maurice Edwards.

In this year we read in the register :

Rebecca Edwardes, the wife of Mr. Maurice Edwards, who was minister of
Gode's word in this Church was buried upon ixth day of September.

Besides (though there is nothing to prove him rector here), we
find :—

1613. Buried was John Sands parson.
1620. John Gimpton instituted.
1624. William Williams, buried here. William Williams, the minister of
Snodland, was buried ye xxiind day of December.
1630. William Medburst.
1631. Thomas Garraway.

There are one or two entries in the registers of this rector.

1648. Rebecca Garraway, the daughter of Thos. Garraway, minister of God's
word, was baptised on the 6th day of August.
1658. Samuel Garraway, and child of Thomas Garraway, Rector of Snodland,
buried on ye 5th day of July.
1666. Thomas Garraway, Rector of Snodland, was buried Feby. 21st.

In his day, the fifth bell was hung in Snodland tower. In Mr.
Garraway's time in the year 1659, Snodland was returned as being
in the gift of the bishop of Rochester and worth £20. It was then
the best living in the valley.

1666. Luke Proctor.

This incumbent is also buried here, for we read : " 1673. Luke
Proctor, Rector of this parish, was buried Jany. 18th."

1673. John Thomas.
1681. John Walwyn.

The entries of this rector's family are numerous.

1683. Mary, the daughter of John Walwyn, Rector, was bapt. Jany. ye 4th.
1684. John, the son of John Walwyn, Rector, was bapt. May 10 ; in which
year king James ye 2nd began his reigne.
1685. Elizabeth, the daughter of John Walwyn, Rector, was baptised July 21.
1688. Herbert, the son of John Walwyn, Rector, was baptised Jany. 17.
1690. Catharine the daughter of John Walwyn Rector was baptised Aug. 25.
1693. Robert, the son of John Walwyn, Rector, was baptised Jany. 26.
1694. Bridget, the daughter of John Walwyn, Rector, was baptised October 25.
1709. John, ye son of Thomas Walwyn, was baptised Jany. 15.
1691. Herbert, ye son of John Walwyn, Rector, was buried in woollen Aug. 14

1708. Robert, ye son of John Walwyn, Rector, was buried in woollen March 24.
1712. Elizabeth, ye daughter of John Walwyn, Rector, was buried in woollen
May 27th.

Neither Mr. nor Mrs. Walwyn, though they are mentioned on
their monuments as lying in the church, are inserted in the registers.
Mr. Walwyn, as elsewhere stated, has left two valuable notes in the
Snodland books.

1712. Thomas Wacher.

The Rev. S. Bickley, M.A., who was rector of Offham, acted as
curate of Snodland during this rector's time, as appears by the burial
register.

Feby. 22. Died the Rev⁴. Mr. Bickley, Curate of this place and Rector of
Offham, where he was buried.

As Mr. Wacher is not mentioned in the registers it would perhaps
appear that he was not a resident. The church plate dates from
this period.

1748. Lewis Hughes.

This incumbent is also not mentioned in the parish registers, and
also appears to have been non-resident, as we find in the burial
register :—

1750. Buried the Rev. Mr. Herndale, Curate of Halling, also Rector of Birling,

who probably served Halling and Snodland together, and in,

1762. May 24, Phoebe Tirrell, daughter of the Rev⁴· Mr. Tirrel buried.
1793. Thomas Barnard (not mentioned in the registers).
1800. George Robson (not mentioned in the registers).
1804. Henry Dampier Phelps buried at Snodland Aug. 4, 1865, aged 88,
after being sixty-one years incumbent of the Parish.
1865. James Gaspard Le Marchant Carey.

In his days the last three bells were hung in the tower. He was
afterwards rector of Boreham, Essex, and then archdeacon of
Colchester.

1874. John George Bingley was appointed rector ; previously he had been
for ten years rector of St. Leonard's, Colchester. One of his
children was christened here.

Besides the clergy mentioned above, we have also this entry :—

1736. The Rev⁴· Anthony Dennis, clerk and rector of Wouldham, and Mary
Villiers, of the same parish, with a licence, married April the 10th.

The Snodland registers are dated from 1560. They simply commence with the words, "The Register Boke of Snodland."
The earlier of these entries in the baptismal book is characteristic
of many in the beginning of the register : it may be noticed that the
child's surname is added after the christian, while it is left out after
the father's name, and the mother's name is omitted.

1560. Gyles Andrew. sonne of Francis, was baptised ye 6th March.
1564. Elizabeth Leuse (Lewis), ye daughter of John of Hamyll, was baptised 23rd April.

This is interesting as correcting the mistake Ham hill, showing us that it was Ham mill and not Ham hill. This entry also points to the antiquity of the name.

1575. Samuel Godden, ye sonne of Thomas, of Paddlesworth, bapt. 17 Sept.

This entry, with others, shows us that the church of Paddlesworth had already become not generally used; though there were incumbents for the next sixty years.

1603. The first year of the reign of King James, Nevill Godden, the son of Edward Godden of Paddlesworth, baptised 28 April.

About this time we have many Latin entries.

1600. Robertus Goldinge, filius Thomæ Golding, baptisatus fuit 29 Febr.
1619. Jane, the daughter of Goodwife Chapman, was baptized 25th day of March.
1684. John. the son of John Walwyn, rector, was baptised May 10, 1684 ; in which yeare King James ye 2nd began his reign, viz., Feby. 6th.
1703. Catherine, daughter of a poor travelling stranger, was baptised May ye 9th.
1742. June 27, Almeria, daughter of a stranger, born at ye Bull Back.
1748. Nov. 6, William, the son of a stranger, born at Wingates.
1762. July 31st, James, son of James and Mary Best, was privately baptised, being dangerously ill. Publickly admitted Aug. 22.
1783. From this place the duty of threepence for every child bapt. or christened is rece⁴· according to the Act of Parliament, by William Lewis. clerk.
1788. May 11, Lydia, daughter of Richard and Lydia Solley, born at ye poor house.
1794. October 2. The duty of 3ᵈ· for every baptism ceased according to an Act of Parliament.
1819. July 19, Henry. son of John and Frances Baker, higgler.
1819. Nov. 19, Charles James and Mary Ann Barnfield, of Pomphrey, in Paddlesworth, smuggler.

The entry of smuggler as a calling in life is very curious.

1846. Aug. 16, Elizabeth Anne, daughter of James and Jane Lee, Snodland, traveller. The child was born at Queenborough ; the father knew not where he belonged to, his parents being travellers. Alas ! H. D. Phelps.

1559. THE BOOK OF MARRIEDGES.

There are no entries for the years 1594, 1595, 1596, 1599, 1600, 1601, 1670.

1641. Edward Cloake of East Kent, married to Mary Netter of Paddlesworth. on the 26th of October.

1561. John Usher of Hoborrowe, was buried 16 October.
1563. A waterman or sayler, whose name was unknown, who had a wife, as he said, and children at Rye, was buried July 5.
1563. Cuthbert Ersh, a young man of London, was buried, 30 September.
,,   16 October, John Hardware, a nurse child, of London, was buried.

There are a great many nurse children buried about this date, principally from London.

1563. Richard Bambridge, a goldsmith of ye town of Rye, was buried 3 March.
1569. One Cornelius, a poor labouring man, was buried 7 January.
1569. Mary, ye daughter of one Austin of London, was buried 8 March.
„ August 21, William Carr, a tayler, was buried.

The mention of the calling of a man so long ago is not usual.

1570. Robert Crane, a single man, was buried March 18th.
„ October 4th, an olde maide called Phillip was buried.
1573. August 7th, also Sparrowe, an olde wydowe, was buried.
1574. June 15th, Theoball Hammon, a French, dwellinge in London, was buried here.
1578. May 20, a child yt was born at Swynborn's house of a woman yt asked for lodging was buried.
1584. Feby. 2, Father Border was buried.
1585. Oct. 27, Joane Swinborne, alias Downe, widow, was buried.
1588. June 2, Deborah, the daughter of John Powlter, miller, who was drowned in the mill pound at Holborough, was buried.
1588. Mary, the daughter of Thomas Pillkington, was buried September 8th.
1589. Jany. 28. An infant of John Hammons who died unbaptised was buried.

Only two Burials in 1500, one in 1597, none in 1599, 1600, 1639, 1640, from 1641 to 1658, one in 1695, and one in 1700.

1591. A man child of William Blies, which departed as soon as it was borne.
1591. Rose, the daughter of Wm. Pallmer, of Redcriffe, by London, taylor, brought up by Abraham Collier, of this parish, was buried 26 June.
1592. Nem. (Naomi) Tiksall, being drowned the day before at the mill, by goodman Leedes his house, was buried.
1601. Elizabeth Angier, a poor maide, was buried.
1602. William Samon died and was buried 27 June.

This style of entry very common in the register at this time.

1603. Mother Becher, servant to Thomas Cordery, died and was buried July 1st.
1606. Nicholas Gregorie, a harvest man come from Cheswick, in Middlesex, was buried xxviiith August.
1609. The 22nd of May, buried was mother Chittenden.
1618. Robert Amies, the son of Goodman Amies, was buried the third day of September.
1625. Old mother Hughes, ye wife of Richard Hughes, was buried the 29th of March.
1626. Elizabeth, ye relicte of James Spenser, was buried the 7th of February,

(The earliest record of the word relict I have found.)

1636. A stranger at Snodland mill was buried the 10th of June.

This seems to point to Snodland having had a paper mill pretty early, as the trade is mentioned about this date.

1659. A child of Farmer Peosk, was buried, July 23. (The child's name is not given.)
1665. The wife of Mr. May, of Sen Margaret, was buried October 23.

12

Sen, for Saint, tells us very plainly of the education of the parish clerk.

1667. Thomas Martyn, householder, was buried.

The phrase householder is found frequently in registers at this period.

1670. William Goteere, aged by his own computation one hundred and eleven years, was buried.

This man and a woman buried at East Malling, and a woman of Offham, are centenarians recorded of these parishes. Goteere lived in the reigns of Elizabeth (perhaps of Mary), James I., Charles I., and Charles II., as well as during the Commonwealth, and must have seen some of the greatest changes in the civil and religious world that could have been seen in a lifetime.

1671. Martin, the widdow, the relict of Thomas Martin. late deceased, August 1.
1672. Feby. 7, John Swift, householder, died here and was buryed at Mepham.
1696. Nathaniel, a stranger, was buried in woollen Dec. ye 4th, 1696.
1703. Sept. 1st, Mary Simson, a poor travelling woman, was buried in woollen.
Memorandum: that her husband went away immediately after her interment, but whither was unknown to us.
1704. Mary Henfield, labourer, was buried in wool, Sept. ye 13th.
1705. James Smith, paper-maker, was buried in wool, Jany. the 5th.

We have here another record of paper being made at Snodland for 200 years.

1708. John, a stranger, who died at Groves, was buried in wool, Oct. ye 2nd.
1729. Katherine, the wife of Edward Owlett, and last descendant of the family of the Pounds, late of the parish. buried March 10th.
1737. July 11. Buried Elizabeth, wife of Henry Taylor, on Punish Hill.

Holly Hill was known till about 40 years ago, as Punish Hill from the old family of Povenash, who had an estate here. In their day the hills from Ryarsh and Birling to Halling appear to have been parks, showing that the predecessors of the present landowners admired what they seem to neglect—the lovely views of this smiling valley.

1750. August 12th, Buried a man his name unknown a stranger.
——. August 17th, Buried the wife of ye above mentioned stranger.

From entries like these we see how little trouble was taken to identify persons, and this will account for the number of persons who were missed in the last century.

1760. Margaret Brown, a child belonging to the foundling hospital, No. 1104 ;

one of many such entries, showing that Snodland was a foundling parish, which is still the case with many Kentish parishes.

1781. Dec. 30, Mary Dartnell, of the smallpox.

We have more cases in 1790 and 1793.

1783. October 3, Thomas James, a stranger, drowned in the Creek.
From this place, the duty of three pence for burials, recd. according to Act of Parliament by William Lewis, Clerk.

1786. Hannah, a stranger, supposed to belong to Marden, appeared to be near 70 years of age ; by the Parish.
1793. April 26th, John, son of John and Sarah Dartnell, aged 5 years ; killed by accident in a sand hole at Birling.
1791. October 2. The duty of 3d. for every burial ceased according to Act of Parliament.
1794. October 8. A stranger, his name not known, appeared to be a seafaring man, and about 30 years of age.
1796. July 24th, John Taylor, aged 30 years, forced overboard by the foresail of Mr. Bensted's barge, and drowned near the mouth of the creek.
1796. July 5th, William Whitfield, aged about 65 years, found drowned in the river.
1807. August 9th, Joseph Cook, drowned near the mill.
Here note that the way which Mr. May hath to ye court Lodge meadows was first claimed as property by him, and in the year 1741 was first allowed upon consideration of a road through his land leading to ye glebe under the hill.
1813. Thomas Mills, drowned in the Medway, March 1st, aged 31.
1852. A woman, unknown, left on the brook by the high tide, evidently had been in the water a long time.
1863. March 11. A man unknown found drowned in the Medway. So in 1875 and 1879.
1873. July 28, Caroline Miller, wife of a captain of a barge, found drowned. August 28, Israel May, P.C., found murdered, age 37 years.

The history of his death is already recorded.

1887. Henry Mayger, killed by accident by machinery, aged 51, September 5th.

PADDLESWORTH CUM DODE OR DODECIRCE.

These parishes at the present time have been joined in one name as Paddlesworth, and have been for civil purposes attached to Snodland. Many books speak of Paddlesworth as a chapelry of Birling, and Dode as a chapel of Meopham. This is perfectly incorrect, as already stated. The old parish of Paddlesworth being ecclesiastically distinct, and Dode or Dodecirce, *i.e.*, Dodekirk, a chapel of that parish. The old ruin of Dode is a Norman church ; it has by some mistake of late years been called by many Buckland. This name it borrowed simply from the manor on which it stands, which was in days gone by the manor of the parish. Dode is mentioned both in Domesday and in Textus Roffensis. Dr. Harris tells us he could not find out its whereabouts.

The ruin of Paddlesworth was built into a barn, and the north porch removed ; the chancel arch is Early English. The church was dedicated in the name of St. Benedict. The incumbents of these parishes were styled rector of Paddlesworth cum capella of Dowde or Dode. Those we have discovered were :—

John Rowe.
1319. Walter de Chesterfield, on the death of Rowe.
1327. Richard de Lonekyn.
—— Robin Rothberry.
1398. John Dunce.

1400. William Tipper.
1405. John Brewster, in place of William Tipper resigned.   Robert Clifford
    is the patron.
1415. Thomas Jade, vicar of Euston, changes with John Brewster.
1460. Thomas Merbury.   Before this vicar of Halling.
1462. William Belthorp.
1464. William Merbury ; also rector of Leybourne.
1467. William Codling, vicar of East Malling, changes with Merbury.
—— Father Tippen.
1506. John Walker, on the death of Tippen.   Robert Watton, patron.
1509. John Parkyn, on the resignation of Walker.
1533. James Roberts, on the death of John Parkyn.
1540. David Welling, on the death of James Roberts ; previously rector of
    Ditton.

By another authority we learn William Baker succeeded Roberts.

1571. Edward Danes.
1581. Robert Paynter, M.A.

In the year 1599, we read in the Rochester registry :—

" Take care that no one admit, institute, or induct to the rectory or the
parish church of Paddlesworth, Rochester diocese. Benjamin Sarryer, Clericus."

In defiance, however, of this statement we have in

1600. Robert Chambers, instituted to Paddlesworth, Lord Watton being the
    patron.
In 1623, Edmund Aldley was appointed by the king.

How long he held the living we do not know, but there are no
more incumbents mentioned as being instituted ; and as we find
Paddlesworth was returned in the gift of Sir J. Watton, and the
value *nil* in 1659, he was probably the last incumbent.

The name Paddlesworth we are told by Harris was originally
Paulsford ; if so, the reason perhaps of the decay of this living was
the giving up of the pilgrimages to Canterbury (as the old church
stood on the Pilgrims' Road, and if they turned off from the foot of
the hills to cross the river here, the pilgrims probably paid their
votive offerings for a safe passage in this church), and there being
afterwards no resident squire.

The parish registers have been lost, but in the register of Birling
we have the record of a marriage that took place here.

1687. George and Joan Ray, married at Paddlesworth.

This seems to point to service being held here later than the last
incumbent. These Rays or Wrays and a family of the name of
Godden, are mentioned in the registers of Snodland, always as of
Paddlesworth ; the Goddens are first mentioned : we give their
names.

1575. Samuel Godden, ye sonne of Thomas of Paddlesworth, bapt. 17 Sept.
1592. John, ye sonne of Edward Godden of Paddlesworth, 12 December,
    baptised.
1593. Thomas, ye sonne of Edward Godden of Paddlesworth, 17 October,
    baptised.

- 1596. Antony, ye sonne of Edward Godden of Paddlesworth, 23 April, baptised.
1598. Bridget, ye daughter of Edward Godden of Paddlesworth, 1 November, baptised.
1601. Edward, ye sonne of Edward Godden of Paddlesworth, 13th January, baptised.

As these appear to be insertions, it is possible that they were copied into the Snodland register from the Paddlesworth one.

1603. Nevill Godden, the son of Edward, was baptised, 28 April.
1608. Elizabeth, the daughter of Edward Godden, was baptised the third day of April.
1654. Martha, the daughter of John Godden of Paddlesworth, gent, was baptised the 27th day of February.
1615. Edward Godden of Paddlesworth, aged . . ., buried on Sunday, August 25, in ye church.
1670. Nevill, the son of Nevill Godwin, of Birling, gent., was buryed Sept. 3rd.
1785. Sept. 17th, a stranger man died at Paddlesworth, middle aged, had but one eye ; supposed to be a native of Ireland : by the parish.
1801. Windsor, stranger from Paddlesworth, Xtian name unknown, 19th October.

These entries suffice to show us that Paddlesworth was well known as a distinct place from Snodland or Birling, and indeed, it must be considered as such, not only historically, but also ecclesiastically. The next oldest parish to Snodland as regards its register is

LEYBOURNE.

The church is dedicated in the names of SS. Peter and Paul. Though it has been much repaired, there are good samples of the ancient Norman and early English buildings; the tower is a modern erection. The heart shrine of the De Leybournes has been spoken of elsewhere. On the font may be seen the places where the taper was fixed, and the salt or chrysm deposited in pre-Reformation times.

1276. We find Peter, rector, and John, the chaplain (probably one of the clergymen of the chantry endowed by Sir Roger), declared to have trespassed on the lands of Ralph Ruffyn, who appears to have transferred certain tenements to the de Leybournes, which they afterwards settled upon the church.
1279. Thomas Bacun; also rector of Langley, in order that he might maintain Sir Roger's chantry.
1311. Walter de Lecton, or de Leghton. We find him also parson in 1314, and he is, perhaps, the same as Walter, the parson of Leybourne, who, in 1347, paid 16s. 8d. to make the Black Prince a knight.
John Ashlying, patronage in the hands of St. Mary, Graces.
1391. Thomas Smyth, vice John Ashlyng.
1423. John Clifton changes with John Burchbacke, rector of Elmley.
1425. John Browning.
1437. John Cowper.
1438. John Lee succeeds John Cowper.
1441. William Midelton.
John Clifton.

1464. William Merbury succeeds John Clifton, he was also rector of Paddles-worth cum Dode.

1495. William Millys, a Cistercian monk.

He was one of the court that sat on a commission to enquire into the patronage and income of the prebend of the High Mass of the High Altar at the monastery of West Malling in 1493. In the deed, he is described as rector of Leybourne, though not then instituted. If he is the same that is buried in West Malling church, he died in 1497, but in the Rochester register we read that Thomas Sewell succeeded William Millys on his death, in 1510, and also the same from another authority.

1510. Thomas Sewell.

1526. John Larke, on the death of Sewell.

1545. Hugh Woodward.

In the Rochester register this clergyman is emphatically spoken of as belonging to the Church of England, probably to show he was the first incumbent of the diocese who had been instituted, who belonged to the reformed faith.

1559. Hugh Williams.

He appears to have been also rector of Ditton, in his days the register was commenced.

1582. William Mounte.

In his days the curious second bell of Leybourne was given; he was probably presented by Robert Godden.

1582. Thomas Lovelace.

1602. William Drury (Sir John Levison presented him).

At the beginning of the register we learn :—

Memorandum that Sir William Drury, Parson of Leybourne, gave out the Boke of Articles upon Sunday, being the fifth day of December before myne induction, Anno Dom. 1602, Anno Regni Dominæ nostræ, Elizabethæ Reginæ 45, in the presence of those whose names are hereunder written Pme. John Astun, Robert Olyver, George May. Robert Millys, Edward Dounes, William May, Robert Olyver, son of the above.

We have entries of several of this incumbent's family :—

1607. April 7.   Thomas Drury, son of William Drury,* Rector of the parish of Leybourne, born the second day of April, and bapt. the 9th of the said month.

1608. John Drewry, the son of William Drury, baptised March 16.

1613. Richard Drewry, buried at Offham April 16.

1614. Richard Drewry the son of William Drewry, born 9th of August, baptised.

1614. December 14, Richard Drewry, junior, buried the 14th of December.

1616. Francis Drewry, the son of William Drewry, born 23rd of April, being St. George his day, baptised the sixth of May.

1621. March 6.   William Drury, born and baptised : the day following, he dyed and was buried.

1630. Nov. 17.   Jane Drury, wife of William Drury, rector of Leybourne, buried.

---

* In this as in many other cases the spelling of names is according to the register which often gives two or three ways.

In the year 1616, the first Leybourne bell was hung.

1640. On the death of William Drury, John Codd obtained the living of Leybourne.

This rector has also left behind a few memorials of his incumbency of the parish. He was presented by the Recorder of Rochester, Henry Clerke.

1643. Elizabeth Codd, the daughter of John Codd, rector of Leybourne, and Gertrude his wife, baptised at Rochester July 27th.
1643. Upon the 29th of December 1653, a daughter of John Codd and Gertrude his wife, buried in Rochester Cathedral, being dead born.
1649. Baptised was Gertrude the daughter of John Codd, rector of this parish, and Gertrude his wife, buried May 7.
1650. John Codd, born 10 July, bapt. 17.

Mr. Codd was also prebend of Rochester, 1660; he resigned Leybourne and was instituted to St. Margaret's, 1662. In his days the return of Leybourne (1659) was value £17 13s. 4d. Patron Sir F. Clark.

1662. John Lorkyn.

Mr. Lorkyn was also rector of Wouldham, prebend of Rochester and archdeacon. In the burial register we read:—

" Mr. John Lorking, prebend of Rochester, and rector of Leybourne, died at Leybourne Jan 8, and was buried at Roch. Jan. xi., 1666. He was presented by Sir Francis Clark.
1667. Nathaniel Hardy, Doctor of Divinity, presented by Sir Francis Clark.
1675. Meric Head.

He succeeded Dr. Hardy, as he tells us in the commencement of the register.

" Meric Head succeeded Dr. Hardy in the Church and Parsonage of Leybourne. He was presented by Sir Francis Clark, and inducted the twentieth of February 1674—5, and the next day being Sunday, performed all the offices and duties required by the Act of Uniformity, which were necessary to his full settlement in the living."

His name appears once or twice in the register.

June 3, 1682. Buried was Sarah Head, the daughter of Mr. Head, and Elizabeth his wife.
March 12, 1687. Merick, alias Meril, Head, Doctor of Divinity, was buried March 12, 1687.
1687. William Gotier.

This rector is several times noticed in the registers:—

Nov. 3, 1687. William Gotier of Leybourne, Rector, and Frances Robins of Town Malling, were married by licence.
Sept. 9, 1688. Mary, daughter of William Gotier and Frances his wife, was baptised. Buried Jany. 7, 1689.
June 3, 1690. William, the son of William Gotier and Frances his wife was baptised.
1690. Henry Ullock, D.D., Dean of Rochester 1689—1706.

He gave the cup to the parish of Leybourne, on it is inscribed:
" The gift of Henry Ullock, D.D., Dean of Rochester, and Rector
of Leybourne in Kent, 1691." We find in the register :—

> 1706. June 28th. Henry Ullock, D.D., dean of Rochester and rector of
> Leybourne was buried.
> 1729. Oct. 31st. Buried Mrs. Margaret Ullock, widow and relict of H. Ullock,
> dean of Rochester and rector of Leybourn. in linen.
> 1706. Samuel Spateman.

This rector's family is largely entered on the registers. The patron
of the living was Captain William Saxby.

> 1710. Samuel, ye son of Samuel Spateman, rector of Leybourne, and
> Elizabeth his wife, was baptised ye 26th of November.
> 1711. Joseph, ye son of Samuel and Elizabeth Spateman, was baptised ye 20th
> of January.
> 1713. April 28th. Margaret, ye daughter of Samuel Spateman, rector of
> Leybourne, and Elizabeth his wife, was baptised. Buried March
> 3rd, 1713-14
> 1714. Mary Greenway, gentlewoman, sister to Mr. Spateman. buried 26 Feby.
> 1714. 15th April. Elizabeth, ye daughter of Samuel Spateman, rector of
> Leybourne, and Elizabeth his wife, was baptised.
> 1716. Ye 30 April. Mary, ye daughter of Samuel Spateman, rector of Ley-
> bourne, and Elizabeth his wife, was baptised.
> 1717. July 21. Joseph, ye son of Samuel Spateman. rector of Leybourne, and
> Elizabeth his wife, was baptised.
> 1718. March ye 29. Joseph Spateman, infant, was buried.
> 1718. August 31. John, ye son of Samuel Spateman. rector of Leybourne, and
> Elizabeth his wife, was baptised.
> 1720. July 24. Thomas, ye son of Samuel Spateman, rector of Leybourne, and
> Elizabeth his wife, was baptised.
> 1720. October 1st. Buried, Samuel Spateman, rector of Leybourne.
> 1721. July 4. Buried Thomas Spateman, infant.

On the death of Samuel Spateman, Robert Hall was inducted as rector, as we
read from a memorandum in the registers.

> Mem. Robert Hall, A.M. was inducted into the parsonage of Leybourne, the
> 3rd of December, 1720.

He was also buried here after a short incumbency, as we read :—

" November 9. 1723. Mr. Robert Hall, late rector of this parish."

He was presented by Captain William Saxby.

> 1723. George Whitworth, instituted on the death of the last.

Related probably to the Whitworths of the Grange, in whose gift
Leybourne then was.

> 1727. Francis Hooper, D.D. whose name is on the fly leaf of the register, was
> then inducted ; he appears to have died in 1758. Presented by Francis
> Whitworth, Esq.
> 1758. George Burvill ; presented by Sir Charles Whitworth. Knt.
> 1797. Charles Cage was rector of Leybourne fifty-two years : he was also
> instituted to Bredgar in 1794.

There is a memorial to him in the church; he was buried here as we read :—

1849. Jany. 30. Rev. Charles Cage, rector of the parish of Leybourne, aged 79 years. He was presented by the first Sir Henry Hawley, Bart.

The first bell is dated 1826.

1849. Henry Charles Hawley instituted.

There is a memorial window to him in the church. His family being also the squires of Leybourne, will be given at length elsewhere. Presented by Sir Joseph Hawley, Bart. In his time the church was completely restored.

1877. Charles Cusac Hawley was instituted on the death of the last; presented by Sir Henry Hawley, Bart., the present rector.

The other clergy mentioned in the Leybourne registers are :—

1718. Charles Brown, clericus, curate of Town Malling, married Mrs. Lydia Elliston, of ye same, by a licence, Jany. 31.
1750. March 6. Buried John Shaw, curate.
1781. Baptised, May 21, Francis Letitia, daughter of James Thurston, clerk, and Mary his wife.

In the commencement of the Leybourne registers, besides what we have already stated, we read: "The register of the christenings of Laiborne beginning Anno Dom. 1560, annoque Dominae nostrae Elizabethae secundo" (in the year of our Lord, 1560, and the second year of our lady Elizabeth). It must be remembered the old registers, which have been destroyed, were carefully copied. The first entry is :—

1560. June 5. Joane King,

and runs down to 1584; then we read :—

Here the boke was very imperfect, therefore I leave out the yeare.

The next entry is :—

George Bredham, gent., and Mary Goddin.
Here the boke was imperfect, and therefore I leave yeare and day unwritten.

1594. Anne Morris, widow of Jasper Morris, was buried upon Ascension Day.
1641. Martii 9. Married were Edward Larking and Anne Chexitty, single persons of Brenchley, by licence from Rochester.
1656. Nov. 12. Elizabeth Aiherst, the youngest and most hopeful daughter of Mr. Aiherst of Leybourne Castle, widow, was buried in the church.

Another person named Dane is described as of Leybourne Castle, in 1655.

1680. Sept. 4. John Morton, brazier and traveller, was buried.
1699. Thomas and John, sons of John Bowden and Isabelle his wife, inhabitants, according to his report, of St. John St. near East Smithfield in London, were born at Lunsford in the parish of East Malling, but were baptised here at Leybourne, Sept. 3r., there being not at this time any incumbent at East Malling.

1701. December 30. Richard Buttenshaw and Elizabeth Wood, were married with the bishop's licence.

The licences from the bishop's court, are mentioned in very formal ways in the registers at this period.

1703. July 19. Richard Buttenshaw of Shorne, and Elizabeth Burvill of Leybourne. having a licence to be married there, were married there then accordingly.

From this, it does not distinctly appear whether they were married at Shorne, or Leybourne.

1705. Feby. 19. John Hills and Elizabeth Edmunds, both of ye parish of Trosscliff, where the banns were thrice published, were married.

This is an early notice of banns.

1705. Dec. 25. David Price and Mary Fleet, both of ye parish of Stepney, in ye county of Middlesex, the banns having been thrice published in ye said parish church, as appeared by a certificate under ye curate's hand, were married by Mr. Harper.
1708. Sept. 26. Elizabeth, daughter of Francis and Margaret Brook of Colebrooke Passinger, baptised.
1709. Sept. 20. James Stone of Trossley, and Dorcas Huggins of this town, were married in ye parish church of Leybourne, having been thrice asked at both places.
1709. Oct. 9. David Ray of East Mallin, and Ann Sethardcn of this town, were married in Leybourne church, by banns thrice published in each parish.
1710. Oct. 10. William Mercer of Speldhurst, and Elizabeth Burgesse wid. of Tonbridge, were married by a licence had from ye archbishop's court, Oct. 10, 1710, in ye parish church of Laybourne, Oct. 12.
1712. Sept. 5. On ye same day, buried mulier ignota (an unknown woman), and infant.
1714. Nov. 22. Richard Buttenshaw, aged about 84, buried at Leybourne.

This is an early entrance of the age.

1719. Aug. 23. Sarah Beavour of Birmacham, traveller, buried.
1719. Sept. 6. Buried a traveller from ye Grange.
1731. July 13. Buried John Anthonio, a blackmoor.
1737. Mar. 13. Baptised . . . the daughter. . . .

There is a similar entry just below, " Sept. 24, 1738."

1750. James Butler, ye son of John and Replenish Butler, born ye 18th, buried ye 29th, of April.
1754. Kemsley Margaret, wife of Francis Kemsley Hoppers, and Catharine Kemsley their daughter, buried 22 Sept.
1766. May 17. Mary Lloyd, housemaid at the Grange, buried.
1787. March 14. Buried, a woman from Mr. Newman's outhouse, unknown.
1792. Sept. 23. Buried, a woman and stranger from Mr. Saxby's of ye name of Bellcbridge.

The next oldest registers, after Leybourne, are those of

ADDINGTON.

The church which is dedicated in the name of St. Margaret, has some relics of Norman times, but it was added to to the westward, and the tower built in the mediæval period, which time is farther marked by the fine wood carving over the church porch. There are traces of where the rood loft was once placed. There is a handsome reredos in this church. The monuments in the Watton chapel are worthy of notice. The register dates from 1562. The incumbents appear to have been as follows :—

1326. Laurence de Polle.—Patron Roger de Leschekere.
1349. Wymundus Conyntone.—Patron Roger atte Eschekere.
1349. John Mount.—Patron Roger atte Eschekere.
1349. Robert de Cuxton.—Patron for the next two turns Sir Nicholas Dagworth.
1350. Richard Gerveys.

In about twelve months Addington had four rectors at the period of the Black Death, which we have spoken of elsewhere ; and which not only swept off many ecclesiastics, but also left another indelible mark upon church history, by causing a cessation of church building.

John de Wynchecoumbe, also precentor of St. Paul's, London, and perhaps rector of Snoreham, Essex.
1356. John de Lexeden, on resignation of the last.—Patron for next four turns John de Colonia.
1356. Thomas Drapier.
1358. Simon de Tonebregge, alias Goman ; previously vicar of West Greenwich.
1361. John atte Ffelde ; previously vicar of Excete, diocese Chichester.
1396. John Graunger ; also prebend of the high mass at the great altar of the conventual church, of the nunnery of Malling 1396, and dean of Malling 1400.
John Marshall ; also vicar of St. Nicholas, Rochester, and rector of St. Michael's, Lewes.
1411. Thomas Clerk, changed with John Marshall.
1416. Edmund Webley, previously rector of Estburgate, diocese Chichester, exchanged with last.
1418. Simon Stokk exchanged with Edmund Webley. The Wattons patrons till 1513.
1435. Robert Bradly ; previously vicar of Welcomstowe.
1438. Thomas Chaworth, on resignation of the last.

His brass on the east wall of the church, records him as a clerk in the King's Chancery, and cousin to Elizabeth, wife of Robert Watton, Esq. It also says he was rector of Long Melford, diocese Ely.

1447. Thomas Skelton.
1451. Thomas Dyne, on resignation of the last.
1453. Robert Watton, a Minorite friar, on the death of Dyne.
1455. Robert Stroke, on the resignation of Watton.
1456. Alexander Broun, on the resignation of Stroke.
1494. Richard Smith.
1495. Thomas Goodale.
1502. William Layfielde, on the resignation of Goodale.
1505. John Houghtone, on resignation of Layfield.
1514. Robert Houghtone, on the death of John Houghtone. Presented by Richard Welbeck.

1533. James Goldewell, on the death of Robert Houghtone. He was vicar of Dartford.
1549. Richard Taylor.
The next four presentations are by the Wattons.
1551. Robert Goodaye.
1559. Robert Salsberry (or Salisbury).

He was prebendary of the fifth stall of Rochester Cathedral. He was collated to the rectory of Trottescliffe, on the deprivation of Bartholomew Bowsfell in 1554, and to Ryarsh in 1572. Trottescliffe he had to resign to its old rector in 1560. It would thus appear that this gentleman was a true vicar of Bray, managing to keep his preferments in Mary's and Elizabeth's times. His record during these reigns being, rector of Trottescliffe and Addington in Mary's time, and rector of Addington and vicar of Ryarsh in Elizabeth's days. He was possibly a son of Dean Salisbury, of Norwich.

1583. Henry Syliard; presented by the Queen. Also rector of Igtham, where he is buried.
The third and fourth bells of Addington date 1602.
1615. Edward Drayner, previously vicar of West Peckham. Presented by Thomas Watton, Esq.
There is an entry in the baptismal register:—
1624. Martha, ye daughter of Edward Drainer, parson, of this psh., the 20th day of June, was baptised.
1635. John Smith. Presented by John Smith, senior, of East Malling.

The second bell of Addington dates 1635. The first bell is probably older than all, but has no date. We find his name in the register:—

Baptised, Phœbe, the daughter of John Smith, parson, of this psh., the 3rd day of January, 1638.
1643. The 7th day of April, baptised Abel, the son of John Smith, minister of Addington.
1660. Peter Davies; presented by William Watton, as also the next two incumbents were.

The cup and paten-cover date from this incumbency. He is mentioned in the registers:—

1663. John, ye sonne of Peter and ffrances Davies, rector of this parish, was baptised Feby. ye 14.

On page 30 his baptismal entry, with several others, is given again in the burial register.
Also we read:—

1679. John Davies, of St. Bridget's and St. Bride's, son of Peter Davies, late rector of Addington, was buried October 14th.

He too, himself was buried here, as appears from the register:—

1673. Mr. Peter Davies, Rector of Addington, was buried in the chancel 22nd October.

Mr. William Polhill, who preached his funeral sermon, succeeded him in the said rectory, being inducted November 19th, 1673. His widow also is buried here, as we read :—

1716. Mrs. Frances Davies, widow, Dec. 3rd.
1673. Mr. William Polhill.

Thus he always names himself, though the ignorant clerk corrupts this into Polly. He was also rector of Offham, where he was buried.

1675. Robert Topp, on the death of William Polhill.

He was previously vicar of All Hallows, 1600. He is mentioned as being rector when a register was bought.

Also we read :—

1677. Maximilian, the son of Robert Topp, rector of Addington, and Elizabeth his wife, was born at Offham, October 15. Bapt. Oct. 18, 1677.
1679. Maximilian, the son of Robert Topp, rector of Addington, and Eliza-beth his wife, was buried March 29th.
1687. Andrew Frederick Forneret ; presented by James Hickford.
1689. Abraham Lord, on the death of Forneret ; presented by William Watton. Presented to West Malling in 1695; he held both livings together till 1698.
1698. Samuel Atwood ; he was instituted in 1701 to the rectory of Ash. Presented by William Watton.
1701. Robert Worlidge, on resignation of Atwood. Presented to vicarage of Ryarsh in 1676.
1702. John Boraston.

He presented a flagon to the church in 1721. There is a paten of the same date. The bell frame is dated 1732. He is buried in the churchyard, where his monument may still be seen, on which we find it recorded that he was, " a fellow of University College, Oxford, a man truly learned, charitable and religious; of a temper mild, cheerful, and humane ; for which while living he was beloved by all that knew him, and when he died was by all lamented." His burial register runs :—

1741. The Rev$^{d.}$ Mr. John Boraston, rector of the parish, buried in ye church-yard, June 14.

He was the last presentation by a Watton.

1741. Thomas Buttanshaw.

Was one of a family long resident in this district, of whom the last about here was the Rev. John Buttanshaw, curate of West Peckham, fifty years ago, and chaplain, afterwards, of Barming Asylum, several of whose sons are incumbents in different parts of England. He was curate of Trottescliffe, he was a minor canon of Canterbury, rector of St. Peter's, vicar of Westgate, and vicar of St. Stephen's, or Hackington, Canterbury. He and his wife are buried here. We read in the registers :—

1761. Mrs. Jane Buttonshaw, wife of the Rev$^{d.}$ Mr. Thomas Buttonshaw, rector of this parish, buried Oct. 26.
1768. Reverend Thomas Buttonshaw, rector of this parish, buried Aug. 26th.

He was presented by Sir Roger Twysden, Bart., in right of his wife.

1768. Daniel Hill, vicar of East Malling.

He was vicar of East Malling from 1762 to 1805, and was also vicar of Yalding. He was buried at East Malling. He was presented by Sir Roger Twysden, Bart., in right of his wife.

1805. Peter Elers.

He had been curate of Trottescliffe, Birling, and Addington. He was buried in the churchyard. In the register we read :—

1820. "Rev. Peter Elers, Nov. 14, aged 62 years." Presented by Leonard Bartholomew, Esq.
1821. Thomas Bowdler.

He was instituted to Ash and Ridley in 1811. He resigned in 1834. He was presented by Hon. John Wingfield Stratford.

1834. George Robert Paulson, on resignation of the last.

In his incumbency the church was restored by the widow of the Hon. John Wingfield Stratford, in the year 1856. He died August 14th, 1869, and was buried here, as we read in the registers :—

George Robert Paulson, rector, 71. 20 August, 1869.

His brother also is buried here :—

1857. John Thomas Paulson, Commander in the Royal Navy, June 4, aged 58 years.

And his widow :—

Fanny Paulson, aged 50 years. 28 Feby., 1870.

Mr. Paulson was presented by Hon. John Wingfield Stratford.

1869. James Newton Heale, previously vicar of Swindon, Staffordshire, and afterwards vicar of Orpington, Kent. He was presented by J. Wingfield Stratford, Esq.

The handsome reredos was erected in memory of John Wingfield Stratford, Esq., by his widow, at Christmas, 1881.

1883. Julian Guise, on the resignation of the last ; previously vicar of Lea, Gloucestershire, presented by Edw. John Wingfield Stratford, Esq., the present rector.

Besides those mentioned above we have the following clergy :—

1620. Robert Wheeler, Doctor of Divinity, and Mary Clerke, were married by a licence, ye 8th of February.
Thomas, ye son of Thomas Pyke, vicar of West Malling, and Elizabeth his wife, was born on Saturday ye ninth day of April, and was baptised on Tuesday ye third day of May, in ye yeare of our Lord God 1698.

The registers of Addington commence in the year 1562, without any preface. The first entry is :—

1562. Thomas Booreman, son of John Booreman, on the third day of March was baptised.

A little lower down we read, "Memorandum, that in 1575 was omitted these five names following"—which are then given. The next is a remarkably early entry of the name of the officiating clergyman.

1610. James, ye sonne of Arnold Curlen, the 23rd day, was baptised by Mr. Hooper, parson of Offham.

1641. Baptised, the . . . son of . . . flower.

The omissions show the clerk filled in from memory.

1610. A maid called Joan Wood was buried, from Anthony Godden's, x^th of December.

1622. The fourth day of November, a poor child of a wayfarer, name unknown, was buried.

1565. Joan Godden, the servant to Mr. Tilden, the 4th day of November, was buried.

1567. John Martin, a frameyer, the seventh day of March, was buried.

1569. Robert Stilt, alias Round, the fourth day of February, buried.

1582. The eleventh of March, a gentlewoman of rich birth from Mr. Daus, was buried.

The strangeness of this entry consists in the clerk knowing this lady to be of *rich birth* yet he does not know her name.

In the years 1578, 1579, 1580, 1581, 1583 and 1599 no marriages are entered. The marriages under Cromwell's Act render the registers of Addington most peculiarly interesting.

1651. Married, John Storey and Susanna —— of Wrotham, the 19th of June.

1654. John Henge and Margaret Sladden were married, their banns being first published 3 several daies, the 8th day of June.

1656. Thomas Hatch and Margaret Hatch were married by Justice Maddun of Boxley, and by the minister of Addington, the twenty-fourth day of Sept.

1657. John Kendon of Stanstead and Mary Woollett of Meopham, were married the 15th day of September, 1657, having three several market days their banns published in Rochester.

1657. William Shileren and Helen Stimpson, both of Tonbridge, were married the 31st day of January, 1657, their banns being 3 several daies published in the market of Tonbridge, by the J(us)t(ic)e of the peace.

1673. Richard Fen and Catherine Noale were married by banns, 18th December, though here inscribed. (This entry is inserted between 1657 and 1659.)

1659. John Dennis of Boxley, in Kent, and Mary Boorman of Ryarsh in the said county, their intended marriage being 3 several market days published in Maidstone, and no exception against them, were married the 7th day of April.

1659. John King and Susannah Prior, both of Wrotham. having had their proposed marriage 3 several Lord's days in the congregation of Wrotham published, were solemnly married at Addington the 13th of October.

1660. Thomas Wood and Sarah ffenne, of ye parish of Ash, having a licence, were married February 7th.

The second register is headed, "This register booke was bought at the charge of the parish of Addington, May 25, 1675, Robert Topp being then Rector."

1691. William Woorsley, born at Tenterden, was publickly baptised in ye parish church of Addington (being of years of discretion), January 4th, 1690—91.
1698. Elizabeth, ye daughter of Mary and James Watts, jun., of this parishe was borne and baptised October ye 7th.
1752. Joshua Cole, an adult, aged about 44, bapt. May 24.
1752. John, son of William and Mary Nettlefold, born Oct. 10, bapt. Oct. 22. Pencil note : Died 1846 in Malling Union, (a case of 94 being *bonâ fide* reached).
1771. William de Coffee, a black, baptized January 24.
1773. Ann, daughter of Henry and Ann Jeffreys, named Feb. 20, brought to church May 15.
1687. William Terry, of Trotterycliffe, widower. and Mary Lamb, of Addington, were married, the banns being first lawfully published, May 5th.
Married An°· Domini 1748, Thomas Draper, a parishioner of the parish of St. Anne's, Boar Street, Soho Square, London, by serving an apprenticeship to John Greenway, a currier, and Elizabeth Meads, a traveller, were married by banns, August 18.
1732. Buried, Sarah, a stranger, Feb. 16th.
1757. William Barton, a foundling infant. buried June 23.
1834. Priscilla, daughter of Philip and Anne Rowlands.

In the column for the abode of the deceased is entered " Vagrants," and for their profession is entered " Vagrancy," by Rev. Geo. Paulson, who was then rector.

The remaining register that dates from the sixteenth century is that of

East Malling.

The church, which is dedicated in the name of " St. James," is one of the finest in the neighbourhood; but though very ancient, there is little to be traced in it of Norman or earlier times, compared with some of the parishes of which we are speaking. The records of the incumbents are fairly complete for nearly 700 years.

1311. Sir William Nicholas, dean of Shoreham, in the time of Edward I., is mentioned in the Kentish Fines.
1323. Thomas de Leghton, on the death of the last incumbent, as we learn from the Lambeth registers.
1363. John Lorkyn.

He looked successfully after the loaves and fishes of his living, as he obtained from the abbess of West Malling the grant of certain tithes for the augmentation of his salary, and also added the chapel of Newhythe and its income to the living.

Simon Blake resigned.
1370. John Kempstan ; changed with
1371. Robert de Gaynesburgh, vicar of St. Clements', Sandwich ; changed with
1373. William Chamberlayne, vicar of Lullingstone.
John Aston ; changed with
1401. Nicholas Grene, vicar of Altisdon, dio. Chichester ; changed with
1410. Richard Smyth, vicar of Ffreuyngham (Farningham) ; changed with
1415. John Wyndesor or Windsor, vicar of Welcombe (London diocese).

1435. Robert atte Kyrke, on the death of Windsor.
Thomas ——.
1439. William Codlyng, changes with Thomas ——.
1467. William Merbury, vicar of Paddlesworth, near Snodland, changes with
Codlyng.
1522. Richard Adams, prebend of the high mass in the monastery of West
Malling.

Was vicar of East Malling as is shown by his brass in East
Malling church, but he is not noticed in the archiepiscopal registers
at Lambeth, which are very incomplete from 1435—1556.

Roland Rice deprived.
1556. John Wells ; presented by the serene princes Philip and Mary, king
and queen of England.

I suppose this entry of Cardinal Pole's register, to be one of the few
public documents that can be found, acknowledging Philip what he
was not, but what his too faithful wife laboured to make him—king
of England.

1571. John Wheler, on the death of John Wells.

In his days the registers start, in which are preserved in a wonderfully continuous chain, the incumbents of East Malling. He was
buried here according to the registers.

1576. Buried was John Wheler, clerke, vicar of East Malling, the 8th of
November.
1576. Launcelot Sympson ; on the death of John Wheler.

There are several entries in the registers, referring to himself
and his family.

1591. Married was Robert Chambers, clarke, and Anne Hartridge, the 8th
day of June, baptised was Launcelot, the son of Launcelot Simpson,
clarke, the same day.
1594. Baptised was Bridget Simpson, the daughter of Launcelot Simpson,
clarke, ye 30 day of May.
1596. Baptised was Mary Simpson, the daughter of Launcelot Simpson,
clarke, the 20th day of August.
1597. Buried was Judith Simpson, the daughter of Launcelot Simpson, the
5th day of April.
1627. Buried was Launcelot Simpson, vicar of East Malling, Dec. xxvii.

In his day the paten was given to the parish.

1627. Robert Whittle.

There are many entries referring also to this vicar and his
family :—

1630. Married Nov. 25, were Robert Whittle, vicar, and Bridget Holmden.
1631. Baptised, August 28, was William, the son of Robert Whittle, vicar,
baby born seventeenth day.
1632. Baptised, February 26, was Thomas, the son of Robert Whittle, vicar,
being born the thirteenth day of same month.

13

1634. Baptised, February 17, was George, the son of Robert Whittle, clerk, and Bridget his wife, born the fourth day.
1637. Baptised, January 16, was Robert, the son of Robert Whittle and Bridget his wife, being born January 1st.
1638. Buried, June 25th, was Robert, the son of Robert Whittle and Bridget his wife.
4663. Married, July 26, George Sampson and Frances Whittle.
1667. Buried, Dec. 9, George, the son of Robert Whittle, vicar.
1668. Buried, March 1st, Bridget the wife of Robert Whittle, vicar.
1671. Frances Sampson, widow, daughter of Robert Whittle, vicar of East Malling.

This entry is inserted without dates, between the entries of Oct. 8th and Nov. 7th in this year.

1673. Buried, March 10th, Thomas Whittle, the son of Robert Whittle, vicar.
1673. Baptised, July 11th, Robert, the son of Thomas Whittle.
1679. Buried, July 16th, Robert Whittle, late vicar of East Malling.
1679. Buried, Nov. 10, William Whittle, minister of Luddesdown.
1732. Jany. 21. Baptised was Thomas, ye son of Robert Whittle and Judith his wife.
1734. Sept. 27. Baptised was Mary, ye daughter of Robert Whittle and Judith his wife.
1735. Jany. 23. Baptised was Judith, ye daughter of Robert Whittle and Judith his wife.
1737. December 7. Baptised was George, ye son of Robert Whittle; buried Oct. 3rd.
1738. April 1. Baptised Thomas, ye son of Robert and Judah Whittle, Jany. 8, 1738.
1739. Buried. Robert Whittle also, September 25th.
1811. Married Feby. 15th, John Allen and Sophia Augusta Whittle.
1679. John Crosse.

So is the name in the registers; but on one of the bells, five of which are dated 1695, it is spelt Grosse; there are several baptisms of his family.

1681. Baptised, 9th of October, Frances, the daughter of John Crosse, vicar, and Elizabeth his wife.
1683. Baptised, 25th day of May, was Philip, the son of John Crosse, vicar, and Elizabeth his wife.
1684. Baptised, 20th day of July, was John, the son of John Crosse, vicar, and Elizabeth his wife.
1685. Baptised, Oct. 13th, was Harry, the son of John Crosse, vicar, and Elizabeth his wife.
1686. Baptised, Oct. 20th, was Elizabeth, the daughter of John Crosse, vicar. and Elizabeth his wife.
1688. Baptised, Feby. 6th, was Anne, daughter of John Crosse, vicar.
1689. Baptised, July 26, Sarah, daughter of John Crosse, vicar.
1693. Baptised, May 19, Jane, daughter of John Crosse, vicar.
1705. April 26. Buried was Jane, ye daughter of Mr. Crosse.
1705. July 14. Buried was Frances, daughter of Mr. Crosse.
1701. Richard Berrow.

There are the entries of two of his children :—

1701. Nov. 22. Baptised was John, ye son of Richard Berrow, vicar of this parish, and Catherine his wife.

1703. Jany. 23. Baptised was Millicent, ye daughter of Richard Berrow, vicar of this parish, and of Catherine his wife, born on ye 22nd.
1 05. Thomas Hill. He was buried here.

There are a number of entries of his family.

1706. Jany. 8th. Baptised Elizabeth, ye daughter of Thomas Hill, vicar, being born Dec. 28th.
1707. Feby. 25. Baptised Thomas, ye son of Thomas Hill, vicar, being born 13th of same month.
1708. July 2. Baptised Anne, ye daughter of Thomas Hill, born ye 28th same month, about noon.
1710. Feby. 25. Baptised Richard, son of Thomas Hill, being born ye 24.
1712. May 12. Baptised Edward, ye son of Thomas Hill, vicar, being borne April ye 3rd.
1713. March 13. Buried Edward, son of Thomas Hill, vicar.
1714. March 1st. Baptised, born Feby. 17th, Susan, daughter of T. Hill, vicar.
1718. Sept 16. Buried Mr. Thomas Hill, vicar of East Malling.
1718. Thomas Cockman.

There are no entries of this vicar's family. He obtained a dispensation to hold Trottescliffe in 1724 with East Malling, he died in 1741. The East Malling flagon and alms dish date from this time.

1742. William Perfect.

Of his family, we have this information from the registers :—

1748. Buried Mrs. Dorothy Perfect, the vicar's daughter.
1756. Baptised Oct. 19, was Sarah, the daughter of Mr. William Perfect and Mrs. Elizabeth his wife, by his grandfather the present vicar. Witness my hand, WILLIAM PERFECT.
1757. Buried the Reverend Mr. William Perfect, vicar, June 8th.
1803. Buried Mrs. Perfect, 58.
Thomas Gowland Sherret and Sarah Perfect, married July 24, 1757.
1757. Richard Jacob.

He was also vicar of New Romney. There are two entries of his children :—

1758. Philip, son of Richard Jacob, vicar, and Anne his wife, Sept. 1st.
1760. July 25th. Mary, daughter of Richard Jacob, vicar, and Anne his wife. Buried, Sept. 9. Richard Jacob, A.M., vicar, 1762.
1792. Feby. 28th. Mary Jacob, widow of the late vicar, 73.
1762. Daniel Hill.

He was also rector of Addington, 1768-1805, and for some time vicar of Yalding. There are entries in the register that show this pluralist vicar was resident here.

1782. Jany. 29. Buried, Susanna, wife of Daniel Hill, vicar.
1793. Jany. 10. William Hill, A.M., rector of Wickham Bishop in Essex, and canon of the Cathedral Church of Wells, buried.
1796. July 25. Lætitia Hill, widow, daughter of Daniel Hill, vicar, 46, buried.
1805. Feby. 26. Daniel Hill, vicar, in the 94th year of his age, buried.
1784. May 25. William Hill, clerk, and Lætitia Hill.
1805. Samuel Francis Godmond.

There are several entries of this vicar's family, into whose hands the presentation had then fallen.

1807. Baptised, April 5, Ellen Elizabeth Jackson. daughter of Rev^d· Samuel Francis, and Anne Godmond, born, Jany. 28.
1808. Baptised, Nov. 2, Mary Anne, daughter of Rev^d· Samuel Francis and Anne Godmond. born Jany. 18.
1828. Buried, Nov. 5th. James William Godmond, aged 17 years.
1839. Isaac Singleton Godmond, clerk, June 8th, aged 34 years, buried.
1845. October 2nd. Rev^d· Samuel Francis Godmond, clerk, vicar of East Malling, aged 72 years.

In his days, the East Malling bell No. 1 was added.

1845. C. F. Godmond held the living for two years. He was son of the Rev^d· Samuel Francis Godmond ; upon his resignation,
1847. William Lewis Wigan obtained the incumbency, the advowson of which had been purchased by the family.

There are five of his family mentioned in the registers previous to his own burial, which we find in the registers thus :—

1876. January 13. Buried William Lewis Wigan, vicar of the parish of St. James. R.I.P. 58.

, He greatly ornamented the church, and in his time the Newhythe chapel was built, of which we shall speak more particularly below, which was erected in the year 1854.

1876. Septimus Wigan.

He was previously vicar of Fring, Norfolk, 1861-66 ; chaplain of Faversham Almshouses, 1866-67 ; vicar of Tettenhall, Stafford, 1867-74. His wife and daughter are buried here. He is the present vicar.

Besides these, there are the following entries of clergy in this Register :—

Buried 1784, Dec. 17, Thomas Hartley, clerk, rector of Winwick in Northamptonshire, 77.
1795. Margaret Ramsay, daughter of Reverend Richard and Sarah Warde, was born, Jany. 31, bapt. Feby. 1.

Mr. Warde, in the year following, was appointed rector of Ditton.

1814. Jany. 13. Rev. John Henry Norman, of Harrietsham, and Elizabeth Norris, married.
1881. April 21. Rev. Frederick William Reade, and Harriet Dadson, married. Besides these we have the christening of 3 sons and one daughter, of the Rev. William Frederick and Dom Mary Woods, and of two daughters of the Rev. Henry Amherst and Constance Eleanor Orlebar, and two sons of the Rev. Octavius Charles Legge and Jane Monkhouse Wilkinson.

The registers of East Malling begin, " Burials, Marriages and Christenings in the Parish of East Malling, 1570."
One of the earliest entries is in 1570.

Baptised was Robert, the sonne of George Thomas, without any swathing, the 9th of January.

1571. Buried was Johan Pyne, the daughter of one Pyne of London, the 2nd of June.
1582. Buried was a nurse child the 14th day of April.
1588. Buried was mother Smith the 14th of September.
1597. Buried was one who said he was borne in Hartfordshcere, the 29th day of August.
1598. Buried was Thomas Foster, a Sussex man, who died soon after he was released out of prison, 28th February. (The prison we suppose was Maidstone.)

We have after this, several entries of poor girls and poor women.

1628. Buried, Jany. 18, was old Will Johnson.
1631. Buried, March the 19th, was Francis Lewis, a vagrant, who died as he was passing to Lullingstone, in Kent.
1631. Buried October 15th, a chrisom child of Richard Pierson.
1631. Buried October 24th, was Alice the wife of Thomas Parsley, and her chrisom child, of the said Thomas Parsley (cum aliis).*

N.B.—The register of the marriage of George Fowler and Harriet Bassett, which occurs in the year headed 1726, was a forgery perpetrated by individuals, who it is hoped will be brought to justice. This wicked insertion is obvious from its outward visible sign, viz., after the register of Oct. 28th. "Baptised John, the son of John Phipps," is inserted " May ye 21, married George Fowler to Hannah Bassett."

On the page between 1641 and 1642, we find this curious insertion, and again, in the year 1726, we find the insertion again. It has evidently been written there over a previous entry which had been erased, and is of later date than the entries before and after it, and written in blacker ink.

1647. At the end of the entries for this year is an interpolation of much later date.

Born at the parish of Goudhurst, was Abraham Walter, the son of John Walter, the ii. day of August.

In this century though the register is never blank, still in 1624 there are only four entries, in 1625 fifteen entries, in 1648 only three entries, in 1649 six entries, in 1650 eight entries, and in 1651 four entries.

1653. Married, December 27, Thomas Worlidge, hoyman,† and Susan Norton after publication of the intention thereof three preceeding lord's dayes.
1654. Buried July the 17th, was an apprentice of Thomas Ward, yt was drowned.

The clerk does not seem to have troubled much about the unfortunate youth's name.

---

* A chrisom child was a child under a month ; so named from the "oil" or "chrisom" used formerly in anointing at baptism. The white vestment used in Baptism was also named a chrisom.
† Hoys appear to have been frequent at Newhythe, at this date (though now we have only barges), by several entries of this style in the register.

1657. Baptised John, James, and Mary, the three children of —— Bathe on
—— February 1657.

The next entry is the burial of this triplet.

There are no marriages in the years 1655, 1656, 1658, 1659 and
1661.

1695. March 10, married John Adams, and Temperance Davies.

In this we have the survival of one of the quaint old Commonwealth Puritan christian names, some of which, as Charity, Mercy,
Constance and Prudence, have lasted even down to our own times.

1706. July 30. Buried a strange boy, that was drowned at Newhythe.
1706. Dec. 17. Buried old Goody Mills. (A not uncommon entry.)
1707. Jany 13. Buried Eliza, ye wife of Goodman Curteis.
1708. Jany 14. Buried old Goodman Wood. (These entries are frequent.)
1739. Feby 20. Married John Foster, of ye parish of Brentsley, and Mary
      Wingate, after ye banns had been thrice published, and she had
      assured me her former husband was dead.
1753. Buried Dec 4, (aged 105), ye widow Baker.

There is a stone to this old centenarian, let into the wall of the
church on the outside of the west door.

1755. Feby. 23. Baptised ye son of John Shepherd, of Larkfield, being about
      a quarter old when baptised.
1760. March 8th. Maria Anne, daughter of John Charlton, and Anne his wife,
      was rec^d· into the congregation, being privately baptised three years
      before.
1760. October 9th, two strangers buried.
1764. Nov. 10. Augustina Barham, an infant from the Foundling Hospital.

This shows that East Malling like the rest of the parishes round
took the foundlings from London in the eighteenth century.

A paper in the early registers tells us that in 1776, when Rev.
Daniel Hill was incumbent, there were 953 inhabitants, and no
dissenters: five men were furnished to the national militia. This
last entry shows us that the early movement for forming militia
was listened to in our valley.

1777. Nov. 24. A man unknown found dead in the Medway.
1789. June 12. John, a cooper, unknown, about 50.

Here is an instance of very recent date, of how it is we find so
many people called after trades; not only is it because some were
so named, originally, but also that persons whose names had been
forgotten were named after their trades when nought else could be
remembered at all.

1798. Nov. 8. Elizabeth Ann, d^r· of Thomas and Mary Spicer, nearly five
      years old, baptised at Lynn in Norfolk, but fully so here.

In this entry we trace an early example of that curious idea of
the poor that a person privately baptised has not been really
christened, which they signify by calling it half-named.

We have also this insertion :—

N.B.—The tablet erected to the memory of the Revd. D. Hill, A.M., was at ye joint and equal charges of Lady Twisden (widow of the late Sir Roger), Mrs. E. Norris, Charles Smith, Esq. (who was many years a resident of the rectory house in this parish), and Miss Smith, his sister, now of West Malling then (1828) of the rectory, and founder of the alms houses in Mill Street : the intention of this praiseworthy act originated with the latter lady.

> 1820. Robert Waggon, March 12, 67 years ; 33 years clerk of this parish, the duties of which office he fulfilled with credit and general respect.

The chapel of ease to East Malling called Newhythe is half-way between the hamlets of Newhythe and Larkfield, and was erected in 1854. The various curates have succeeded each other as follows :— Rev. R. Dimock, 1854—1872, Rev. S. Wigan, 1874, Rev. F. H. D. Ness, 1876, Rev. W. F. Woods, 1881, and Rev. O. Legge Wilkinson, 1888. Mr. Wigan informs the author that Newhythe has never been held as a curacy-in-charge, but the clergyman for the time being has been always looked upon as only a curate of East Malling.

The next oldest register, after East Malling, in the valley is that of

ALLINGTON,

which, however, is sixty years later, and with the two neighbouring parishes of Aylesford and Ditton, dates from the seventeenth century; it commences in the year 1630. The parish church is dedicated in the name of "St. Laurence," but it has been completely rebuilt only a few years ago; with the exception of the tower, which is probably fifteenth-century work, there are no traces either of the Norman or the Perpendicular in the building. So much has the church been altered that even the old inscriptions of Thorpe's days have disappeared. The incumbents we have found to be as follows :—

> 1132. Robert de Donam.
> 1279. Odo. This vicar engaged in a suit with Gregory de Elmham, whether Longsole belonged to Allington or Aylesford.
> 1318. Jordan De Sale ; presented by Henry de Cobham.
> 1322. Thomas de Clare.
> 1326. Simon Ladnsville.
> 1358. Richard Broot.
>   Richard Bricton.
> 1361. Henry atte Chambre, in succession to Richard Bricton.
>   Richard Grigg.
> 1394. William Levinge, by change with Richard Grigg. He was previously vicar of Bearsted.
> 1398. John Essex, on the resignation of William Levinge.
> 1402. John de Warcham.
> 1404. John May.
> 1404. John Crip. Also vicar of Beauchamp.
> 1422. William Sprote.
> 1431. John Disse.
> 1451. Simon Drake.
>   John Wyllys.

1461. Richard Yogesby on the resignation of John Wyllys.
1514. Robert Saunders, a monk of Boxley.
    Robert Hedcorn.
1530. Richard Taylor, on the resignation of Robert Hedcorn; presented by
    Sir Henry Wyatt.
1576. Thomas Ely.
1582. Robert Carr; presented by the Crown.

It would appear that the old cup of the church belongs to his time, as it dates from 1599.

1622. William Carr.

Instituted, according to the Rochester registers, to the " Parish church of Allington Castle." In his days the registers commence.

1636. Edmund Jackson.
1636. Richard Thomas.

This clergyman is buried in the churchyard, as we learn from a tombstone :—

" Hic jacet Richard Thomas, magister in artibus utriusque academiæ nuper pastor hujus ecclesiæ, qui obiit Feby. 8, 1656."

The entry is not to be found in the register. In his days the bell was hung. He mentions himself as having christened Jacob Ashley, in 1651.

1656. John Collins, was presented, on the death of Thomas, by Isaac Astley.

In the year 1659, the living of Allington was declared worth £6 16s., and the patronage was then in the hands of Lord Aylesford. Collins was also rector of Bearsted.

1677. Edward Darby, mentioned in the register.
1711. John Richards, succeeded Edward Darby.

He is mentioned in the registers.

1712. John Richards, succeeded his father.

He soon after obtained also the living of Teston, and after that he obtained the living of Nettlestead also, when he resigned Allington. Hasted tells us that he died " distracted " in 1761.

1714. Richard Spencer.

In his days the paten-cover, dated 1726, was probably presented. This rector was presented by Lord Romney.

1757. Edward Weller.

We have an entry of his performing a marriage in the year 1769.

790. Jacob Marsham, D.D., son of the 2nd Baron Romney; was canon of
    Windsor, and canon of Rochester. ·

Amongst his children were the late head of Merton College, Oxford, and the vicar of Shorne, in this county.

1831. George Frederick John Marsham.

He is the first rector that appears to have been resident for many years. We have a number of entries finishing with the burials of his wife, and himself, who died both of them at an early age :—

Baptisms.

1838. Feby. 15.   Catherine Elizabeth, daughter of George Frederick John (clerk) and Elizabeth Marcia Marsham.
1840. July 2.   Marcia Elizabeth Maria, daughter of George Frederick John (clerk) and Elizabeth Marcia Marsham. .
1842. Feby. 4.   Frances Penelope, daughter of George Frederick John (clerk) and Elizabeth Marcia Marsham.
1843. Sept. 10.   Elizabeth Isabella Sophia, daughter of George Frederick John (clerk) and Elizabeth Marcia Marsham.
1845. Dec. 11.   George Jacob, son of George Frederick John (clerk) and Elizabeth Marcia Marsham.

1845. December 26.   Buried, George Jacob Marsham.
1848. May 21.   Annie Harriet, daughter of George Frederick John (clerk) and Elizabeth Marcia Marsham, baptised.
1848. May 27.   Buried, Annie Harriet Marsham.
1849. April 23.   George, son of George Frederick John (clerk) and Elizabeth Marcia Marsham, baptised.
1849. April 26.   Elizabeth Marcia Marsham, aged 38 years, buried.
1852.   George Frederick John Marsham, aged 45, buried Feby. 5.

Mr. Marsham was vicar of Halling, as well as rector of Allington.

1852.   Edward Brown Heawood.

Of whose family there are several entries. Mr. Heawood is the present rector.

Besides the clergy mentioned above we have in the register :—

Jonas Provost, of London, cleric, and Susan Clarke, of Maidstone, widow, were married 10th July, 1648.

Rodulphus Mabbe, Generosus Magister Cantab., clericus, et vicarius Ecclesiæ de Gran (Grain ?) sepultus fuit 28th August, 1649.

1631. James Wilson, of Boughton, in the county of Kent, clerk, and Mary Rayner of Hollingbourne, in the county of Kent, spinster, were married together in the church upon the fourth day of March, Anno Domini 1631, by licence from the facultie.
1736. The Reverend Mr. Edward Crank, of the parish of Goudhurst, and Mrs. Mary Philcox, of the parish of Horsmonden, were married May 1st.

We have several curious entries in the registers as under. The book is prefaced : " The registers of the Christenings, Marriages, and Burials in the parish of Allington, near Maidstone, in the county of Kent, began in Anno Dni. 1630."

We have a number of baptisms, etc., from Maidstone :—

1649. Thomas, the son of Thomas More, dead of Maidstone, and Elizabeth his wife, the 20th of September.
1722. Samuel, a black servant to Mr. Goatley, was baptised.

At the end of the baptismal register we read : " The account of the inhabitants of Allington is taken by Robert Fauchon, overseer, on the 27th of July, 1811. Thomas Britten, parish clerk at the time. Males 31, females 24 ; total 55."

The number of marriages was wonderfully augmented by strangers from 1648-1653, owing to Cromwell's Act ; probably on account of the passage boats on the river, and the sequestered position of the church, which gave opportunity for runaway matches. A singular thing about these weddings is that nearly all the bridegrooms are described as widowers : the number of weddings almost reached the probable population one year ; they were 12 in 1648 ; 18 in 1649 ; 32 in 1650 ; 38 in 1651 ; 29 in 1652 ; and 25 in 1653. After this we find no marriages in 1654, 1656, 1657, 1659, 1683, and during the whole periods from 1686—1695, 1699—1713, and from 1736—1744.

October ye 22,1753. William Fraser, of the parish of St. Laurence Poulteney, London, and Mary Evans, of St. Mary Abchurch, London.

This marriage was solemnised in the parish church of Allington, in the county of Kent, this eighteenth day of October, 1769, between John Fauchon, of the parish of Northfleet, in the county of Kent, and Susannah Russel, of this parish, by me E. Weller, rector. John Fauchon, Susannah Russell. Witness, Henry Butt, John Russell. This marriage was registered in this book, a new one not at this time having been provided.

Will Field, of the Hermitage,* was buried the twenty-fifth day of April, 1639.

John Fletcher, of Little Buckland, was buried the 14th of March, 1650.

Whereas I am credibly informed, upon the oath of Robert Fauchon, of Allington, in the county of Kent, farmer, that on the seventeenth day of June instant Elizabeth Barrett, a child of about ten years of age, was found drowned in that part of the river Medway lying in the said parish of Allington, that the said child had been missing for a week past, and appears to have been drowned by accident : these are therefore to certify that you permit the body of the said child to be buried, and for your so doing this warrant and authority is given under my hand and seal the 18th day of June, 1803. GEO. MILLS, Coroner.

Henry Russell, found dead in the River Medway, in this parish, Oct. 5th, 1859. Age about 50.

John Rawson, drowned in the river Medway in this parish, by accident, Jany. 27, 1862.

The next oldest register is that of

AYLESFORD.

The church, which is dedicated in the name of the apostle St. Peter, has been thoroughly renovated. The building is of the fourteenth century.

The various incumbents of Aylesford, we have found, are as follows (those previous to the Reformation were presented by the monastery of St. Andrew's, Rochester ; those after that period by the chapter) :—

1145. Jordan.
1285. Gregory de Elham.

This vicar had a dispute as to whether the Hermitage of Longsole belonged to Aylesford or Allington. It appears in the end to have been incorporated in Allington parish.

* This is the last record of the Hermitage of Longsole, if we except the name Hermitage woods that still survive.

—— Thomas de Borstall.
1326. Galfridus de Cowling ; also vicar of Chilham in this county.
1327. John Orset.
1329. John Acholt.
1336. Robert de Haldenc succeeded John Acholt.
—— Richard Baker.
1394. John de Battiscombe, on the resignation of John Baker.
1397. William Gorynge, on the resignation of John de Battiscombe.
1404. John Long.
1422. John Stubbercroft disputes Longsole with Thomas Wilson of Ditton,
   and William Sprote of Allington.
——. William Handton, *alias* Stringer.
1425. William Battisford, instead of Handton.
1425. Philip Arngorm, *pro* William Battisford.
—— Richard Bride.
1432. John Hill succeeds Richard Bride.
—— William Redysdale.
1451. Thomas Carter.

Mentioned also as vicar in 1459, 1462, and 1464.

—— John Rosc.
1521. Henry Fletcher, on the death of John Rose.

He afterwards became vicar of West Malling, and prebend of the
High Mass of Malling Abbey, whereupon he resigned Aylesford.

1524. Robert Blacus, on the resignation of Fletcher.
1572. Thomas Shastebrook.
1574. William Giles.
1593. Henry Barnewell ; was also rector of Barming, and afterwards, in 1605,
   archdeacon of Rochester.
1608. George Smith.

The first Aylesford cup and paten are dated 1628.

—— Henry Eryngton.

In his days the registers commence. He was buried here, as we
learn from them :—

1654. Henry Eryngton, Esq., vicar of this parish, was buried 20th September.

The fourth and fifth bells of Aylesford date from 1652.

1654. Joseph Jackson.

Is called minister of Aylesford in the first leaf of the registers.

1655. Daniel Aldene.

Is mentioned as buried here. We read in the register :—

1666. Daniel Aldene, gent., minister of this parish, and one of the surrogates
   of the diocese of Rochester, and brother of Dr. Edward Aldene,
   chancellor of the said diocese, was buried the first day of September.

The seventh bell (1661), and the eighth bell (1666), belong to the
incumbency of Mr. Aldene.

1666. Thomas Tilson.

In 1679 he was appointed rector of Ditton, and held the two

livings together till his death in 1702. There are many of the
family of this vicar mentioned in the registers as well as himself.

1671. Hugh, the son of Mr. Thomas Tilson and Joane, his wife, was baptised
the xith day of October.
1673. Sarah, the daughter of Mr. Thomas Tilson, vicar, and of Joane, his
wife, was borne Dec. 6th, and christened ye 16th of ye same Dec.
1674. Thomas, the son of Mr. Thomas Tilson and Joane, his wife, was baptised
the 16th day of December.
1677. Maria, the daughter of Mr. Thomas Tilson and Joan, his wife, was
baptised Sept. 26th.
1678. Martha, the daughter of Mr. Thomas Tilson, baptised Aug. 23.
1680. Caleb, the son of Mr. Thomas Tilson and Joanna, his wife, was baptised
May 25.
1680. Caleb Tilson was buried June 24th.
1680. Mrs. Joanna Tilson was buried Aug. 30th.
1698. Mr. George Luce of St. Margaret's, Westminster, and Mrs. Sarah Tilson
of Aylesford, were married July 13th.
1702. Mr. Thomas Tilson, vicar of this parish, was buried July 26th.
1702. Thomas Tilson, son of the above.

Was instituted like his father to both Aylesford and Ditton.
He presented a paten to the parish, on which is inscribed :—

·· Tuum est domine tibi reddo, T. Tilson, Vic. Aylesford, 172¼ " ; on the
flagon is, " Sumptu parochiæ de Aylesford, et Thomæ Tilson, conjunctim, A.D.
1711 " ; and on the Alms dish, "The gift of Lady Taylor to the Parish of
Aylesford.   T. Tilson, Vic. 172⅓."

We have a wedding here in his day of one of his family.

1711. Robert Drew and Mary Tilson were married the 1st day of June.

He was buried here.   There is a monument to him and his father.
The entry in the register is :—

1749. Revᵈ· Thomas Tilson was buried Feb. 17th.

The sixth bell of Aylesford was also added in his days ; it is dated
1708.

1749. John Lawry.

He was also rector of Lee, in Kent, which he held together with
Aylesford.

1773. Charles Colcall.

He was vicar of Ashburnham, Bucks, as well as rector of
Aylesford, and petitioned that he might hold these livings together,
they being not more than forty miles apart, and was permitted to
do so.

1784. William Eveleigh.

He was also rector of Lamberhurst.   There are several entries
of his family.

1794. Susannah Rebecca, daughter of William and Susannah Eveleigh, born
March 31st, baptised 8th of May.

1794. Susannah Eveleigh, Aug. 30th, buried.
1794. John Eveleigh, October 25th, was buried.
1795. John Eveleigh, May 30th, was buried.
1798. William George, son of William and Susannah Eveleigh, born Sept. 4, 1797, baptised Jan. 6.
1799. John, son of William and Susannah Eveleigh, born May 13th, baptised July 21st.
1801. James, son of William and Susannah Eveleigh, born Aug. 29th, baptised September 1st.
1801. James Eveleigh, September 7th, buried.
1803. Thomas Chase Eveleigh, Feb. 12, buried.
1803. William George Eveleigh, August 31, buried.
1805. Susannah Eveleigh, Jan. 19, buried.
1808. Susannah Jane, daughter of Rev. William and Susannah Eveleigh, born 18 June, baptised July 8.
1809. William George, son of William and Susannah Eveleigh, born 12th of March, baptised Aug. 23.
1823. August 1, married Henry Fage Bilson of St. Margaret's, Rochester, and Anne Hanwood Eveleigh.
1830. William Eveleigh, LL.B., born July 23, 1757, died Oct. 29, 1830 ; 38 years vicar ; aged 73 years ; buried.
1834. Susannah Eveleigh, Feby. 27 ; 68 years ; buried.
1831. John Griffith was vicar.
1832. William Tolbutt Staines.

He is buried here. We find in the registers :—

1840. William Tolbutt Staines, vicar of Aylesford, Oct. 1st, aged 65 years.
1840. E. G. Marsh.

In the marriage register we read :—

1856. March 26. Thomas Abbott and Anne Caroline Marsh.
1862. Anthony Grant.

He was previously rector of Romford and archdeacon of Rochester and St. Albans. His daughter's marriage is to be found in the register.

1878. Cyril Fletcher Grant. Honorary canon of Rochester, 1890.

There are several entries of the vicar's family in the register. Other clergy mentioned in this register are :—

1665. William Jole, rector of Ditton, gent., and Mrs. Catherine Aundey of West Malling, were married by virtue of a licence out of ye court of Rochester, the 5th day of May.
1710. Richard, the son of Mr. Hill, minister of East Malling, was buried here at Aylesford, the 4th day of October.
1718. Mr. Edward Harrison, clerk, Otham, and Abiezer Betts of the same, were married Dec. 4.
1744. The Rev^d. Thomas Marshall Jordan and Susannah Woodgate, both single persons of Aylesford, were married April 24th.
1735. Mary, daughter of the Rev^d. Mr. John Williams and Mary, his wife, was baptised Feb. 13th.
1737. John, son of the Rev^d. Mr. John Williams and Mary, his wife, was baptised Jan. 28th.
1746. Mary, daughter of the Rev^d. Thomas Marshall and Susannah Jordan, was baptised Oct. 1st.
1782. The Rev^d. Mr. James Nance, buried Jan. 21st.

1784. Rev^d· Joseph Milner,* D.D., of Preston Hall, buried Aug. 5th.
1790. Rev^d· Richard Warde * and Sarah Ramsay of Gillingham, married
    Feb. 4th.
1791. Sarah Catharine, daughter of Richard Warde, clerk, and Sarah his
    wife, born 12th Dec. 1790, baptised Jan. 15th.
1843. Rev^d· George Lockyer Perry, April 15, buried ; aged 73 years.
1855. Feb. 12. William Kiteley, clerk, M.A., and Kate Mary Ann Burroughs
    Nash, were married.

Also the baptism of several children of Rev. A. H. Cheshire.
There is no preface to the Aylesford registers. Those entries
that concern Cromwell's Acts have been previously mentioned. As
in the Allington registers, so we find in the Aylesford registers a
remarkable number of marriages; only at rather a later period, from
1661-1668, are they most numerous here. Most of the bridegrooms
are described as widowers. There are a great number of weddings
again about 1701; and in 1735 there are entered no less than 40;
and the marriage of persons, both alien to the parish, we find for
many years.

Another curiosity in this register that we notice is that, for
the first ten years of the register, births are entered instead of
baptisms, e.g.—

1653. Francis, ye daughter of John Hills and Jane, his wife, was born ye
    viii^th day of March.
1662. Elizabeth, the daughter of George and and Mary Burde, baptised the
    30th day of January, being the first baptised in the new font after
    the iniquities of the tymes had broken down the old one.

This is the most remarkable testimony to the destruction wrought
in the churches by the soldiery in the Civil War. Possibly the
children entered above as born, were not baptised, but received into
church.

George Battie, a man who was drowned in the river, or some said his name
    was Thomas Batt, was buried the xxiv^th day of April, 1654.

This entry gives us a remarkable example of the fondness of Kent
people for adding a " y " to words : we hear still rosey for rose, posty
for post, and flinty, that is flittymouse, for a flittermouse or bat.

1680. A stranger, who was found drowned, was buried, Dec. 6.
1709. A baker was buried here. Feby. 26.
1709. Anne, daughter of John Dawson and Mary his wife, was born the 25th
    day of August, but was never baptised ; by reason they profess them-
    selves to be of that erroneous sect, of the Dippers or Anabaptists.

We have the same entry in 1711 of Patience, daughter of the
same couple.

1711. Richard, the son of Richard Cogill and Anne, his wife, departed this
    life, here at Aylesford, ye 7th day of February, but was buried at
    Raynham, the 10th day of ye same February.
1711. Thos. Normond, a servant that lived at Mr. Porteous', buried 5th day
    of April.

————————

* Also rector of Ditton.

1711. Mr. Stalebrass, schoolmaster of the parish, was buried 7th of November.

John Bessant, son of Nicholas Bessant, marriner, being drowned by casualty was buried the xvi^th day of June, 1658.

John Sampson, a stranger, died at the Lady Culpeper's, was buried the 4th of September, 1659.

Thomas Day, being shot by a soldier, who with three others were stealing bonds on the warren, was buried the ix^th of March, 1660.

Henry Gorham, and John Allen, the one a bricklayer, and the other a carpenter's apprentice, going into ye river at Forman's forstall to wash them, being upon the xxv^th day of June, 1661, were both drowned, and were buried in two several graves in this churchyard, the xxvii^th day of ye said June, 1661.

The word " forstall " is in common use in East Kent to this day, but not so far west as Aylesford, for the fore part of a farm.

Peter Dyne, an apprentice to Robert Kembsley of this parish, by falling from a horse, or being thrown, or strooke, or trod upon by the horse, so bruised and wounded that he died thereof, was buried the xxvi^th of July 1661.

1664. Dorothy Birthall, senior, an antient maid, aged about 75 years, buried on the 12th February.

A travelling man who sold earthen pots and other earthenware, being found dead in Thomas Smith's barn, was buried in the said Thomas Smith's orchard the said seventh day of February, 1665.

It is possible this orchard was one of the sacred grounds belonging to Tollington, Cossington, or Eccles chapels; if not, it is strange this interment, which would be then most irreligious, should be entered on the Aylesford registers.

Richard Kemsell and his child, being unbaptised, was buried the one and twentieth day of October, 1666.

John Philpot, a stranger, being taken blind at Rochester, the nineteenth, as was expressed in his pass then dated and given under the hand and seal of the city of Rochester, aforesaid, to convey the said John from officer to officer to Snargate, in the county and shire of Kent, his former place of abode, was brought hither ye xx^th, and died ye xxi^st, and was buried here the three and twentieth day of November.

John Nethersole, a man that dyed at Mr. Duke's at Cossington, was buried ye xxvi^th day of February 1670.

1729. Two children of William Groombridge was buried.

1750. Thomas, son of David and Honour Phillips, was baptised, Aug. 13. Note:—
The parents of the child belong to Lampeter ponsteven (upon Severn) in the county of Cardigan, in the principality of Wales, and was born in the parish of Maidstone, in Kent.
According to the late Act of Parliament made in the 24th year of his majestie's, King George the Second's, reign, the date begins the first day of January, 1752.

1756. John Henrick Heinman, corporal in Hodenburg's regiment, and Count Captain Phalenberg's Company, was buried, March 18th.

We have several instances of regiments named after their leaders about this time, in the different registers.

1791. Mary Freeman, an infant, daughter of George Freeman, at the Lower Blue Bell, was buried Feby. 22.

1792. Samuel Kemsley, an infant, from the Upper Blue Bell, was buried May the 3rd.

Edmund Poynter, gardener to the Dowager Lady Aylesford, who was unfortunately drowned in the river Medway, by the upsetting of a boat in a gale of wind, on Thursday Feby. 2, 1792, was buried Feby 7, 1792, aet. 29. *N.B.*—The body of Stephen Somers, who was unfortunately drowned with Edmund Poynter, by the upsetting of a boat, on Thursday Feby. 12, 1792, aet. 23, was never found.

This is erased and the following added :—

The body of Stephen Somers, was found March the 14th, and buried March the 15th, 1792.

F. Pelctca, a private in Captain Coleman's independent company, was buried March 14, 1793.

The last parish whose records begin in the seventeenth century is

DITTON.

The church, which is a good specimen of Norman work, is dedicated in the name of St. Peter. It is nearly filled with monuments to the Brewer and Golding families.

The names of the rectors discovered are as follows :—

1317. Adam (dictus ad aquam Maydestone, *i.e.*, surnamed Maidstone-water).
1326. Walter de Roya ; presented by the priory of Leeds.
—— Bartholomew, rector de Ditton.
1356. John Roe.
1371. Henry Shibbard ; also vicar of Marden.
—— John King.
1402. John Sapnethorn, vice John King.
1422. Thomas Wilson ; disputed Longsole, with the Vicar of Aylesford, and rector of Allington.
1423. Robert Blakstolbe, in place of Thomas Wilson.
—— John Florence.
1442. William Howday, in place of John Florence.
1444. William Sampson.
1444. Thomas Thorp, in place of William Sampson.
1449. John Solom.
—— Hugh Hudson.
1501. Laurence Skoye, on the death of Hugh Hudson.
1510. Thomas Greane, on resignation of Laurence Skoye.
1526. John Bechynge, on the death of Thomas Greane.
1533. David Welling, on the death of John Bechynge ; afterwards rector of Paddlesworth cum Dodeciree.
1533. William Kemp. The last presented by the Convent of Leeds.
1546. Nicholas Archebolde, in place of William Kemp.
1554. George Attke, in place of Archebolde, probably deprived. Hasted here inserts William Clough, 1553.
1565. Thomas Bayard, in place of George Attke.
1577. Hugh Williams ; also rector of Leybourne.
1579. Edmund Godyn, or Godwin, succeeds Hugh Williams.
1608. William Prewe ; presented by Richard Shakerley, Esq., on the death of Godyn.
1638. John Smith.

The No. 1 bell of Ditton belongs to this incumbent's time. The first entry in the register is : " 1663. Elizabeth, the daughter of John Smith, baptised 12 August."

But whether this is the daughter of the rector it were difficult to tell. In his days Ditton was worth £11 15s. 0d., and was in royal patronage.

1663. William Jole.

In this rector's time the registers were commenced. There are two memorials to him in the church. He and his family are noticed in the registers. He was married, as we have already shown, in Aylesford Church 1665, on the fifth day of May.

1665. Daniel, son of William Jole, rector of Ditton, baptised March 8.
1670. Anne, daughter of William Jole, rector of Ditton, baptised Feby. 1. Buryed Feby. 15.
1672. Thomas, son of William Jole, rector of Ditton, baptised, August 27, born, August 31.
1675. William, son of William Jole, rector of Ditton, baptised, Jany. 15, born December 26th, 1674. Buryed April 26.
1676. Mary, daughter of William Jole, rector of Ditton, baptised Dec. 6th, born Nov. 18th.
1667. Katherine, the wife of William Jole, was buryed Sept. 14th.
1670. Anne Jole, the daughter of William Jole, was buryed Feby. 15.
1675. William Jole, son of William Jole, rector, was buryed April 26th.
1678. William Jole, rector of Ditton, dyed September 19th, buried 21st in ye chancel.

1678. Joseph Smith.

This rector's incumbency was very short ; as we read in the following year :—

1679. Mr. Joseph Smith, rector, was buried May the second.

The following may be his widow's burial :—

1691. Mrs. Martha Smith was buried, June 1st.
1679. Thomas Tilson presented to Aylesford in 1666.

He was allowed to hold the two livings till his death in 1702. The silver cup which has on it : " The gift of Mary Brewer, to the Parish of Ditton in Kent, for ye use ·of the Church, Jany. 4, 1689," was given during his incumbency.

1702. Thomas Tilson, son of the last rector ; presented and inducted to both Ditton and Aylesford this year.

In the beginning of the register we read an entry made in his incumbency, August 1, 1711 :—" That every acre of Woodland in the parish of Ditton by immemorial custom pays tithe to the rector." He gave a silver paten to the church on which is inscribed, " Túum est domine tibi reddo donum. Thomas Tilson, Rector, 1735." It will be remembered he made a like present to Aylesford. The second bell of Ditton was also given in his day :—

1750. John Oare.

Presented by the Earl of Aylesford, according to Hasted.

1757. Charles Bowles.

In his days the description of the glebe lands, and an inventory of

14

church goods, which we have given elsewhere, were made. He died in 1768, and was buried in the chancel of West Malling, near the vestry door. Presented by the Earl of Aylesford.

1769. Joseph Butler.

He was also rector of Burham. He took the name of Milner, by which he gained the property of Preston Hall. He was buried in Aylesford church, where there is a monument to him as Joseph Milner, D.D., of Preston Hall. Presented by the Earl of Aylesford.

1784. Samuel Bishop.

Head Master of Merchant Taylors' School. He was also rector of St. Martin's, Outwich, in the city of London, which he held together with Ditton. He was presented by the Earl of Aylesford.

1796. Richard Wanle.

He perhaps was curate at Aylesford, as he was married there in 1790, and had a daughter christened there in 1791. He was presented by the Earl of Aylesford.

1840. William Hamilton Burroughs.

Though not buried here, there is a tablet to his memory over the vestry door. The Earl of Aylesford was patron.

1856. John Young Stratton.

The present rector. He has busied himself to accommodate and civilise the hoppickers, with much success; and is secretary to several organisations for this purpose. The living is now in the gift of the Earl of Aylesford. The rector's children were baptised here.

There are the following curious entries in Ditton :—

1712. Robert Norris and John Day's maid were married October 2nd.
1723. Feby. 21, Mrs. Tomlin, wife of old John Tomlin of Ditton, dyed ye 17th day of February, was buried ye 21st, at East Malling.
1720. Moses. a stranger, was buried September 4th.
1758. Dec. 26. Thomas, son of Thomas Harris and Mary his wife, was privately baptised.
1759. Jany. 12. The above Thomas was admitted into ye church.
A.D. 1767. July 5. Thomas, son of Thomas Derham and Susanna his wife. strangers from Bristol, baptised. *N.B.* The child was born in Trooper's field, in ye parish of Aylesford.
1790. Sarah Dod and Stephen Dod were buried September the 22nd. The above persons were unhappily burnt to death, by setting fire to a barn of Mr. John Golding's during hopping.
1793. July 24th. William Starthup, an infant, drowned in a pond of Mr. Thomas Golding's.
1797. Susan, daughter of Richard and Fanny Baber, was born the 11th, and baptised September 17th, they were hopping at Mr. J. Golding's.
1764. Francis Jersey, of the parish of St. James, Westminster, bachelor, and Ann Bridges of this parish, married in this church by a prerogative licence the 22nd of March, in the year one thousand seven hundred and sixty-four, by me, Charles Bowles, rector. This marriage was solemnised between us. Francis and Anne Jersey, in the presence of George and Sarah Luck.
1812. A man, name unknown. who was found drowned, buried Dec. 26th.

Two registers do not commence till the eighteenth century. One of these,

WEST or TOWN MALLING,

begins in the year 1700. The church dedicated in the name of the Virgin has some old Norman work in the chancel, and the tower is also an old building ; but the body of the church, as has been previously stated, feil down at the end of the last century, and is rebuilt in the Georgian style. There is in the church a picture of the Last Supper, over the west gallery. The royal coat of arms, of the reign of James II., is considered by connoisseurs a fine specimen of art.

The incumbents, so far as we have discovered them, are as follows :—

1339. Robert de Beulton, presented by the Abbess of Malling.

A dispute having arisen concerning his tithes, he obtained the lesser tithes for his income, and also personal tithes in what were then known as Holyroode Street and Tan Street ; in return, the vicar was to find everything for the use of the church, bread and wine for the sacraments, processional tapers and lights for the chancel, accustomed ministers, rochets, surplices, unconsecrated napkins, vessels, basons, and green rushes to strew the church if necessary.

1348. Richard Benson, in place of Robert de Beulton.
1360. Richard Gresham.
1392. John Watson. He was also prebend of the High Mass of Malling Abbey.
1395. William Baron, instead of John Watson.
1399. Robert de Geulton.
1399. Thomas King ; also vicar of Lynstead, near Sittingbourne.
1400. John Caldewell.
1402. John Reynolds.
1413. Simon Dawes ; also rector of St. Mary Magdalene, Canterbury.
1426. William Rose or Rotse ; also rector of Norton, Kent.
1440. John Pure.
1452. John Rose ; buried in West Malling church.
1455. Daniel Everard succeeded John Pure. He was also prebend of the High Mass of Malling Abbey.
1500. Thomas Nevill. He was also prebend of the High Mass of Malling Abbey, in which office he succeeded John Whitmore, but it does not appear that this last was the intermediate vicar of the parish.
—— Thomas Smyth.
1515. William Lawson, succeeded Thomas Smyth.
1517. John Bamborough, succeeded William Lawson. In 1522 he was also appointed prebend of the High Mass.
1524. Henry Fletcher. He succeeded John Bamborough, as vicar of West Malling, and prebend of the High Mass. He was previously vicar of Aylesford.
1571. Milo Carrards ; presented by Hugh Cartwright, Esq.
1574. Thomas Brande ; presented by Hugh Cartwright, Esq.
1577. Nicholas Grier ; presented by Hugh Cartwright, Esq.

The Delft-ware stoup is Elizabethan.

1581. Thomas Thomson ; presented by Hugh Cartwright, Esq. A presentation
was made upon his holding the two livings of West Malling and
Teston.

1610. Christopher Wray or Wragge ; presented by Sir Robert Brett.

1630. Mr. Robert Throckmorton.

1637. William Gibson, vicar of West alias Town Malling, on the death of
Throckmorton. Patron, Sir John Rayney.

The fourth, sixth and seventh bells belong to this incumbent's
time. In his days the returns were made to the see of Rochester,
and we learn that Town Malling was £10 0s. 0d. a year, and in
the gift of Fitzjames.

1662. Samuel French.

In his incumbency the third bell was hung ; according to Hasted,
he was ejected in this year : probably he was an interloping
Puritan minister.

1662. Joshua Allard.
——— Samuel Ellwood.

In the register we read: " Dec 9, 1701, Mary Ellwood, relict of
Samuel Ellwood, vicar of this parish."

1695. Abraham Lord.

He was appointed, in 1689, vicar of Addington ; he held both
livings together till 1698. In his incumbency the fifth Malling bell
was hung.

1698. Thomas Pyke, previously curate of Addington.

In his days the register commences. There are several entries of
this vicar's family in the register :—

Thomas, ye son of Thomas Pike, vicar of West Malling, and Elizabeth his
wife, was born in ye parish of Addington, on Saturday ye ninth of April, and
baptised ye third day of May, next following, which was in the year of our
Lord God, one thousand six hundred and ninety eight.

Mary, ye daughter of Thomas Pike, vicar of West Malling, and Elizabeth
his wife, was born on Wednesday ye third of Jany., and baptised ye day
following, being ye fourth of the same month, Anno Dom. 1699—1700.

Mary, ye daughter of Thomas Pike, vicar of West Malling, and Elizabeth
his wife, was buried on Thursday, March 28, 1700.

1703. Mary, ye daughter of Thomas Pike, vicar of West Malling, and
Elizabeth his wife, was baptised on Thursday, March 10.

1704. March 3. Thomas Pike, late vicar of the parish.

1704. Robert Scudamore.

There are several notices of this vicar in the register :—

1707. June 1st. Revd. Robert Scudamore, Esq., vicar of this parish, batchelor,
and Margaret Wild, of ye same psh., spinster, married by licence.

1712. April 30. The Revd. Scudamore, Esq., and vicar of Town Malling, was
buried, aged thirty-five.

Another member of his family is probably entered :—

1809. **May** 2. John Scudamore, of the parish of Maidstone, and Charlotte
    Elizabeth Downman, married.
1712. Simon Babb.

This vicar is entered in each register as rather noteworthy in the
words: "The year that Mr. Simon Babb came to Town Malling."
We do not, however, find anything remarkable during his incumbency.
He is buried here, as we learn from the registers:—

1730. October 26. Mr. Simon Babb, vicar of this parish.
1730. Charles Brown.

This vicar is mentioned in the baptismal register, but no entry
is made of himself or his family:—

1748. James Webb.

His burial is mentioned in the register:—

1759. Sept. 26. Ye Rev^d· James Webb, curate of this place, and rector of
    Trotreclief.

Though called curate, it would appear that he was really vicar.

1759. Robert Style; resigned in 1770.
1770. Richard Husband. He tells us in the register, he was inducted to ye
    vic^o· of Malling, in this year, Dec. 29.

In his days the body of the church fell down; a brief was issued
to rebuild it in 1779, and it was reopened in 1781. Mr. Husband was
rector of Stowting, near Hythe in this county, which he held with
West Malling: he is buried here, and his monument is in the
church. In the burial register we read:—

1813. Sarah Husband, Malling, Feby. 27, 83. William J. Coppard.
1813. Richard Husband (late vicar, Malling), April 1, 78. William J. Coppard.
1814. George Fern Bates. He was also vicar of South Mims, Middlesex,
    which he held together with Town Malling.

Mr. Bates and his wife, and his father and mother, are all buried
here, as can be learnt by the tablet in the church. He was a great
benefactor of the schools.

1842. John Henry Timins, on the death of Bates.

Mr. Timins has just completed his jubilee as vicar. During his
time much has been done towards the improvement of the church;
as the reseating of it and placing in it of an organ by the Lucks,
in memory of the late Mr. Luck, and the building of a more commodious vestry, and the renewing of the chancel, and the hanging of
the first, second, and eighth bells. The Abbey chapel has been also
repaired and used once more for Divine service. Several of his family
are mentioned in the registers, which include the marriage of his
daughter.

1868. Sept 9. John Adolphus Boodle, and Alice Elizabeth Timins.

Mr. Boodle was twenty years curate of Malling, and thirteen
years Diocesan Inspector of Canterbury, when he was appointed to

his present living of Boughton-under-Blean. Two handsome altar lights with a suitable inscription have been placed in the church, which with other presents and suitable addresses were given to the vicar this year, 1892, to commemorate his jubilee. Also a carved oak lectern, the work of a carving class conducted by Mrs. Parry, widow of the late Bishop Suffragan of Dover.

The tithe of Tan Street and Holyrode Street, now Frog Lane, and Swan Street upon trade had continued till Mr. Timins came here, and augmented the living of Malling by £15: this the present vicar never levied, and it has consequently fallen into desuetude. The living is now in the hands of the trustees of J. Lawson, Esq.

Entries of other clergy in this register are :—

1718. Nov. 7.   Samuel, the son of the Rev. Mr. Samuel Bickley, rector of Offham, and Mrs. Mary his wife, baptised.
1721. July 18.   Thomas, son of the Rev. Mr. Bickley and Mary his wife, baptised.
1750. June 11th, was buried the Rev. Mr. John Willis, M.A., for many years an inhabitant of this parish.
1760. Charles Bowles of this parish clerk. and Catharine Weekley of this parish, Spinster. married in the church by licence this eleventh day of February in the year one thousand seven hundred and sixty, by me, Dennis, minister.   This marriage was solemnised between us, Charles Bowles, Catherine Bowles, in the presence of B. Hubble and Anne Hubble.
1768. June 3rd.   The Rev. Mr. Charles Bowles, rector of Ditton.
1783. Oct. 26.   Thomas, the son of the Rev. James Thurston, vicar of Ryarsh, and Anne his wife.
1784. The Rev. James Pritchard, the curate of West Malling, left the place March 14th.
1790. Jany. 25.   Charles Thomas, son of Rev. James Thurston and Mary his wife, baptised.
1797. Aug. 5.   The Rev. Dale Lovett, aged 75, buried.
1800. Jany. 30.   Mrs. Catherine Bowles. widow. aged 79, buried.
1814. April 29.   Elizabeth, daughter of William and Elizabeth Bowles, baptised.
1815. March 9.   Robert Pye, Esq., and Elizabeth Bowles, married.

These last appear to have been both of the family of the Bowles of Ditton.

1814. Dec. 4.   Elizabeth Emma, daughter of Rev. John Henry and Elizabeth Norman, baptised.
1839. April 7.   Rev. John Liptrott, rector of Offham, and vicar of Ryarsh, buried.
1849. April 17.   Rev. William Lewis Wigan (vicar of East Malling) and Caroline Ramsey Akers, married.

Besides the vicars of Town Malling, the undermentioned have been chaplains of Malling Union since the Union chapel was built in 1872.

1873—1875. Rev. John Manus. chaplain.   In his time the register commences.
1875—1880. Rev. John Stuart Robson.
1880—1889. Rev. Henry Frederick Rivers.   Since then vicar of St. Faith's. Maidstone.
1889. Rev. Cecil Henry Fielding.

The present chaplain ; author of this book. His daughters' baptisms are entered in the baptismal register of West Malling.

The following entries of West Malling are worthy of notice, besides those that we have already spoken of :—

1701. April 10. Richard Knowles, a carpenter, buried.

A very early entry of the trade or calling of any one.

1701. July 28. Jane Lane was buried on Sunday.

We often find the entry of Sunday in the Malling register at the early period, as if it was unusual to bury on that day : if so, it is a pity that this has not been adhered to.

1701. Aug. 16. Anthony Gilly, a trooper, buried.
1702. March 28. Anthony Fochard, a trooper in Duke Schomberg's horse, buried.
1703. April 11. Wid. Godden was buried on Friday.
1707. Aug. 13. John Burton of the parish of Eightham (buried).

This entry gives the old name of Ightham, which means eight homes.

1708. We find only a solitary burial entered in this year.

1710. Richard Huye, an apprentice, buried April 12.
1710. April 29. Ja. Bromfield of this parish, bat., vel. Bumfield, and Eliza Pretty, of ye same, spinster, were married, the banns being first lawfully published.
1711. May 11. Henry, the son of John and Elizabeth Ridge, was baptised. Note :—This was Goodwife Gransbury's son, by a former husband, and not christened till after she came to Malling, and then the said boy was ten years of age.
1712. May 25. Rich. Bignal, singal man, and Dorothy Davis, widow, both of this parish, were mar⁴·, ye banns being first lawfully published.
1713. April 7. Wid. Stephenson's child buried.
1713. July 26. Tudor Lamb of London, smallpox, buried.

This and two or three other entries of the same in Snodland register are the only mentions of this dire complaint in the district.

1714. Aug. 24. A soldier, belonging to my Lord Orkney's Regiment of Foot, buried.

We have inserted these and other similar notices as matters of history, as they show us how the regiments were distinguished before numerals or territorial designations were given to the divisions of our army.

1714. Aug. 22. Baptised, a hopper's child.

The earliest notice of hopping in this parish.

1714. Oct. 19. Moyse, Sir Felix Wild's man, buried.
1715. Nov. 29. Old Googer.

This is as short an entry of burial as is to be found anywhere.

1716. Jany. 13. John ——, Mr. Lovegrove's apprentice, buried.
1720. May 12. Anne, daughter of Anne Addison and William, baptised.

In this entry the wife seems to have quite eclipsed the husband.

In the year 1727, there is entered only one baptism; two in
1728; and three in 1729. In the first of these years there is only
one burial, and none in the other two years.

> 1734. Oct. 29. Jeanne Anne Marguerite, fille de Guillaume Boysier, et
> Marguerite sa femme, Marchands de Genes en Italie, fit baptizee jour
> d'Octobre 18, O. S., par moi, Chas. Browne, Vicaire de cette paroisse,
> a Genes.

(Jane Anne Margaret, daughter of William Boysier and Margaret, his wife,
merchants of Genoa in Italy, was baptised 18th day of October, Old Style, by
me, Charles Browne, vicar of this parish, at Genoa.) *Memorandum.*—The
vicar and churchwardens are desired to permit this to be entered on their new
registry, wherever this shall be transcribed, it being a very great inconvenience
to the French Protestants, when they cannot prove that their children have
been baptised by a Protestant clergyman.

> 1741. Aug. 23. Mary, daughter of Patrick and Eleanor McDonald, Irish,
> baptised.
> 1743. Aug. 24. Martha, daughter of George and Mary Davis, Scotch, baptised.

These two entries rather point to where the Home Rule notion
sprung from; this fondness for distinguishing the inhabitants of the
different parts of the British Isles from each other.

> 1737. Aug. 21. A hopper's child, buried.
> 1739. Aug. 29. Jane Mortimer, a hopper, buried.

After this we have frequent mention of hoppers.

> 1743. Dec. 23rd. A poor woman from the workhouse (buried).

The first mention of this institution in any of the records we have
been examining.

> 1744. May 11. Robert, son of Robert Watson, a soldier, baptised.
> —— May 27th. John, son of John Grover, farrier to a troop of horse,
> baptised.
> 1747. December 24th. Peter, son of John Isaac, a soldier in Pulteney's
> Regiment of Foot, buried.
> 1748. March 6th. An infant, daughter of a soldier in General Pulteney's
> Regiment, buried.

In this last entry the regiment has quite extinguished the indi-
viduality of the father.

> 1748. Feby. 27. A man, name unknown, died through the inclemency of
> the season.
> 1749. August 13th. Anne Hind, a stroling player, buried.

Strolling is often used for travelling.

> 1750. January 7. Elizabeth Costin, servant at New Barnes, buried.

1756. There was a great mortality in this neighbourhood during
this year: we have 64 deaths in West Malling; and a number more
in the neighbouring parishes.

> 1757. A soldier, in the seventh regiment of foot, buried.

In this year we have several entries from the Foundling Hospital,
and the same occur afterwards.

> May 8th. John Burnet, an infant son of a soldier, buried.
> April 1, 1758. Margaret Campbell, a soldier's child, buried.

These entries of soldiers must engage our attention; because, for some reason, they are very frequent in Malling for the size of the place.

1760. About this period we have the word "late" inserted in the register with the bride's name, in this parish, thus :—

Nov. 13, 1760. Edward Harcourt, and Elizabeth Harcourt, late Compton.

1765. Anne, daughter of Robert and Sarah Dunbar, a dragoon belonging to Sir John Mordaunt's regiment, baptised.

1770. October 6. A woman, Mr. Maplesden's hopper, name unknown, buried.

1773. October 23. A hopper of Mr. Stewart's, buried.

These two entries are curious, showing that these people had lost their identity in their master :—

Dec. 11. Alice, daughter of Gilbert and Eleanor Graham, of the 50th Regiment, baptised.

Dec. 18. Grace, daughter of Thomas and Grace Upstone, of the 50th Regiment, baptised.

1780. Jany. 14th. Grace Upstone, an infant, of the 50th Regiment,* buried.

1779. June 14th. John, son of Thomas and Susannah Mecham, of the Buckinghamshire militia, was baptised.

1784. Jany. 22. A stranger woman, name unknown, found dead in the cage.†

1786. Dec. 3. John Clow, black servant to Mr. Perfect, buried.

1794. Jan. 17. John Doidge, soldier of the South Devon Militia, buried.

1797. Aug. 31. Bartholomew Davis, late organist, buried.

This is as early an entry of an organist in a country church as we think is likely to be found. No wonder Town Malling can be complimented on its good choir !

1799. Dec. 26th. Joseph Barlow, soldier in the Royal Wagon Team, buried.

1801. Jan. 8. Henry Piggot, from Eaden Bridge, aged 38, buried.

This entry gives us an intermediate way of spelling the place: Eatonbridge being frequent some years ago, it is now always spelt Edenbridge.

1801. May 27. John Wood, beadle, aged 60.

This record is interesting as recalling the old parish beadle, whose rule in Bumbledom Dickens has so admirably portrayed.

1820. Mary Walter (by order of the coroner), near New Barnes, May 14, aged 38. G. F. Bates.

There is no reason given for the cause of the coroner's order.

We have several entries from 1820 to 1830 of people from Coxheath Workhouse; why, it is not very clear.

On the 17th day of August of 1796, Charles Peto of this parish, a bachelor,

---

* The 50th Regiment, which was originally raised in Kent, has again been designated the West Kent Regiment. It must be glad to have lost its number, which gained it the sobriquet, "The dirty half hundred."

† The lock-up or cage formerly stood on the piece of waste ground that lies on the side of the hill between Ryarsh Lane and High Street. It was pulled down some years ago.

and Mary Green of this parish, spinster, were married by the vicar R. Husband. It is declared to be null and void, because the bride being a minor had not obtained her father's consent.

But the next entry declares her lawfully married because his consent had been then given. In these days, when there is so much laxity in the marriage laws, it were good if the old rule and canon of the church, as regards minors obtaining their father's consent, were strictly adhered to.

> 1824. Rev^d. Jacques François Stuart de Lenneville. French priest, formerly of Notre Dame de bon Report, chaplain of Roman de Pavilly, rector of Champigny near Melun. June 28, 1824, buried.

Evidently one of the refugees of the French Revolution.

> 1831. Jane Tilley, a soldier's child.

A very unfinished entry for its date, there is only added, " May 5, infant. John Scott." (Burial register.)

> 1832. Timothy Hogarty, a traveller from Barming, seized with cholera in the London road through this parish, Sept. 18. 35. John Scott, curate.
> 1839. Henry Mountfort, Lunatic Asylum, Aug. 26. 21. S. F. Godmond.
> *N.B.* He was an inoffensive, amiable man, much respected in the neighbourhood, and was an inmate of the asylum 32 years.

It now only remains for us to speak of

HALLING,

the registers of which parish are very recent, beginning as late as 1705; the older records down to that date having been destroyed or lost. The church, which is dedicated in the name of St. John the Baptist, has been much pulled about; and though we know there was a church here in Saxon and Norman times, there are little or no remains of either. The bishop's palace has been nearly swept from the face of the earth; and the chapel of Saint Laurence is built into a cottage. The incumbents of Hamo de Heth's favourite parish, as far as we can find, are as follows : —

> —— Michaelis.
> 1317. Hugh Girton. Patrons, Strood Hospital.
> —— John Argent.
> 1327. Thomas Lardner.
> —— John de Wileshyr.
> 1329. Robert de Dercham.
> 1330. John Champneys.
> 1338. John de Ripara.
> 1359. Nicholas Plumele.
> —— John Erpingham.
> 1391. Robert Clerk succeeds John Erpingham.
> 1391. John Penysthorp, in place of Robert Clerk.
> 1393. Thomas Bekonsfield follows John Penysthorp.
> —— Stephen Porchet.
> 1429. Thomas Pende, chaplain of Hoathe or Hothe, near Reculver.
> 1442. William Hammond.
> 1445. Thomas Ratcliffe.

1447. Thomas Carlton.
—— Thomas Merbury.
1465. William Martyn.
—— Robert Cass.
1500. John Cotton.
1513. Richard Clark, in place of John Cotton deprived.
1515. Thomas Snydall on the death of Richard Clark.
1534. Robert Johnson, on the death of Snydall.
—— Thomas Bedlowe.
1554. Launcelot Gylhawke, in place of Thomas Bedlowe deprived.

Gylhawke was evidently one of the papists introduced by Queen Mary; as Thomas Bedlowe we find was restored, and the mention of Gylhawke is passed over.

1567. Walter Hait, in the place of Thomas Bedlowe.

He is also called Heath. He was vicar of Shorne in 1575. He resigned Halling for St. Margaret's, Rochester, in 1587, when he was made prebend and archdeacon. He resigned St. Margaret's for Goudhurst in 1589, and Goudhurst for Cuxton in 1589. Dean and Chapter of Rochester, patrons.

1587. William Ledes or Leeds; is mentioned as being here in 1608, and by another authority in 1630.
1638. John Bath.

Halling, in 1659, is valued at £7 13s. 9d., in the gift of the Chapter of Rochester.

—— Bailey.

In the year 1675, the five bells of Halling were hung.

1688. Robert Beresford.
1705. William White.

The registers commence in this vicar's incumbency.

1723. Ralph Bishop.
1729. John Price.

Mentioned in the registers.

1739. William Pattison.
1769. Robert Fountain.
1777. John Leach.

He was married here according to the registers :—

John Leach, clerk, widower, and Susannah Fuller, spinster, both of this parish, were married 7th day of July, 1778.

His daughter was christened here :—

Feb. 25, 1779. Matilda Ann, daughter of John Leach, vicar of this parish, and Susannah his wife.
1791. William Dyer.
1818. Samuel Browne.
1825. William Henry Drage.
1843. George Frederick Marsham. Also rector of Allington, where he resided.

1852. Joshua Nalson.

His wife and he are buried here :—

1867. Harriet Nalson, May 31, aged 71 years.
1885. Joshua Nalson, The Rectory, Halling, July 2, aged 81 years.
1885. Frederick Goldsmith—made dean of Perth, Western Australia, 1888.
1888. George Plumptre Howes, the present vicar.

There are very few things worthy of note in the short record which the Halling register contains. We, however, give the following.

Besides the clergy above mentioned we have :—

The Revᵈ Mʳ Major Nourse, of the parish of Shorne in the county, a batchelor, and Isabella Hill of the same parish, spinster, were married by licence, March 2, 1747.

Mr. Nourse was for some years vicar of Higham, where he lies buried. Other entries are :—

1729. Feby. 16. A daughter of Thomas Fuller buried.
1730. Jany. 25. Diana and Thomas, twins, were baptised.
1730. October ye 25. Mary, an infant, unknown, buryed from Peckman's.
1731. A stranger, unknown, died in Mr. Ray's barn, Feby. ye 7th.
1732. William Beecham and Thomas Bailey were both drowned and buried together, March ye 10th.
1750. Delona, daughter of a stranger unknown, Nov. 26, baptised.
1736. Matthew Lofte and Anne Cox, were married at St. Nicholas, Rochester. by John Price, vicar of Halling, June ye 14th.
1736. November ye 13th. Gregory Everest and Rachel Humphries, were married at St. Nicholas, Rochester.
1734. John Cheeseman, servant to Mr. Hatch, April 21, buried.
1737. A stranger, Stephen, who died at ye Compasses, and Thomas Fowler, were buryed August 24, by John Price.
1777. November 1. Garland Partridge (buried).
1780. August 2. A boy about 17 years old, unknown, drowned in the Medway.
1804. May 2. A stranger, drowned in the Medway, buried.
1806. October 3. A person unknown. drowned in the Medway, buried.
1809. April 21st. A man, unknown, drowned in the Medway, buried.
1820. April 15th. A man, unknown, drowned in the Medway, buried.
1888. —— Harris, Dec. 15, 2 years. G. P. Howes. This child was found dead in the woods, after having been lost sixteen days.

CHAPTER XII.

WE have now gone through the various clergy, and the curious entries to be found in the different registers. Our next task will be to go through the principal families to be found in this district; after which we shall give the names of other gentry to be found here, then those of yeomen and farmers so specially entered, and conclude this account of the registers with names in the district and the various occupations mentioned.

The first family to speak of in the neighbourhood, are the Lords of Abergavenny, whose seat was for so many years at Birling, in the register of which we find many of them mentioned as christened, married, and buried. In connection with them we must also mention their ancestral window in Nettlestead Church.

NEVILL.

1576. Frances Ladie Abergavenny, wyf to the Right Honourable Lord Abergavenny, was buried the xth day of Sep., with the garter king Harold's devise.

1586. The Right Honourable Sir Henry Nevill, Knight, Lord Abergavenny, was buried honourablie ye xxist of March in the yeare above written.

1588. Mr. Edward Nevill, was baptised the 4th day of June, in the chapell at Comfort.

This entry is most interesting as preserving the old name of the seat of the Nevills, which is recorded by Harris, but has now been quite forgotten in name; though some of the old house and walls at Birling Place are undoubtedly of this period.*

1602. Margaret Nevill daughter of Edward Nevill, Lord of Abergavenny, was buried the xth day of October.

1610. Edward, the second son of the Right Honourable Edward Nevill, Lord Abergavenny, was buried the first day of November.

1616. Lady Rachell, wife to the right Honourable Edward Nevill, Lord Abergavenny, was buried the viiith day of October.

1617. Lady Elizabeth, first wyffe to the Right Honourable Sir Harry Nevill,

---

* Comfort is mentioned again in this entry: "Ferdinande Ashbee, gentleman that died at Comfforte House, was buried the xiiii. day of July 1681."

221

but not wyffe to Sir William Sedley, Knight and Baronet, was buried ye fifteenth day of August.

1617. The Right Honourable Edward Nevill, Lord Abergavenny, was buried the third of November, ut supra.

1628. The Right Worshipfull Sir Thomas Nevill, sonne and heir apparent to the Right Honourable Sir Henry Nevill, Knight and Lord of Abergavenny, was buried the 7th day of May.

1641. The Right Honourable Henry Lord Abergavenny, was buried the 24th of December.

1649. Jany. 7. Sir Christopher Nevill, buried.

1649. July 10. The Lady Katharine Nevill, widdow, was buried.

1662. The Right Honourable John Lord Abergavenny, was buried ye 23rd of December.

1666. The Right Honourable George Lord Abergavenny, 14th day of June, was buried.

1727. The Right Honourable George Nevill, Lord of Abergavenny, was born June. 24th. His Majesty King George the Second, His Grace Lionel Duke of Dorset, and her Grace, the Duchess of Newcastle, being sponsors (at St. Margaret's, Westminster).

Nothing is said, we must notice here, as to the christening, though undoubtedly the noble sponsors, and the church, are intended to convey this to the mind.

1828. March, 28. Henry Nevill, buried.

1828. March, 28. Augusta Nevill, buried.

1855. Jany. 14. Ludoviek Edward, son of Hon. Edward Vesey and Isabel Mary Francis Bligh, baptised.

1856. July 27. Llewellyn Nevill Vaughan, son of Hon. Thomas Edward Mostyn, M.P., and Henrietta Augusta Lloyd Mostyn, baptised.

1857. May 15. Henry Richard Howell, son of Thomas Edward Mostyn, Esq., M.P.. and Henrietta Augusta Lloyd Mostyn, baptised.

1861. May 16. Houble. Thomas Edward Mostyn Lloyd Mostyn buried.

1863. Jany. 17. Constance Emily, daughter of Hon. Ralph Pelham, and Louisa Marianne Nevill, baptised.

1868. August 25. William Nevill, Earl of Abergavenny, buried.

1872. May 21. Cicely, daughter of Hon. Ralph Pelham and Louisa Marianne Nevill, baptised.

1872. May 27. Caroline Nevill, Dowager Countess of Abergavenny, buried.

1873. July 22. Emily Georgiana, daughter of Hon. Ralph Pelham and Louisa Marianne Nevill, baptised.

The monument from Nettlestead is as below :—

" Here sleepeth in the Lord with certain hope of resurrection, the body of borne Lady Mary, Barroness Despencer and Burwash, who departed this transitory life into an eternall, upon the 28th day of June anno dom. 1626. and the 72 year of her age. She was sole daughter and heir of Henry Barron of Abergavenny, all which three barronies were derived to him from Elizabeth. sole daughter, and heyre of William Beauchamp, barron of Abergavenny and Earl of Worcester, his great grandmother ; and from Isabella Despencer her mother. This noble lady was wife only to one husband, Sir Thomas Fane. Knight of Badsell in Kent, who left her a widdowe, at the age of 35 years, and he left this life the 28th day of February, Anno Dom. 1589, in the 52 year of his age. He was buried first at the Church of Tudely, in Kent, and now, in obedience to the command of her last will, her executors and executrix, her humble daughter-in-law, Mary Countess dowager of Westmoreland, who errected to her memory this monument in the year 1639, hath translated his body to accompany hers until the general day of resurrection."

On brass plates above the monument :—

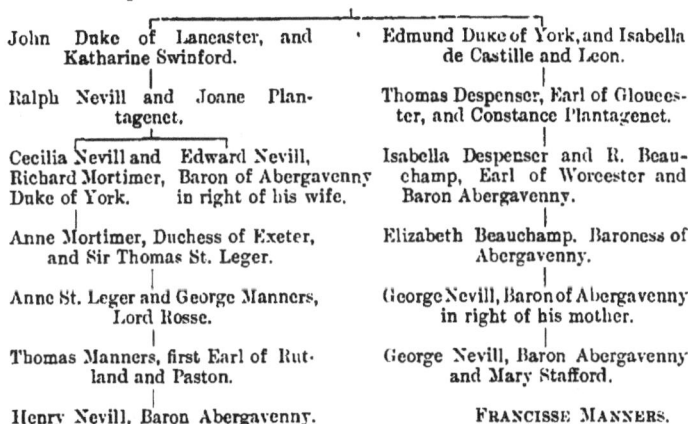

John Duke of Lancaster, and Katharine Swinford.	Edmund Duke of York, and Isabella de Castille and Leon.
Ralph Nevill and Joane Plantagenet.	Thomas Despenser, Earl of Gloucester, and Constance Plantagenet.
Cecilia Nevill and Edward Nevill, Richard Mortimer, Baron of Abergavenny Duke of York. in right of his wife.	Isabella Despenser and R. Beauchamp, Earl of Worcester and Baron Abergavenny.
Anne Mortimer, Duchess of Exeter, and Sir Thomas St. Leger.	Elizabeth Beauchamp. Baroness of Abergavenny.
Anne St. Leger and George Manners, Lord Rosse.	George Nevill, Baron of Abergavenny in right of his mother.
Thomas Manners, first Earl of Rutland and Paston.	George Nevill, Baron Abergavenny and Mary Stafford.
Henry Nevill, Baron Abergavenny.	FRANCISSE MANNERS.

FINCH (AYLESFORD).

1700. The Right Honourable William Lord Dartmouth, and the Honble Anne Finch, were married July 11th.

Mary Finch, daughter of the Rt. Honble the Lord of Guernsey, was buried the 21st day of August, 1783.

The Right Honble Heneage, Earl of Aylesford, was buried August 8th, 1719.

1731. Mary, daughter of the Honble John and Eliza Finch, was baptised Jany. 25th.

1734. Elizabeth, daughter of the Honble John Finch, Esq., and Mary his wife, was baptised July 30.

1736. Savile, son of the Honble John Finch, and Mary his wife, was baptised Sept. 22.

1736. John, son of Honble J. Finch, buried May 27th.

1739. Elizabeth, daughter of Honble John Finch, buried April 25th.

1819. The Honble Charles Finch, 66 years, buried December 24th.

1827. The Honble Seymour Finch, July 13, aged 18 months, buried.

1848. Hon. Lady Mary Elizabeth Finch, Oct. 27, aged 82, buried.

1861. Rt. Honble John Finch, Nov. 30, aged 68, buried.

Besides the above, we have two other entries of the name ; one at Aylesford :—

1658. Mrs. Ann Finch died at Cobtree, in the parish of Allington, was buried here at Aylesford, the seven and twentieth day of July :

and the other at East Malling :—

1605. Buried was Edward Fynch, gentleman, the 23rd of January.

In the east window of the north chapel of Aylesford church, we find a window in memory of " The Earls of Aylesford and their wives." Above are figures which have inscriptions, " In my Father's house are many mansions," " I am the Resurrection and the Life " ; below are in the three lights, first, C. F., W. C. F., M. F., H. F.,

J. F., K. F., C. F., D. F., C. P.; then C. A., M. A., C. A., L. A.,
A. S. A.; and H. F., M. F., J. C., L. C. W., S. F., J. H. F., S. F.,
C. G. M. F., G. A. F., with the dates 1719, 1757, 1777, 1812, 1859,
1871, the dates when the earls succeeded. The window is dedicated :
" To The glory of God, and in memory of dear ones gone to rest."

On the north side of the church is a window, which has an
inscription, with figures : " Why seek ye the living among the dead? "
" Behold, I send the promise of my Father upon you." The window
is dedicated :—

" In memory of Georgiana Agnes, daughter of William Lord Bagot, the wife
of the Honble· Charles W. Finch, B. May 22, 1852 : D. April 12, 1874.

The other members of noble families found in these registers
are :—

1881. Oct. 6. Married Captain Charles Robert Pratt, son of 3rd Marquis of
Camden, and Florence Maria Stevenson,

at Aylesford. And at West Malling :—

1741. Feby. 18. The Honble· Charles Stewart, Esq., Vice-Admiral of ye White,
buried in linen, and 50 shillings paid.
1779. Feby. 25. Charles Stewart, Esq.

In addition to these we find two families of baronets. The first
of these was created in Charles II.'s day, though springing from
a still older honourable family. This is the family of

TWISDEN (BRADBOURNE, EAST MALLING).

1653. Baptised, September 3, Isabella, the daughter of Thomas Twisden, Esq.
1655. Buried, Oct. the 30th, Isabella, the daughter of Thomas Twisden,
Serjeant at Law.
1669. Thomas Twisden, the eldest son of Roger Twisden, and Margaret his
wife, was borne in the parish of St. Giles-in-the-Fields, the 10th of
November, 1668, being Tuesday about 8 o'clock in the morning, and
baptised the next day after, in the same parish ; but it is desired to
have his birth and baptism to be registered in this parish of East
Malling the 13th of April.

A copy of St. Giles register is on a slip of paper, with the
marriage of Roger Twisden and Margaret Marsham, not later than
Feby. 1668.

1670. Baptised, June 11th, Roger, the son of Roger Twisden, Esq.
1672. Baptised, October 1st, John, the son of Roger Twisden, Esq.
1675. Buried, April 9th, Francis Twisden son to Sir Thomas Twisden, Baronet.
1677. Jane Twisden, daughter of Roger Twisden, Esq., and Margaret his
wife, born in London, in the parish of St. Giles-in-the-Field, Oct. 31st,
and baptised November 6 by Dr. Stillingfleet.
1679. December 23. Buried, Heneage Twisden, Esq., third son of Sir Thomas
Twisden, Knight and Baronet.
1680. November 28. Buried, William Twisden, Esq., son of Sir Thomas
Twisden, Knight and Baronet.
1681. Baptised, the first day of June, was Elizabeth, the daughter of Roger
Twisden, Esqre·, and Margaret his wife.

1681. Buried, 9th of November, was Matthew Tomlinson, Esq.
1681. Married, February 27, were Sir Thomas Style, Baronet, and Margaret Twisden, the daughter of Sir Thomas Twisden, Knight and Baronet.
1682. Baptised 12th day of November, was Francis, son of Roger Twisden and Margaret his wife.
1682. Buried, the ninth day of November, was Sir Thomas Twisden, Knight and Baronet, one of the Judges of the King's bench.
1683. Buried the ninth of March, was Francis, the 5th son of Sir Roger Twisden, and Margaret his wife.
1683. Buried the 18th day of May, was John, the third son of Sir Roger Twisden, Baronet, and Margaret his wife.
1700. Feby. 7. Buried was Thomas Twisden, Esq., second son of Sir Thomas Twisden, Knight and Baronet.
1700. Nov. 22. Buried was Mr. John Twisden, ye sixth son of Sir Roger Twisden, Baronet.
1702. Oct. 2. Buried was Dame Jane Twisden, wife of Sir Thomas Twisden, Knight and Baronet.
1703. March 5. Buried was Sir Roger Twisden, Knight and Baronet.
1704. Jany. 2. Baptised was Thomas Twisden, Esq., firstborn son of Sir Thomas Twisden, Baronet, of ye parish of St. Martin's-in-ye-Fields, in London, being born the same day.
1704. June 17. Married were Richard Newdigate, Esq., of Lincoln's Inn, in ye city of London, and Elizabeth Twisden, daughter of Sir Roger Twisden of this parish.
1705. April 4. Baptised was Roger, ye second son of Sir Thomas Twisden, Baronet, being born ye same day.
1706. March 31. Baptised, William, ye third son of Sir Thomas Twisden, born ye same day.
1707. March 17th. Buried Mrs. Jane Twisden, daughter of Sir Roger Twisden.
1709. April 27th. Baptised John, ye fourth son of Sir Thomas Twisden, Bart., being born ye same day.
1719. Sept. 17. Buried John Twisden, fourth son of Sir T. Twisden, Bart., and Anne his wife.
1724. Feby. 5. Buried Mrs. Twisden, widow.
1728. Sept. 19. Buried Sir Thomas Twisden, Baronet.
1729. Oct. 25. Buried ye Lady Anne Twisden.
1733. Feby. 27. Buried, Mr. William Twisden, son of Sir Roger Twisden, Bart.
1737. Sept. 12. Buried, Mr. John Twisden.
1737. Nov. 7. Born in ye parish of Addington was Roger Twisden, ye first-born son of Sir Roger Twisden, Bart., of ye parish in ye year of our Lord God, 1737.
1738. Dec. 7. Buried, Jane and Elizabeth Twisden, infants.
1740. Oct. 16. Baptised, Thomas Phillip, son of Sir Roger Twisden, Bart., and Elizabeth, his lady.
1742. Dec. 13. Buried, Thos. Twisden.
1743. Jany. 20. Baptised, John Papillon, son of Sir Roger Twisden, Bart., and Elizabeth, his lady.
1762. April 22. Philip Papillon, Esq., buried.
1762. Dec. 5. Elizabeth, relict of Philip Papillon, Esq., buried
1772. Jany. 2. William Twisden, Esq., second son of Sir Roger Twisden, Barronett, buried.
1772. March 16. Sir Roger Twisden, Barronett, buried.
1772. Sept. 29. William Twisden, brother to the late Sir Roger Twisden, Bart., buried.
1775. March 11. Dame Elizabeth Twisden, relict of the late Sir Roger Twisden, Bart., buried.

15

1779. Feby. 22. Sir Roger Twisden, Baronet, and Rebecca Wildash, of the
  Parish of Chatham, married.
1779. Oct. 13. Sir Roger Twisden, Bart., 42, buried.
1780. Jany. 6. Rebecca, daughter of Sir Roger Twisden, Bart., lately deceased,
  by Dame Rebecca his wife, baptised.
1782. April 8. Sir John Papillon Twisden, Baronet, and Elizabeth Geary, of
  Great Bookham, Surrey, married.
1784. John. son of Sir John Twisden and Dame Elizabeth his wife, born
  Sept. 28, baptised Oct. 9.
1810. Feby. 23. Sir John Papillon Twisden, Bart., æt. 57, buried.
1811. Feby. 25. Sir John Twisden, Bart., of this parish, and Catharine
  Judith Coppard, spinster, of Gillingham, married.
1812. John, son of Sir John Twisden, and Catharine Judith Lady Twisden,
  was baptised Jany. 18.
1815. Dec. 26. Dowager Lady Elizabeth Twisden, Bradbourne, 61 years, buried.
1819. April 20. Dame Catharine Judith, wife of Sir John Twisden, Bradbourne, 29 years, buried.
1833. Feby. 12. Dame Rebecca Twisden, relict of Sir Roger Twisden, from
  Hunton, aged 74 years, buried.
1843. June 20. Mrs. Ann Twisden, the wife of Captain Twisden, R.N.,
  Bradbourne, aged 71 years, buried.
1853. June 29. John Twisden, Bradbourne, aged 85 years, buried.
1871. Jany. 31. Amy Rebecca Twisden, aged 62 years, buried.
1879. March 21st. Elizabeth Twisden, aged 78 years, buried.
1880. Feby. 18th. Charlotte Harriet Twisden, aged 70 years, buried.
1881. Nov. 29. Mary Matilda Twisden, aged 70 years, buried.
1887. May 25. Thomas Edward Twisden, aged 68 years, buried.

(West Malling.)

1721. Aug. 7. Mrs. Jane Twisden, buried in linen.

Hawley (Leybourne).

1777. Dec. 26, James Hawley, Esq., M.D., buried.
1777. October 20th. Christened, Dorothy Elizabeth, daughter of Henry
  Hawley, Esq., and Dorothy, his wife.
1783. Dec. 11. Mrs. Dorothy Hawley, buried.
  August 22nd, 1780. Baptised, Harriet, daughter of Henry Hawley, Esq.,
  and Dorothy his wife.
1789. Jany. 26. Frances Ann, daughter of Henry Hawley, Esq., and Ann his
  wife, baptised.
1790. August 26th. James, son of Henry Hawley, Esq., and Ann his wife,
  baptised.
1791. November 8th. Sarah Hawley, ætatis 70, buried.
1808. Feby. 9. Catherine Anne. Daughter of Henry Hawley, Esq., and
  Catherine Elizabeth, his wife, born Jan. 4, was baptised Feby. 9th,
  by J. K. Shawe Brooke.
1809. Sir William Brook Bridges of Goodnestone, and Dorothea Elizabeth
  Hawley, married Dec. 16.
1810. John Brook Bridges, of Saltwood, and Charlotte Hawley, married
  Nov. 28.
1826. Buried, Sir Henry Hawley, Bart., Leybourne Grange, Jany. 28, aged
  81 years.
1828. Nov. 14th. Buried, Lady Ann Hawley, relict of the first Sir Henry
  Hawley, Nov. 14, aged 72 years.
1834. April 5th. Sir Henry Hawley, Bart., Leybourne Grange, buried April 5,
  aged 54 years.
1834. July 27. Caroline Elizabeth Hawley, buried July 27, aged 12 years.

1836. April 6. Mary Ann Dorothy Hawley, London, April 6, aged 26 years
    buried.
1848. October 28. Ada Mary, daughter of Henry Charles and Mary Elizabeth
    Hawley, baptised.
1850. November 22. Frances Charlotte Hawley, Clifton in Bristol, 34 years,
    buried.
1851. June 22. Charles Cusac, son of Henry Charles and Mary Elizabeth
    Hawley, baptised.
1853. March 6. Kathleen Augusta, daughter of Henry Charles and Mary
    Elizabeth Hawley, baptised.
1854. December 3. Frederick William, son of Henry Charles and Mary
    Elizabeth Hawley, baptised.
1857. August 22. Ethel Maud, daughter of Henry Charles and Mary Elizabeth
    Hawley, baptised.
1859. May 23. Sir Michael Cusac Smith, Baronet, aged 65 years, buried.
1862. March 2nd. Dame Catherine Elizabeth Hawley, of the parish of
    St. Mary-le-Bone, aged 75 years, buried.
1862. March 22. Baptised, Arthur Cecil, son of Henry Charles, and Mary
    Elizabeth Hawley.
1866. June 28. Mary Anne Angelina Cusac Smith, Norwood, Surrey, June 21,
    75 years, buried.
1871. September 27. Elizabeth Hawley, Reigate, aged 60 years, buried.
1875. April 27. Joseph Henry Hawley, Bart., London, aged 61 years, buried.
1877. February 22. Rev^d. Charles Henry Hawley, rector of this parish
    (Leybourne), 22 February, aged 53 years, buried.
1891. July 21. Winifred Mary Hawley, Fairby Lawn, Boston, July 21, aged
    10 years, buried.

(Ryarsh.)

1839. John Hawley, Sept. 10, aged 78, buried.

Besides the above titled families, we have also the following that
have not been previously mentioned in this work :—

Ashley (Allington).

Jacob, the son of Isaac Ashley of Maidstone, Knight, and Dame Anne, his
    wife, was baptised 13 January, 1651, by me, R. Thomas. He was
    baptised in Maidstone.

Faunce (Aylesford).

1713. Elizabeth, wife of Sir Robert Faunce of Maidstone, was buried here at
    Aylesford the third day of December.
1716. Sir Robert Faunce of Maidstone was buried here at Aylesford,
    Feby. 16.
1723. Mrs. Margaret Faunce was buried Nov. 23.
1739. Jany. 29. George Faunce, Esq. was buried.

Williams (Aylesford).

1700. June 7. Elizabeth, the daughter of Sir Francis Williams, was buried.

Wild, or Wylde (East Malling).

1677. Jane Wild, daughter of Felix Wild, Esq., and Eleanor his wife, was
    baptised in London, in the parish of St. Giles-in-the-Fields, October
    19th, by Dr. Stillingfleet.
1697. Oct. 28. Married was Thomas Sympson and Jane, daughter of Sir
    Felix Wylde, Bart.

(West Malling.)

1710. March 19.   Mrs. Elinor Wylde, the third and youngest daughter of Sir Felix Wylde, Bart., buried.·

1722. October 26.   Sir Felix Wild, Bart., buried.

Whitworth (Leybourne).

1750. Feby. 26, was born Margaret Whitworth, daughter of Charles Whitworth, Esq., of the Grange, and Martha, his wife ; baptised March ye 6th by me, John Shaw, curate.

1751. March 21.   Catharine, daughter of Charles Whitworth, Esq., of the Grange, and Martha, his wife, was baptised.

1752. May 29.   Charles Whitworth, son of Charles Whitworth, Esq., and Martha, his wife, christened.

1753. June 24.   Mary, daughter of Charles Whitworth, Esq., was baptised by G. Richards, vicar of West Peckham.

1754. July 2.   Baptised Francis, son of Charles Whitworth, Esq., and Martha his wife.

1755. October 24.   Richard Whitworth, son of Charles Whitworth, Esq., and Martha his wife, baptised.

1760. April 7.   Priscilla Whitworth, daughter of Charles Whitworth, Esq., and Martha his wife, baptised.

1761. October 16.   Robert Whitworth, son of Charles Whitworth, Esq., and Martha his wife, baptised.

1763. May 3.   Anne Barbara Whitworth, daughter of Charles Whitworth, Esq., and Martha his wife, baptised.

Tufton (Offham).

1625. Richard Tufton, ye son of Richard Tufton. Esquire, was buried ye second day of July.

Twysden (Addington).

1833. June 21st.   Married William Twysden and Elizabeth Polhill.

Goringe (Birling.)

1609. Lucye, the daughter of Sir George Goringe, was baptised the 1st of October.

Besides these titled people, we have the following families of gentry who were squires in these parishes from time to time. From the long period of their holding the manor and the advowson, those that deserve the first mention are the Wattons and the families descended from them—the Bartholomews and the Stratfords.

Watton (Addington), Bartholomew, and Stratford.

1572. Jany. 13.   Thomas Watton and Mary Rutland ye xiii. of January were married.

1580. July 20.   Thomas Watton, Esquire, the twentieth day of July was buried.

1594. June 2.   John, the son of Thomas Watton, Esquior, the second day of June was baptised.

1599. Thomas, the son of Thomas Watton, Esquior, the —— of September was baptised.

1621. Thomas Watton, son of William Watton, was baptised ye 22nd day of March.

1622. Elizabeth, the daughter of William Watton, Esquior, the 4th day of June was baptised.
1623. Margaret, the daughter of William Watton, Esquior, the 10th day of June was baptised.
1624. William Watton, the son of William Watton, Esquire, the 20th day of June was baptised.
1626. Thomas Watton, ye son of William Watton, was baptised ye last day of December.
1628. Martha Watton, daughter of William Watton, the xv. day of September was baptised.
1645. The 15th day of April Elizabeth Watton, the daughter of William Watton, was baptised.
1661. May 19. Frances, daughter of William Watton, Esq., and Margaret his wife, was baptised.
1679. Margaret, the relict of William Watton, Esquire, was buried Sept. 20.
1680. Aug. 16. John, the son of William Watton, Esq., and Mary his wife, was born Aug. 5, and baptised Aug. 16.
1685. Thomas Watton, gent., was buried May 2nd.
1685. Aug. 9th, Elizabeth Watton, daughter of William Watton, Esq., and Mary his wife, was born ii. July 1685, and baptised the 9th day of August.
1688. Will. Watton and Mary his wife buried Aug. 7th.
1695. Mary, the wife of Captain William Watton, was buried June ye 11th.
1699. Mary, ye daughter of Mr. Edmund Watton, and Mrs. Sarah his wife, was borne and baptised ye 19th of April.
1703. William Watton, of Addington, Esq., April 30th. Affidavit brought May 1st, Reginald Peckham, Esq., executor.
1703. July 7. Elizabeth, daughter of Edmund Watton of Addington, Esq., and of Sarah, his wife, born July 3rd.
1705. Feby. 23rd. Margaret, daughter of Edmund Watton, Esq., and Sarah, his wife, born Jan. 11, and christened Feby. 23rd.
1706. October 1. Ann, daughter of Edmund Watton, Esq., and Sarah, his wife, born and baptised Oct. 1st.
1707. Ann, daughter of Edmund Watton, Esq., buried Aug. 31st.
1707. Dec. 17. Sarah, wife of Edmund Watton, Esq., buried.
1711. Sept. 11. Mrs. Elizabeth Watton, buried.
1715. July 22. Mary, daughter of Edmund Watton and Ann his wife, born July 22.
1715. August 9. Mary, daughter of Edmund Watton, buried.
1717. April 16th. Edmund Watton, Esq., buried.
1719. Jany. 1st. John Krow, son of Philip Bartholomew, Esq., born Dec. 1, baptised.
1727. Sept. 30. Elizabeth, daughter of Leonard Bartholomew, Esq., buried.
1728. June 17. Leonard, son of Leonard Bartholomew, Esq., and Elizabeth,* his wife, born June 15, baptised.
1729. May 23. Edmund, son of Leonard Bartholomew, Esq.. and Elizabeth, his wife, born May 12, baptised.
1730. October 8. Leonard Bartholomew, Esq., buried.
1731. Jany. 28. Jane, daughter of Leonard Bartholomew, Esq., and Elizabeth, his wife, born July 12/30, baptised.
1735. Jany. 23. Jane, daughter of Leonard Bartholomew, Esq., buried.
1737. Jany. 10. Roger Twisden, Esq., and Madame Elizabeth Bartholomew, widow, married.
1737. November 14. Roger, son of Sir Roger and Lady Twisden, born Nov. 7, baptised.
1739. May 29. Mrs. Frances Watton, buried.

* The last of the Wattons.

1743. July 17.   Edmund Bartholomew, son of Dame Elizabeth Twisden, of
    Bradbourne, by Leonard Bartholomew, Esq., buried.
1799. April 25.   John Wingfield, bachelor, and Frances * Bartholomew.
1802. October 23.   Buried Mrs. Frances Bartholomew, wife of Leonard
    Bartholomew, Esq.
1808. March 12. Frances Amelia, daughter of the Honble. John and Frances *
    Wingfield Stratford, baptised.
1810. March 22. Isabella Harriet, daughter of the Honble. John and Frances
    Wingfield Stratford, baptised.
1810. October 27.   Buried Leonard Bartholomew, Esq., aged 82 years.
1811. Feby. 7.   John, son of the Honble. John and Frances Wingfield
    Stratford, baptised.
1827. July 21. Frances Wingfield Stratford, aged 51 years, buried.
1850. August 10.   The Honble. John Wingfield Stratford, aged 78 years, buried.
1863. April 4.   Harriett, the widow of the Honble. John Wingfield Stratford,
    aged 80 years, buried.
1880. May 13.   John Wingfield Stratford, aged 70 years, buried.

WATTON (LEYBOURNE).

1652. May 30.   Thomas Watton, ye sonne of William Watton, Esqre., of
    Addington, and Margaret his wife, was laid here in his own grave.

BARTHOLOMEW AND STRATFORD (WEST MALLING).

1764. Dec. 22.   Humphrey Bartlemew, M.D., buried.
1773. Leonard Bartholomew of this parish, bachelor, and Frances Thornton,
    of the same parish, married in the church by licence, the nineteenth
    day of August, in the year 1773.
1775. Dec. 27.   Baptised Frances, daughter of Leonard Bartholomew, Esq.,
    and Frances, his wife.
1808. July 11.   Married Robert Richard, of the parish of Maidstone, bachelor,
    and Sarah Stratford of this parish, spinster.

WINGFIELD STRATFORD (RYARSH).

1856. Nov. 9.   Leonard Guise, son of John and Guise Wingfield-Stratford.

OLIVER (LEYBOURNE).

This family took its name from William, son of Oliver Quintin,
who in the eleventh year of King Henry VI. purchased some lands
called Hilks, in the parish of Seale.   In the deed he is frequently
called William, son of Oliver, and from this he and his descendants
came to be called Oliver, and dropped Quintin.   The Grange passed
from them by the marriage of their last descendant, whose name
was, strange to say, Juliana, like the last of the De Leybournes, to
the Coverts, from whom it passed in a similar way to the Saxbys.

1561. Nov. 19.   Buried, William Oliver.
1562. March 13.   Married, Thomas Petley and Joane Oliver.
1565. March 8. Married, Richard Wood and Elizabeth Oliver.
1568. November 13.   Joane Oliver, baptised.
1570. November 3.   William Oliver, baptised.
1571. November 30.   Elizabeth Oliver, baptised.
1573. January 2.   Robert Oliver, baptised.
1576. November 2.   Margaret Oliver, baptised.

---

* The last of the Bartholomews of Addington.

1580. Jany. 25. Margaret Oliver, baptised.
1581. August 7. Sylvester Oliver, baptised.
1584. May 17. William Oliver, buried.
1592. Nov. 11. Margaret Oliver, buried.
1601. July 5. Mary, daughter of Mr. Robert Oliver, jun., baptised.
1613. October 23. Robert Oliver, sen., buried.
1623. December 22. Francis Brudonell and Judith Oliver, married.
1624. Edward Brudonell, the son of Francis and Judith Brudonell, baptised the 22nd of Feby.
1627. Mary Oliver, the daughter of Mr. Robert Oliver, jun., baptised the 24th April.
1629. April 7th. Juliana Oliver baptised.
1630. September 22. Elizabeth Oliver, daughter of Mr. Robert Oliver, baptised.
1631. December 24. Mr. Robert Oliver, sen., was buried.
1632. October 25. Robert Oliver, son of Mr. Robert Oliver, baptised.
1634. June 1. Elizabeth Oliver, the daughter of Mr. Robert Oliver, was buried.
1634. Oct. 16. Judith Oliver, daughter of Mr. Robert and Juliana Oliver, baptised.
1638 Oct. 4. Francis Oliver, the daughter of Mr. Robert Oliver, was baptised.
1639. Dec. 16. Thomas Oliver, the son of Robert Oliver, was baptised.
1640. March 2nd. Baptised was Jane, the daughter of Robert Oliver.
1644. Feby. 2. Buried, Juliana Oliver, the wife of Robert Oliver.
1647. May 30. William Oliver of the Castil in Leybourne, was buried.
1656. Nov. 4. Robert Oliver, the eldest and most accomplished son of Mr. Robert Oliver of this parish, was buried in this church.
1656. Dec. 16. An intention of marriage betwixt Mr. Edward Covert, and Mrs. Juliana Oliver, * both of this parish, was published three Lord's days, vid. the 1st day of November, the 8th day of the same month, and the 15th day of the same, after Divine service, and no objection made thereunto, in witness of which I have subscribed. John Codd, registrar of the parish of Leyborn.

The aforesaid marriage was solemnised before me in the presence of Robert Oliver, the father of the said Juliana and John Covert, brother to the said Edward, and Barnham Oliver, brother of the said Juliana, upon the 16th day of December 1656. William James.

1664. August 11. Mrs. Frances, daughter of Mr. Robert Oliver, was married to Mr. John Stowell of Rochester.
1666. Mrs. Mary Oliver, ye eldest daughter of Mr. Robert Oliver of this pish., died Aug. ye 18, and buried 20, ejusdem.
1667. Mrs. Mary Oliver, the mother of Mr. Robert Oliver, died Oct. 22nd, and was buried 29th, ejusdem.
1668. Dec. 29. Mr. Robert Oliver, departed this life and was buried January 1st.
1670. Barnham Oliver, the youngest son of Mr. Robert Oliver, departed this life, September 28th, was interred Octo. 13.
1678. Oct. 15. Thomas Oliver, Esquire, was buried.
1703. Aug. 24. Henry, the son of William Saxby, gent., and Elizabeth, his wife, was baptised.
1705. April 25. Elizabeth, ye daughter of William Saxby, gent, and Elizabeth, his wife, baptised.
1707. Sept. 24. William, ye son of William Saxby, gentleman, and Elizabeth, his wife, was baptised.

---

* She conveyed Leybourne Grange from the Olivers to the Coverts, and her daughter carried it to the Saxbys.

1708. Aug. 20. Thomas Golding, gentleman, and Frances Saxby, gentlewoman, were married by licence, in ye parish church of Leybourne,
on ye 20th Aug. 1708, a licence being first had from Rochester.
1708. November 7th. Thomas, ye son of William Saxby, gentleman. and
Elizabeth his wife. was baptised.
1710. October 12. Mary, ye daughter of William Saxby, gentleman. and
Elizabeth. his wife, baptised.
1711. Oliver Edward, ye son of Willlam Saxby, gentleman, and Elizabeth,
his wife, baptised.
1713. Robert. ye son, of William Saxby, gent., and Elizabeth, his wife, was
baptised, ye 24th of April.
1717. George. ye son of William Saxby and Elizabeth, his wife, baptised
March 25.
1718. April 21. John, ye son of William Saxby, gent., and Elizabeth, his
wife, baptised.
1719. July 18. Elizabeth, wife of Captain William Saxby of ye Grange,
baptised.
1719. Feby. 14. John, ye son of Captain William Saxby, buried.
1725. April 5. Buried, Mrs. Mary Saxby.
1725. Dec. 15th. Buried, Oliver Edward Saxby.
1737. July 17. Buried, Mr. Henry Saxby, of the parish of Ryarsh.
1744. Sept. 2. Buried, Mr Henry Saxby. Affidavit made.
1746. Mrs. Mary Saxby of Town Malling, buried April 5th.
1786. Nov. 2. George Saxby, Esq., buried.
1788. April 27. Christening. Thomas, son of Henry and Elizabeth Saxby.
1791. Feby. 20. Baptised, Mary dr. of William and Susannah Saxby.
1793. April 28. Baptised, Ann, dr. of Henry and Elizabeth Saxby.
1795. March 6th. Robert, son of Henry and Elizabeth Saxby, baptised.
1797. Jany. 20. Michael, son of Henry and Elizabeth Saxby, baptised by
the Rev. Mr. Warde. rector of Ditton.
1798. Jany. 23. Elizabeth Saxby, aged 71, from London, buried.

RYCHARDS (TROTTESCLIFFE).

The chief families of Trottescliffe died out at a very early period,
as will appear by the following extracts; but the parish is the home
of the family of Godden, from which they have spread over the
whole valley. The oldest of these families is the Rychards.

1547. Mrs. Annis Rychards was buried the xvith. day of February.
1548. Mr. John Rychards, gentleman, was buried the iiii. day of February,
ut supra.
1566. Mr. Richard Rychards, was buried the 13th day of April.
1568. Maryon Rychards, gentlewoman, was buried the xiiith. day of January.
1570. Thomas Rychards, gentleman, was buried the xixth day of October.
1575. Richard Rychards, gentleman, was buried the xxvi. day of March.
1579. Margaret Rychards, gentlewoman, was buried the 31st day of March.

ATTWOOD (TROTTESCLIFFE).

1594. Johan Attwood, uxor Nicholas Attwood, was buried the first of May.
1600. Samuel Atwode, the good youth, was buried the 12th day of May.
1649. Mr. Edmond Attwood, aged about 63 years, buryed May ye 7th.
1647. June 10. Elizabeth, daughter of James Atwoode, gent., was baptised.
1648. Mary, ye daughter of James and Mildred Atwood, was baptised the
same day that King Charles the First was beheaded, Jan. 30.
1649. Mr. Edmond Attwood, aged about 63 years, buried May the seventh.
1656. Millicent, ye daughter of Mr. James Attwood, was buried Jany.
the 30th.

1663. Elizabeth Attwood, the daughter of William Attwood and Sarah Attwood, his deare beloved wife, was baptised the 31st day of July.
1669. Edmund Attwood, the honest gent., was buried September ye 13th.
1674. Mr. James Attwood, aged 67 years, was buried October 7th.

We find in 1667 Mr. Samuel Attwood was curate of Trottescliffe. About the same period also we have a family named

MARSH (TROTTESCLIFFE).

1666. December 13. Millesend, the daughter of Mr. Richard Marsh, baptised.
1667. Richard, the son of Richard Marsh, Esq., and Elizabeth, his wife, was baptised the ninth day of December.
1668. James, the son of Richard Marsh, Esq., and Elizabeth, his wife, was baptised November ye 15th.
1669. Edmund, the son of Richard Marsh, Esq., and Elizabeth, his wife, was baptised April 20th.
1671. Elizabeth, the daughter of Richard Marsh, Esq., and Elizabeth, his wife, was baptised Feby. 21st.
1672. Elizabeth, the daughter of Richard Marsh, Esq., and Elizabeth, his wife, was buried March 9th.
1673. Richard Marsh, Esq., buried December 12th.
1673. Mary, the daughter of Richard Marsh, Esq., and Elizabeth, his wife, baptised 3rd of October.
1676. Mary, the daughter of Richard Marsh, Esq., was buried March 26th.

GODDEN (TROTTESCLIFFE).

The records of this family, gentle and simple, nearly fill the register of Trottescliffe. Besides those mentioned as belonging to this parish, we have also a great many of their records in several of our parishes.

1606. August 5th. James Godden, the son of Thomas Godden, of the Court Lodge, was baptised.
1610. The 20th day of May John, the son of Thomas Godden of the Court Lodge, was baptised.
1635. September 5th, Mr. Godden, the curate of Trottescliffe, was buried.
1641. James Godden of Rouses was buried ye 11th day of February.
1645. May 27th. Anne, daughter to John Godden, son of Thomas Godden at the Court Lodge, baptised.
1646. March 4. Thomas Godden at the Court Lodge, aged about 60 years, was buried.
1647. March 9. Mary, widow to the above-named Thomas Godden, at the Court Lodge, was buried.
1649. November 27. Dorothy, ye little daughter of James Godden of Rouses, was buried.
1657. Martha, the wife of Mr. Thomas Godden of the Court Lodge, dyed the 4th of August, and was buryed the 8th day of the same month.
1670. December 29. Thomas Godden, gent., was baptised.
1680. July 5. John Godden, gent. Affidavit was made by Dorothy Rose.
1684. May 27. Thomas Quarrington and Mary Godden, the daughter of Thomas Godden, were married, being Whitmonday.
1700. There was buried in woollen Mr. James Godden. Affidavit was made by Samuel Attwood, rector of Addington, Nov. 6th.
1707. May 28. There was buried in woollen, Mrs. Dorothy Godden, aged 93. Affidavit was made by Mary Quarrington.

1729. There was buried in linen, Mary Godden. Affidavit was made before Thomas Dallison, Esq., one of His Majesty's Justices of the peace, Jany. 30th.

(Ryarsh.)

XXIInd day of January, 1589. Robert Godden, ye sonne of Thomas Godden, of Callescourt in the parish of Ryarsh, married in Seal church to Jane French of ye said parish of Ryarsh.

1585. Thomas Godden, the younger, was buried the 10th day of May.

1585. Henry Godden, the son of Thomas Godden, was buried the xxvith day of September.

1602. The 17th day of February, Edmund Myllis and Alice Godden, ye daughter of Amos Godden, of Trosly Court, were married at Trosly church on Monday.

1607. Thomas Yonge and Marie Godden, ye daughter of Thomas Godden, of Warlwesworth, were married the eleventh day of March.

1609. Thomas Godden of Dartford was buried ye eight and twentieth day of June, anno predicto.

1620. James Godden, the younger sonne of James Godden, Esquire, of Trottescliffe, and Audrey Cripps of the Ville* of Ryarsh were married the two and twentieth day of April.

1672. William Champion and Dorothy Godden, were married the 20th of February, being Shrove Tuesday.

1714. Oct. 7.   Edward Godden of West Malling and Jane Elizabeth Skinner, were married in the parish church of Ryarsh.

(Snodland.)

1575. Samuel Goddin, ye sonne of Thomas, of Paddlesworth, bapt. 17 Sept.

1592. John, ye sonne of Edward Godden of Paddlesworth (was baptised), 12th December.

1593. Thomas, ye sonne of Edward Godden of Paddlesworth, baptised 17th October.

1596. Antony, ye sonne of Edward Godden of Paddlesworth, baptised 23rd April.

1598. Bridget, ye daughter of Edward Godden of Paddlesworth, was baptised 1st November.

1600. Edward, ye sonne of Edward Godden of Paddlesworth, was baptised 13th January.

1603. Nevill Goddin, the son of Edward Godden of Paddlesworth, was baptised 28th of April, the first year of the reign of King James.

1608. Elizabeth, the daughter of Edward Godden of Paddlesworth, was baptised the third day of April.

1615. Edward Godden of Paddlesworth, aged ——, buried on Sunday August 27, in ye church.

(Allington.)

1640.—Hunt, of Maidstone, and Anne Godden of this parish, were married in the parish church of Allington, the twenty-ninth of September, by virtue of a licence out of the court of Rochester.

1731. Mrs. Frances Godden, of Maidstone, was buried, November 18.

1745. Mrs. Mary Godden, buried Dec. 8th.

Walsingham (Ryarsh).

This family lived at Ryarsh for about two hundred years; they

---

* We have several places called by the French word Ville in Kent, *e.g.*, Dunkirk, near Canterbury ; this, however, is the only example in this neighbourhood.

were probably an offshoot of the Walsinghams of Igtham Mote, and Chislehurst.

1567. The second day of November in the year of our Lord there was buried Ann Walsingham, the wyfe of John Walsingham.

1572. The third day of May in the year of our Lord, there was buried John Walsingham.

1577. The xvii[th.] day of July there was christened Edward Walsingham son of Edward Walsingham.

1579. Anno supradicto, Thomas Walsingham, filius Edmundi Walsingham, baptisatus fuit xxiii° die Augusti.

1579. The twenty-sixth day of March there was buried Thomas Walsingham, son of Edmund.

1579. The twelfth of July there was buried John Walsingham.

1580. John Walsingham, ye son of Edmund Walsingham, was baptised xv[th] day of February.

1581. The 17th day of October, in ye yeare above said, was baptised Elizabeth the daughter of Edward Walsingham.

1582. Joane Walsingham, the daughter of Edward Walsingham, was baptised the fifth day of December.

1586. Mary Walsingham, the daughter of Edmund Walsingham, was baptised the first day of May.

1587. Jane Walsingham, the daughter of Edmund Walsingham, was baptised the fifth day of December.

1589. Jany. 22nd. Robert, son of Edward Walsingham, was baptised.

1589. Feby. 10. Robert Walsingham was buried.

1591. Feby. 23. Mary Walsingham, the daughter of Edmund Walsingham, was baptised, anno supradicto.

1592. June 15. Stephen Walsingham, ye son of Edmund Walsingham, wa baptised.

1593. Nicholas Walsingham, filius Edmundi Walsingham, baptisatus fuit on Sunday ye xiii. August.

1597. The third of May. Robert Walsingham was brought out of the Hundred of Eyhorne, and was buried in the churchyard of Ryarsh.

1600. February 7th. Henerie Serjeant and Elizabeth Walsingham were married.

1602. December 23rd. Jane Walsingham, daughter of Edmund Walsingham, was buried.

1608. September 19th. John Bridger, the parson of the parish of Mereworth, and Marie Walsingham, ye daughter of Edmond Walsingham, were married here.

1612. Marjorie, daughter of Edward Walsingham, baptised ye xxv. of June.

1620. April 30. Martha Walsingham, widow, buried.

1632. Anno domini. Nov. 1st. Edward Walsingham and Jane Caysier were married.

1636. July 21. Margaret, daughter of Edwarde Walsingham, baptised.

1639. June 22. Elizabeth, ye daughter of Edward Walsingham, buried.

1642. August 4. Jane, daughter of Edward and Jane Walsingham, was baptised.

1665. Anno domini. Feby. xxii. John Codd and Jane Walsingham, both of this parish, were married in the parish church of Ryarsh.

1675. Sept. 30. Margaret, daughter of Edward Walsingham, baptised.

1677. Feby. 14. Jane, daughter of Edward Walsingham.

1679. June 2nd. There was buried in linnen, Mary Walsingham ye wife of Edward Walsingham, of the parish of Ryersh, gent., informacion being given to Roger Twisden, Esq., by Goodman Bockett of East Malling, mason; and by virtue of a warrant from him, ye said Roger Twisden, five pounds were paid upon ye goods of Edward

Walsingham, one moiety of which was given to ye informer, ye other to ye poor of the said parish.

1682. Jany. 5. There was baptised Mary, ye daughter of Edward Walsingham and Felix, his wife.

1683. August 21st. There was baptised, Anne, ye daughter of Edward Walsingham, and Felix, his wife.

1685. April ye 25th. There was buried in linnen, William Caysier, of ye parish of Ryersh, yeoman, information being given to Serjeant Selby by Thomas Farnott of ye same parish and by virtue of warrant from him, and by virtue of warrant from him, ye said Serjeant Selbie, five pounds were levied upon ye goods of William Caysier, one moiety was paid to the informer, and the other to ye said parish on 29th of May last.

1688. July 13th. There was baptised, Elizabeth, ye daughter of Edward Walsingham and Felix, his wife.

1692. October 17th. John, son of Edward Walsingham, was borne 17th October, and baptised 21st October.

1695. Sept. 19th. Jane, ye daughter of Mr. Walsingham and Felix, his wife, baptised.

1695. December 24. There was buried in woollen, Jane Walsingham, of Ryersh, widdow, affidavit being made by Mary Whiting of ye same parish, before Thomas Pyke, curate of Addington, in the presence of Mary Boeke and Jane Humphries.

1708. Anne, daughter of Mr. Edward Walsingham, July 22, buried.

1713. November 28. Susan. daughter of Mr. Edward Walsingham, baptised.

1714. March 11. Mr. Edward Walsingham. Affidavit, March 15th.

1716. June 8. Anne, daughter of Mr. Edward Walsingham, baptised.

1717. Mrs. Felix Walsingham, Sept. 8, buried. Affidavit, Sept. 15th.

1718. May 15. Mr. Edward Walsingham, buried. Affidavit, May 21.

1720. Nov. 19. Anne, daughter of Mr. John and Rebeceah Walsingham, baptised.

1721. Rebeceah, daughter of Mr. John and Rebeceah Walsingham, baptised Jany. 9.

1722. March 19. Jane, daughter of Mr. John and Rebeceah Walsingham, baptised.

1724. October 22. Mary, daughter of Mr. John and Rebeceah Walsingham, baptised.

1726. Mrs. Rebeceah Walsingham, buried. Affidavit July 17.

1764. Jany. 29. Mr. John Walsingham, from Birling, buried in the church. Affidavit, Feby. 17.

(Birling).

1575. George Walsingham, the son of Robert Walsingham, was baptised the viith of December.

(Aylesford.)

1678. Edward Walsingham of Ryarsh, and Joane Browne of ye same, were married by a licence out of Rochester Court, November 30.

(East Malling.)

1576 Married, was Thomas Tomlin and Mary Walsingham, ye 31st December

(West Malling.)

1719. May 30. William Chapman, gent., bat., and Susannah Walsingham, widow, both of this parish, were married by licence.

1721. November 21. Mrs. Susannah, wife of William Chapman, gent., buried.

1721. December 12. Ann Walsingham, the daughter of the said Mrs. Susannah Chapman, buried.

ADDISON (OFFHAM).

This family is connected with Offham for nearly three hundred years, which gives it a title to our consideration.

1585. Robert Addison, the sonne of William Addison, was baptised the xiiith day of September.

1591. William Addison, the sonne of William Addison, was baptised the xviith day of January.

1593. Margaret Addison, wife of William Addison, buried 13th day of April.

1610. Jane, the wife of William Addison, was buried August 22nd.

1611. Robert Addison and Margaret Jordan were married the 10th day of June.

1611. John Addison and Dorothy Waite were married the 20th day of October.

1612. William Addison, ye sonne of John Addison, baptised ye 22nd day of November.

1612. Henry Addison, ye sonne of Robert Addison, baptised ye 14th day of February.

1614. John Addison, ye sonne of Robert Addison, baptised ye 4th day of December.

1616. John, ye sonne of John Addison, baptised ye 7th day of April.

1616. William Addison, ye son of Robert Addison, baptised ye 8th day of February.

1617. William Addison, ye sonne of Robert Addison, baptised ye 7th day of March.

1618. Margaret Addison, daughter of John Addison, baptised 4th day of February.

1664. John, son of John Adyson, jun., and Margaret his wife, baptised 22nd March.

1666. John Addison, gent., was buryed 4th August.

1673. Dorothy, the widow of John Adyson, buried 28th April.

1694. John Addison, buryed May 19.

1698. Anne Addison, buryed March the sixth.

1703. Margaret Addison, widdow, buryed December 8.

1721. May ye 8th. Buried John Addison.

1743. June 16th. John Smith, of the parish of West Peckham, bat., and Mary Addison of this parish, spinster, were married by licence at Offham church.

1792. Frances, daughter of Friend and Anne Addison, privately baptised Dec. 19th, 1790, admitted into the congregation, July 17.

1793. Elizabeth Margaret, daughter of Friend and Anne Addison, privately baptised May 4, admitted into the congregation June 11.

1794. John Smith, son of Friend and Anne Addison, privately baptised 18th August, admitted into the congregation 30th September.

1796. George Smith, son of Friend and Anne Addison, baptised Jany. 19th.

1797. October 13th. Susannah, daughter of Friend and Anne Addison, baptised.

1801. Frances Addison, wife of Friend Addison, the elder, buried March 17th, St. Patrick's day, aged 62, at Ryarsh.

1832. May 7th. Anne, relict of Friend Addison, buried.

1834. Nov. 3. John Smith, son of Friend and Anne Addison, buried.

1838. April 1. Frederick Addison, buried.

1843. April 30. George Smith Addison, buried.

1851. May 21. Mary Addison, buried.

(WEST MALLING.)

1719. March 11. Mary, the daughter of William Addison and Anne, his wife, baptised.

1720. May 12. Anne, the daughter of Anne Addison and William, baptised.
1731. July 13. Margaret, the daughter of William and Anne Addison baptised.

(RYARSH.)

1824. Friend Addison, sen., Offham, Oct. 20, aged 81, buried.
1826. Friend Addison, Offham, Aug. 12, aged 60, buried.
1864. Frances Addison, Brixton, Surrey, October 1, aged 73, buried.

GOLDING.

We have reserved this family to the last; members of it are to be found recorded in nearly all our parish registers of the district. The oldest entries of this name are to be found in the register of Snodland.

1601. Robertus Golding, filius Thomæ Golding, baptisatus fuit Feby. 29.
1603. Margaret Golding, the daughter of Thomas Golding, was baptised June 3.

The next oldest are from East Malling.

1627. Buried was Elizabeth Goldinge, ye wife of Steven Goldinge, November the xxvith.
1630. Married, October 26, Edmund Gibson and Margaret Golding.

The Goldings became soon after this settled at Leybourne, in the register of which we find them for about sixty years, where they became connected by marriage with the Saxbys.

1669. Thomas Golding, the son of Thomas Golding and Anne, his wife, was baptised, March 14.
1672. October 24. Thomas Golding, infant, buried.
1675. July 15. Alice Ramsay, the sister of Thomas Golding, gent., and wife of William Ramsay, was buried.
1685. Dec. 7. Edward Golding of Leybourne was buried.
1705. October 5. Thomas Golding, Esq., of ye parish of Ditton, was buried: late of Leybourne.
1708. August 20. Thomas Golding, gentleman, and Frances Saxby, gentlewoman, were married by license in ye parish church of Leybourne, on the 20th August 1708, a licence being first had from Rochester.
1709. Francesse, ye daughter of Thomas Goldin. gent., and Francesse, his wife, baptised ye twenty-ninth of June.
1710. June 12. Thomas, ye son of Thomas Goldin, gent., by Frances, his wife, baptised.
1722. May 1. Marie, ye daughter of Thomas Golding and Frances, his wife, was baptised ye first day of May.
1713. Henery, ye son of Frances Goldin, bapt. 29 Novemb.
1714. Margaret, ye daughter of Thomas and Francesse Goldin, his wife, baptised ye 4th of May.
1716. Elizabeth, ye daughter of Thomas Goldin. gent., and Frances, his wife, baptised Nov. 10.
1718. William, ye son of Thomas Golding, gent., and Frances, his wife, baptised 5 June.
1719. Alice Goldin, daughter of Thomas Goldin, gent., and Francesse, his wife, baptised May 8th at home, and certifed in ye church 23rd.
1719. Margaret, daughter of Thomas Goldin, gent., and Frances, his wife, buried ye 12 August.

1720. July 24. Richard, ye sonne of Thomas and Francesse Goldin, baptised.
1720. Oct. 25. Buried, Henry Golding, gent.
1721. March 29. Baptised, Robert, son of Thomas Goldin and Frances, his wife.
1724. October 14. Baptised, Edward, ye son of Thomas Golding, gent., and Frances, his wife. Buried November 18th.

For a little longer period we find them at Trottescliffe.

1685. Feby. ye 8th. There was baptised, Thomas, the son of Henry Golding and Elizabeth, his wife.
1689. April 12. William, the son of Henry Golding, buried.
1706. August 11. There was baptised William, the son of Henry Golding and Amy, his wife.
1797. September 14. Sarah, the daughter of Thomas and Charlotte Golding.
1799. May 13th. Thomas, son of Thomas and Charlotte Golding, baptised.
1801. October 8th. Mary Anne, daughter of Thomas and Charlotte Golding, baptised.
1804. August 19th. Charlotte, daughter of Thomas and Charlotte Golding, baptised.
1804. October 14th. Charlotte Golding, infant, buried.
1805. December 6th. George, son of Thomas and Charlotte Golding, baptised.
1808. March 13th. William, son of Thomas and Charlotte Golding, baptised.
1810. June 24. Elizabeth, daughter of Thomas and Charlotte Golding, baptised.

We have a few entries in the Allington register.

1651. George Sedwick, of Town Malling, and Elizabeth Golding of the same were married the twelfth of January.
1651. Henry Golding of Barming, and Barbara Hodges of Hunton, were married the 16th of October.
1810. Thomas, son of Thomas and Mary Golding, borned Feby. 12, baptise August 12.
1811. Henry, son of Thomas and Mary Golding, borned April (the date is pasted over) and baptise August 11.
1813. January 10. John, son of Thomas and Mary Golding, farmer.
1814. August 21. Anna Maria, daughter of Thomas and Mary Golding, farmer.

We have two early entries in the parish of Addington.

1686. Anna, the daughter of Henry Golding, and Thomasin his wife, baptised March 18.
1729. George, son of John and Susan Golding, baptised August 3rd.

In West Malling there is only one entry.

1710. Jany. 12. Thomas, the son of Thomas Golding, gentleman, and Frances his wife was baptised.

The next oldest entry is the first found in Ryarsh, which contains the following :—

1729. May 11. Oliver, son of Mr. Thomas and Frances Golding, baptised.
1741. Feby. 1. Thomas, son of Mr. Thomas Golding, buried.
1867. June 15. Mary Ann Golding, Norwood, Surrey, aged 67, buried.
1823. Sarah Anne Otte, the daughter of William Henry and Mary Ann Golding, gent., baptised July 22.
1824. James Otte, son of William Henry and Mary Ann Golding, gent., baptised Dec. 24.

1826. Ellen Otte, daughter of William Henry and Mary Ann Golding, gent.,
    baptised June 14.
1828. Mary Ann Otte, daughter of William Henry and Mary Ann Golding,
    gent., baptised Sept. 25.
1831. Emma Otte, the daughter of William Henry and Mary Ann Golding.
    gent., baptised Sept. 3.
1833. Rebecca Frances Otte, daughter of William Henry and Mary Ann
    Golding, gent., baptised April 25.
1835. Frances Twysden Otte, daughter of William Henry and Mary Ann
    Golding, gent., baptised Dec. 22.
1840. Henry Frank Otte, son of William Henry and Mary Ann Golding,
    gentleman, baptised Feby. 3.
1844. Arthur Henry Otte, son of William Henry and Mary Ann Golding.
    gentleman, baptised May 18th.

In Aylesford we have these entries :—

1758. December 11. Married James Taylor of East Malling, widower, and
    Mary Golding, widow.
1778. Nov. 21. John Cook of this parish, bachelor, and Elizabeth Golding,
    of East Farleigh, minor, with the consent of her father, George
    Golding, married.
1816. July 29. Married William Golding and Elizabeth Dyer.
1851. Dec. 21. Richard Golding, of West Malling, aged 3 years.

In Offham we have :—

1797. March 5. William, son of Thomas and Anne Golding. baptised.
1798. November 25. Sarah, daughter of Thomas and Anne Golding, baptised.
1806. May 25. Richard, son of Thomas and Anne Golding, baptised.
1809. January 29. John, son of Thomas and Anne Golding, baptised.

In Halling we find :—

1780. October 6. William Francis and Mary Golding, married.
1789. September 6. Eliza and Anne, daughters of Walsingham and Anne
    Golding, baptised.
1782. May 16. Henry. son of Walsingham and Anne Golding, baptised.
1805. Buried Anne Golding, aged 46, March 31.
1806. April 17th. Buried, Elizabeth Golding, aged 16 years.
1807. June 1. Buried, Walsingham Golding, aged 52 years.
1839. July 22. Catharine Charlotte, daughter of Thomas and Francis Godfrey
    Golding, curate of Yalding, baptised.
1849. September 12. Frances Godfrey Golding, buried, aged 43 years.
1851. Jany. 1st. Married, Edward Chapman Williams and Martha Ann
    Golding.
1864. April 2nd. George Golding, aged 59 years.

But the home of the Goldings for many years was Ditton, the
church of which is filled with monuments to them.

Thomas Golding of this parish, bachelor, and Mary Harris of the same, spinster.
    were married in this church by licence, this 29th day of March, in the
    year one thousand seven hundred and sixty eight, by me, Charles
    Bowles, rector.
1769. July 26. Thomas, son of Thomas Golding and Mary his wife. privately
    baptised, and admitted into the church on August 8 following.
1771. June 10. Mary, daughter of Thomas and Mary Golding, was baptised.
1771. John Golding of this parish, bachelor, and Anne Stimpson of the same,
    spinster, were married in this church by licence on the eighth day
    of July, in the year one thousand seven hundred and seventy one,
    by me, Dr. Lloyd, minister.

1772. Jany. 26. Thomas, son of John and Anne Golding, was baptised.
1772. December 21. William, son of John and Anne Golding, was baptised.
1773. March 29. William, son of Thomas and Mary Golding, baptised.
1773. May 5. Susanna Rogers, daughter of William and Sarah Golding, baptised.
1774. Nov. 2. John, son of Thomas and Mary Golding, baptised.
1775. Jany. 4. Frances, daughter of John and Anne Golding, baptised.
1775. July 12. James, son of William and Sarah Golding, baptised.
1776. April 24th. Samuel Barnes, of the parish of Aylesford in Kent, bachelor, and Mary Golding of this parish, spinster, married by licence.
1775. July 12. Oliver, son of John and Anne Golding, was baptised.
1777. October 11. Frances, daughter of William and Sarah Golding, baptised.
1777. Feby. 12. Oliver Golding, an infant, was buried.
1778. March 18. Elizabeth, daughter of John and Anne Golding, baptised.
1781. Feby. 10. Elizabeth, daughter of Thomas and Mary Golding, baptised.
1782. March 17. Rebecca, daughter of William and Sarah Golding, baptised privately, and admitted into the church the 7th of April.
1782. Feby. 7. Mary Ann, daughter of John and Anne Golding, baptised.
1782. March 17. Stephen, son of Thomas and Mary Golding, baptised.
1784. June 23. Susannah, daughter of William and Sarah Golding, was buried.
1784. July 30. William Henry, son of William and Sarah Golding, baptised.
1786. March 4. William, the son of John and Anne Golding, buried.
1787. March 31. Sophia, the daughter of Thomas and Mary Golding, baptised.
1789. William Golding died June 16, buried June 22, of the Court Lodge.
1795. Jany. 16th. Mrs. Susannah Golding, aged 92 years.
1788. March 10th. Mary Anne, daughter of Thomas and Mary Golding, baptised.
1805. Mary Anne, daughter of John and Anne Golding, was buried November 30.
1807. August 7th. Mrs. Anne Golding, aged 61 years, was buried.
1807. November 19th. Mr. John Golding, aged 80 years, was buried.
1798. Married, John Golding and Mary Manvell, November 20th.
1811. Married, John Manvell and Elizabeth Golding, June 11th.
1816. Buried, Thomas Golding, Esq., Jany. 13, aged 46.
1821. April 2. Buried, Thomas Golding, Esq., of Aylesford, aged 88.
1823. November 10. Mrs. Sarah Golding, aged 83, buried.
1824. May 3. Mary Golding, aged 39, buried.
1824. Laura Eleanor, daughter of John and Mary Golding, baptised June 17th.
1826. Henrietta, daughter of John and Mary Golding, baptised June 1st.
1826. Jany. 25. Mary Golding, of Aylesford, buried, aged 82 years.
1830. Sept. 23. William Golding of Aylesford, buried, aged 57 years.
1834. Feby. 21. Fanny Golding, aged 18 years, buried.
1837. May 29. Mrs. Mary Ann Golding, aged 49 years, buried.
1837. July 17. Mary Ann Golding, aged 26 years, buried.
1840. September 7th. Mary Golding, aged 18 years, buried.
1843. May 15. Thomas Golding, aged 75 years, buried.
1848. September 12. John Golding, aged 11 months, buried.
1849. February 12. Mary Golding, aged 80 years, buried.
1859. February 26. John Golding, Esq., of Ditton Place, aged 85 years, buried.
1865. August 8. Henrietta Golding, Brixton, aged 40 years, buried.
1874. April 13. John Golding, aged 96, Maidstone, buried.
1874. May 8. Elizabeth Golding, aged 53, buried.
1877. March 22. Reginald Golding, aged 21, buried.

In addition, we have the following names of gentry entered in the various registers :—

Adam	Cogger	Hicks	Monson	Shakerley
Adams	Coleman	Hills	Montague	Sharrett
Akers	Covert	Hodge	Moore	Shaw
Alchin	Crispe	Hogg	Morse	Simmonds
Aldeine	Croxley	Holder	Mount	Simpson
Allein	Davis	Honeywood	Neame	Sisley
Allen	Debarry	Hook	Nicholson	Smith
Andrews	Debrasne	Hooker	Norris	Shepherd
Annwokes	Decritz	How	Norton	Skinner
Ashbe	Delafield	Hubble	Omer	Southgate
Atkinson	Dickens	Humphries	Page	Spain
Austen	Dickenson	Johnston	Papillon	Spong
Baker	Dimock	Jones	Paris	Stacey
Banks	Douce	Juke	Parker	Stephen
Barnes	Douglas	Kendric	Parrington	Stephenson
Bates	Dowle	Kenrick	Parry	Stow
Bax	Downman	Kentish	Parsons	Streatfield
Beaumont	Drury	Kibble	Perch	Style
Besbeach	Dudlow	Kidwell	Perfect	Tapley
Best	Elin	King	Petley	Tassell
Bettenson	Elliston	Knowles	Pettit	Taylor
Betts	Etheredge	Koe	Peneranda	Thomas
Bewley	Evelyn	Lambert	Philips	Thornton
Bigg	Everitt	Langston	Pilcher	Thorpe
Blake	Eves	Larking	Polhill	Tomlyn
Bligh	Fance	La Trobe	Polly	Townsend
Bonnard	Faunce	Laurence	Pope	Tremesse
Boodle	Fletcher	Leckie	Pott	Tresse
Bovey	Forster	Leigh	Poynder	Tucker
Bowles	Franke	Lewis	Prall	Tutty
Boys	Franklyn	Littehall	Pratt	Vale
Boysier	Furley	Littleboyes	Pym	Vaulx
Bradley	Gates	Long	Ralph	Visier
Brattley	George	Lowry	Reade	Waghorn
Bredham	Glover	Luck	Reed	Ward
Breton	Godney	Luxford	Renney	Warne
Brodgam	Godwin	Lydall	Rich	Warren
Brome	Goldberg	Lys	Richardson	Weekley
Brooke	Gotier	Mabbe	Rives	Weston
Brownson	Graham	Maddison	Rix	Whetenhall
Butcher	Grandsden	Mainwaring	Roberts	Whitehead
Butler	Gratwick	Mair	Rogers	Wiggenden
Butt	Graves	Maitland	Russell	Wigan
Cary	Green	Mapleton	Sampson	Wildash
Castle	Greenway	Marsh	Sanctilis	Williams
Cator	Hadlow	Martin	Sandell	Willoughby
Chambers	Hailstone	Manley	Saunders	Wimble
Chapman	Hall	Mason	Savage	Wodd
Cheeke	Handasyde	May	Savell	Wollven
Clabon	Harwood	Maynard	Say	Wood
Clarke	Hatch	Middleton	Scudamore	Woollett
Clements	Hayer	Miller	Seager	Worrall
Codd	Hayman	Milner	Seaton	Wyatt
Collings	Hedges	Milnes	Seeley	Wykes
Collins	Hely	Money	Selby	Young
Cork	Hickes			

The following names are given of persons described as farmers and yeomen :—

Allchin	Costen	Gorham	Mantle	Savage
Andrews	Day	Graham	Matthews	Sealey
Baker	Dibley	Gransden	Pack	Sedwick
Broad	Everest	Hatch	Palmer	Sharpe
Brooks	Fauchon	Henfield	Pankhurst	Smith
Brown	Fogg	Hubert	Penury	Stunt
Burgess	Fowle	Huggins	Phillips	Taylor
Capon	Fremlin	Knight	Pound	Tomlyn
Causton	Garner	Knowles	Reaves	Vowsden
Cork	Godden	Letts	Rich	Wray

The different callings under which people are described are as follows. We give them, because some of the terms are quaint or obsolete ; of course, we have not gone, in this case, into the legal registers of the present century :—

Apothecary	Corporal	Joiner and carpen-	Ropemaker
Apprentice	Counsellor	ter	Schoolmaster
Attorney	Doctor of physic	Journeyman shoe-	Seaman
Attorney-at-Law	Dragoon	maker	Servant
Baker	Exciseman	Journeyman tailor	Shoemaker
Barber	Farrier to a troop	King's officer	Soldier
Beggar	of horse	Labourer	Solicitor
Blacksmith	Fellmonger	Lieutenant in royal	Shopkeeper
Bootmaker	Frameyer	navy	Strolling player
Brazier	Gardener	Limner	Supervisor of the
Bricklayer	Gipsey	Lodger	Excise
Butcher	Glazier	Maid } to Mr. A.	Surgeon
Captain	Goldsmith	Man } or B.	Tailor
Captain in Royal	Greyhunter	Maltster	Tallow chandler
Navy	Grocer	Merchant	Tanner
Carpenter	Harvester	M. D.	Titheman
Chemist	Harvestman	Miller	Trooper
Churchwarden	Hopmerchant	Mingo	Vagrant
Clerk of the Parish	Hopper	Officer of Excise	Victualler
(mentioned as	Householder	Organist	Vice-Admiral of the
early as 1562).	Housemaid	Papermaker	White
Clockmaker	Husbandman	Plumber	Waterman, or Sayler
Cooper	Innholder	Police Constable	Weaver
Cordwainer	Ironmonger	Publican	Wheelwright

The various designations of the clergy have been spoken of in our remarks upon the clergy of the different parishes. As regards the above, some, such as apothecary for doctor, attorney or attorney-at-law for lawyer, clockmaker, counsellor for barrister, frameyer, fellmonger, husbandman, limner, and titheman, may be considered obsolete, or nearly obsolete, terms. Greyhunter as a calling is peculiar ; but the unfortunate grey, the old name of the badger, was thought by the ignorant rustic, and his almost equally ignorant master, the squire, to be a very mischievous animal. It is to their folly in getting rid of the badger, the kestrel hawk, and other creatures of like nature, that they were pleased to call vermin, that we owe the swarms of real vermin, in the shape of mice and rats, of the present day. "Mingo" is I suppose a nickname, which, however, is actually entered in the Trottescliffe registers to describe one of the Goddens. Organist in the last century was not a common calling; and Town Malling is to be congratulated upon its advance in civilisation in having possessed one. The early notice of parish clerk in 1562 in one parish, 1642 in another, and 1652 in a third, throws some light upon the early appointment of that worthy. The description of a gentleman as churchwarden, without mentioning anything else about him, is what we should hardly expect to find in the present century.

We now proceed to give a list of names that may be found generally in these registers, some of which appear to be peculiar to this district.

Absalom	Bass	Boorman	Canay	Collison
Accounts	Bassett	Boreman	Carey	Conolly
Adams	Batchelor	Botten	Carpenter	Cook
Adds	Bateman	Botting	Carr	Coomber
Adkins	Baxton	Bourne	Carrol	Cooper
Adney	Beale	Bovey	Carter	Corsen
Akhurst	Beckett	Box	Causthen	Cortop
Aldridge	Beeby	Brabon	Cell	Cosin
Alexander	Beecher	Bradbourne	Challender	Cossom
Algur	Beesley	Bramley	Chamberlain	Cotinge
Allfree	Beetenson	Brand	Chambers	Cousins
Apps	Beffin	Bray	Champion	Coverty
Arnold	Bell	Brent	Champness	Coward
Ashdown	Bellingham	Bresbeck	Chapman	Craddock
Ashenden	Belman	Brewer	Chappel	Craft
Assiter	Beltinger	Brigden	Chary	Cramp
Astin	Bennant	Briggs	Chatfield	Crampton
Atkins	Bennet	Bright	Cheaseman	Crayher
Attwell	Bensted	Brightwell	Cheeseman	Cresswell
Augur	Berry	Brissenden	Cheverell	Cripps
Averill	Berwick	Brister	Chinkfield	Crittenden
Badley	Betts	Bristowe	Chittenden	Croft
Bailey	Beven	Brock	Chowning	Cromp
Baker	Bewley	Brome	Chuning	Crosley
Balcomb	Bicknell	Brook	Clair	Cross
Baldock	Bignall	Brooker	Clark	Crow
Baldwin	Billinghurst	Bryant	Clarkstone	Crowhurst
Ballard	Binskin	Buckland	Cleese	Crundell
Ballwell	Birchall	Buckshire	Cleggett	Cuckoo
Bann	Bird	Budds	Cleland	Curteis
Barden	Blackbird	Buggin	Clout	Cutbush
Barker	Blackman	Bullfinch	Cobb	Cutler
Barkett	Blackstock	Bumford	Coburn	Cyborn
Barnecastle	Blassern	Bunyard	Cockerill	Dabbs
Barnes	Blisse	Burt	Cockett	Dane
Barre	Boghurst	Burton	Cockoo	Danes
Barrett	Bolger	Busain	Codd	Dann
Bartholomew	Bone	Butcher	Coke	Darbye
Barton	Bonnar	Butler	Cole	Hartnell
Barwick	Bonnard	Buttenshaw	Colegan	Davidson
Basnage	Bonnewell	Campbel	Coley	Davis

Dawson	Faulkland	Gooring	Hely	Jordan
Dean	Fenne	Gorham	Hendfield	Joy
Deavereux	Fentiman	Gosden	Hernden	Jupp
Denham	Fereby	Gosling	Herne	Keeble
Denman	Fetheridge	Gostling	Hervey	Keen
Dennett	Field	Goulay	Higgins	Keish
Dennis	Fisher	Gould	Higmore	Kember
Derffe	Fissenden	Goulder	Hilder	Kemp
Dewe	Fiveash	Gray or Grey	Hill	Kempsley
Ditton	Fust	Green	Hills	Kentwell
Dixon	Flaxen	Greengrass	Hind	Kettle
Doidge	Fleet	Gregory	Hinde	Kirby
Dolton	Fletcher	Griffin	Hoad	Knight
Dorling	Floyd	Grignon	Hobard	Knott
Dormer	Foliard	Grimard	Hodges	Lad
Dorrington	Foreman	Grindell	Hogben	Lake
Dorsett	Ford	Groombridge	Holden	Lane
Douglas	Forier	Grover	Hollanby	Large
Dove	Foster	Grundell	Holland	Lash
Dowcett	Foule	Guilder	Hollands	Latter
Downes	Fowcke	Gull	Holliday	Lee
Driver	Fowler	Gybbus	Holloway	Lenham
Duckinge	Franklin	Hadlow	Holmer	Leopard
Dudley	Freeman	Hagar	Holmewood	Lewin
Duffe	French	Haggar	Holness	Lewis
Duffield	Frogmarton	Hainsworth	Homan	Lilley
Dunk	Fuex	Halfhead	Honey	Littlelove
Duper	Fuller	Hall	Hopkins	Lloyd
Durling	Furmager	Hamlyn	Hopson	Lock
Dyne	Fypes	Hammond	Hormer	Lockhart
Eaglestone	Gambol	Hampshire	Howel	Lockyer
Eagleton	Gammon	Hanbury	Howes	Loder
Easeden	Garbet	Handen	Howlet	Lomas
Eason	Gardener	Harcourt	Hoysted	Long
East	Gardenham	Harding	Hubbard	Longhurst
Eastland	Garland	Harpenden	Hubbucks	Longley
Edenberry	Garnes	Harper	Hudson	Lovegrove
Edenden	Garraway	Harris	Huggett	Lovelace
Edmonds	Garrett	Harrison	Hunt	Lucey
Edmouth	Gates	Hartridge	Iden	Luck
Edwards	Gayton	Hartrop	Ifield	Lunt
Eele	George	Harvey	Irish	Luxford
Elgar	Gibbons	Hastlyn	Isaacs	Lyon
Elliott	Gibson	Hat	Ivory	Lys
Elwood	Gifford	Hatch	Jackman	Lyssey
Esquior	Gilbert	Hattchurch	Jacob	Lyssley
Etherington	Gnatt	Hattrop	Jacobs	Maddox
Evans	Goddard	Havens	Janvil	Mair
Evenden	Godfrey	Hawkes	Jarrett	Mann
Evernden	Godred	Haycock	Jarvil	Manvell
Eversfield	Godwin	Hayes	Jason	Margett
Evestone	Goffe	Hayfield	Jefferies	Marr
Ewing	Goldwell	Hayles	Jeffery	Marsh
Exeter	Golton	Hazleden	Jenkins	Marshall
Farne	Gooch	Heade	Jessup	Martin
Farningham	Goodenough	Heath	Jewess	Martyr
Farquhar	Goodhugh	Heel	Johnson	Maser
Farrant	Goodwin	Helby	Johnston	Mason

Masters	Nutley	Pollard	Savage	Summers
Mattocks	Oakley	Porse	Saxe	Surman
May	Oates	Porter	Scotchford	Suton
Maylam	Obee	Potter	Scott	Sutton
Mayne	Oben	Pound	Scudder	Swan
Mayo	Offin	Powell	Scalcap	Swansbrook
Meale	Oliff	Powers	Seagers	Swift
Meall	Olive	Poynter	Sedgewick	Syburn
Medley	Oliver	Pratt	Segest	Tallurin
Medly	Olyve	Pressnail	Sellis	Tanner
Meles	Onion	Price	Sharp	Tearness
Mellingam	Orgar	Pringe	Shephard	Terney
Mendecoate	Orpin	Prior	Shepherd	Terry
Meopham	Osborne	Pycher	Shuter	Thatcher
Mercer	Otman	Pype	Siggust	Thompson
Merdick	Otway	Quarrington	Silmister	Thornhill
Miles	Ouldridge	Quin	Simmons	Thorpe
Millar	Over	Quy	Sinder	Thunder
Milles	Ovel	Ramden	Singleby	Thwaites
Mills	Overy	Randall	Sisley	Tonbridge
Miskin	Owen	Randell	Skelton	Tongue
Mitchell	Owlett	Randoll	Skinner	Took
Mockford	Owsley	Rapley	Slyther	Tornson
Mois	Oyne	Rastball	Smallman	Towne
Momford	Pace	Ray	Smallwood	Towner
Mongrove	Pack	Reader	Smart	Tracey
Monroe	Packham	Reed	Smedley	Tramper
Moor	Paddamore	Reeve	Smith	Transum
Moore	Pagden	Reynolds	Smithie	Tress
Morfy	Page	Rhodes	Smout	Trionner
Morley	Pannen	Richards	Smyth	Tritty
Mortimer	Paramore	Ridgeway	Sneed	Troul
Morton	Parmenter	Ring	Somerset	Tuffe
Mosen	Parray	Ripps	Southeran	Turke
Mosley	Parrayan	Robertson	Southwell	Turner
Mosling	Pasfield	Robins	Sparrow	Tye
Moss	Patsriman	Robinson	Sped	Tyler
Moyse	Pawley	Rochester	Spencer	Ullutt
Moulds	Payne	Roderick	Spong	Upstone
Murdock	Pedge	Rofley	Springfield	Usher
Mylls	Peek	Rolfe	Spurland	Vallance
Nash	Peerless	Roots	Squire	Vallum
Neal	Pelts	Rorsway	Stanford	Vane
Nepeter	Pemble	Rothwell	Stark	Vernon
Netherway	Penrose	Rottenbridge	Startup	Vicker
Newberry	Perry	Rowe	Steddall	Vidgeon
Newman	Pet	Rowley	Stedman	Vigeon
Newnham	Peto	Ruddock	Stevens	Vigian
Newstead	Phelps	Rust	Stevenson	Vine
Nibbs	Phipps	Sadler	Stewart	Viner
Nicholas	Piercey	Sage	Still	Wadge
Nightingale	Piggott	Salisbury	Stimpson	Waghorn
Noakes	Pilcher	Salmon	Stone	Waite
Norditch	Pilsen	Salt	Stow	Waiter
Norman	Plair	Sandell	Streatfield	Wakelyn
Norris	Player	Sanders	Street	Walkley
Norton	Pledge	Sarns	Stretton	Waller
Nox	Politic	Satten	Stubberfield	Walnutt

Walton	Wheasman	Whiting	Wilmet	Woodyer
Warburton	Wheaten	Whitney	Wilson	Woollett
Warren	Wheeler	Whittle	Wingrove	Worlidge
Waters	Whiffins	Wicking	Winter	Wrey
Watts	Whiskin	Wilde	Witby	Wright
Wave	White	Wildish	Wolfe	Wrothwell
Wayman	Whitebread	Wilkins	Wolley	Yeekley
Webb	Whitehead	Wilkinson	Wolveridge	Yeoman
Weeks	Whiteland	Willett	Woodcock	Yeomans
Wellard	Whitenall	Willis	Wooden	Yorre
Wells				

Any one who has the patience to examine the above list will find many curiosities we might enlarge upon.

Amongst these we should call attention to the names of " birds " as being a designation of many families in the district ; for example, we have Blackbird, Bullfinch, Cockerill, Cuckoo, Crow, Dove, Gull, Knott, Owlett, Reeve, Nightingale, Sparrow, and Swan ; and no doubt many similar things will strike the reader. In some of these registers we find the notice of the money gathered by briefs. In days gone by the way of gathering money for different charitable purposes, was to read the order commanding the same by lawful authority to be collected in the time of public worship ; and we still see that this was authorised to be done during the Offertory Sentences before the Prayer for the Church Militant, in the Communion Service, by the rubric in our Prayer Book. As some of these briefs contain very interesting allusions, we mention them. From registers we have examined, previous to writing this work, we have been struck by the attempt to introduce the Church of England into Poland and Russia, as we have several entries of briefs for helping the churches in Lithuania, and Courland, where Protestant ; we have also briefs showing that assistance was given to private persons, and in aid of sufferers by fire. Besides these we have aids towards helping, amongst others, these parishes in Kent :—

Brenchley	Gillingham	Benenden
Clyffe	Northfleet	Tonbridge
Gravesend	Woolwich	Yalding
		St. Margaret's Cliffe.

In the last two parishes it is distinctly mentioned the purpose is for rebuilding the church : in the register of Cranbook we find that 1s. 6d. was raised by brief for rebuilding West Malling church.

Besides, we have a record of 2s. 6d. raised in the same way in Snodland, for repairing St. Andrew's Harbour in Scotland ; three several sums of 13s. 4d., 13s. 2d., and 25s. for the relief of poor slaves in Algiers ; * and assistance given to the French refugees in 1699.

* An interesting witness to Moorish corsairs making themselves felt in England in the seventeenth century.

In Snodland we again find a curious record of some cattle plague :—
" Collected here upon the cowkeepers' brief August 1st, 1715, three shillings ;"
and of other catastrophes the following :—

" For sufferers by thunder and hail in the County of Stafford, 12th October.
1720, one shilling."
" For sufferers from an inundation at Upchurch in the county of Kent,
damages £4,290. Paid 4s. 8d."
" For the oyster dredgers of the river Medway the sum of 2s. 6d."
" For Folkestone Fishery, 1s. 8d."

Besides we have briefs on behalf of Strasburg in Alsatia, Dutch
Berg, Robi and Villarin in the valley of Luzerne for sufferers by
inundation.

All these show that collections in church are no new things, as
some churls would try to persuade the public, but that the church
even in its most sleepy time was aiding those abroad, and at home ;
as these records stretch from the middle of the seventeenth to the
middle of the eighteenth century. There is one record in Addington
register which is very curious, and we therefore give it particular
notice.

" 1660. Paid for ye fighting brave 6s. 0d."

Whether this entry refers to money paid to assist Royalist or
Parliamentarian troops, or some foreign wars, it were difficult to
determine, as it is just at the end of the Commonwealth.

These briefs, no doubt, would furnish useful jottings for the confirmation and the explanation of history, if examined in the various
parts of the country ; but unfortunately in many cases the accounts
have been lost.

Though not connected directly with the registers, many parishes
have very curious old churchwarden accounts, in which we find that
money was paid to enable the clergy and churchwardens to attend
confirmations and visitations, and to provide the poor with Bibles
and prayer books, and also with the ordinary furniture necessary
for their homes ; and in many cases money was actually paid for the
production of so many sparrows' heads, and for the destruction of
hawks, weasels, badgers or greys, and other creatures considered
vermin.

In concluding this chapter the author would state that, were the
curious old records preserved in our parish chests cherished and
studied as he has studied them, he has no doubt that much valuable
information upon the lives, manners, customs and names of our
forefathers would be gathered ; enough to fill twenty volumes far
more interesting than much of the literature of the present day,
and giving some idea as to how deservedly the Church of England
has received and maintained the name of the nation, which she bears
as her distinction, as one branch of the Holy Catholic Church of
Christ.

CHAPTER XIII.

AS we have had occasion to mention one of the Kentish sayings, it may be interesting to our readers to examine into those we have collected. There are some that refer to the county.

(a) "Kent is famed for hops, fair maids, and civility."

(b) Alluding to the wealth of the inhabitants, we have :—

> "A knight of Cales,
> A gentleman of Wales,
> And a laird of the North Countree ;
> But a yeoman of Kent,
> With his yearly rent,
> Will buy them up all three."

(c)
> "Kentshire,
> Hot as fire."

(d) "As great as the devil and the Earl of Kent."

(e) "Kent red veal and white bacon."

(f) "Lythe as a lass of Kent."

(g) "St. Tyburn of Kent."

(h) "Neither in Kent nor Christendom."

(i) "A man of Kent, and a Kentish man." A man of Kent is one born between the Stour and the Sea; all others are Kentish men. Another opinion says that a Kentish man is one born in Kent, but not of Kentish parents, while a man of Kent is one whose parents as well as himself are Kentish.

(k)
> "Essex styles, Kentish miles,
> Norfolk wiles, man beguiles"—

an allusion to the Kentish labourer's mile being about one and a half.

(l) A Kentish man speaks disdainfully of persons from other counties, as coming "out of the shires," and of "silly Sussex"; while Surrey is a sobriquet for a fool.

(m)
> "Father to the bough,
> Son to the plough."

This is an allusion to Kent being under the law of gavelkind, by which the son came into his father's property, even though the parent were executed for high treason.

(*n*) The word Kentish is frequently used as an adjective with a peculiar significance, as " Kentish ague," " Kentish cherries," and " Kentish pippins " refer to special sorts. " Kentish longtails " are wild oats ; " Kentish cousins " has a somewhat similar meaning to Scotch cousins. A " Kentish stomach " means a strong stomach.

(*o*) " Wealth and no Health," " Health and no Wealth," " Health and Wealth." The cold and dreary marshes of the coast, and round the mouths of the Thames and Medway, producing excellent pasture, but in former days being very bad for the ague, were designated " Wealth and no Health " ; the high chalk hills, with their barren soil, but bracing air, were renowned for " Health and no Wealth " ; the sandstone ranges and the Weald, with their smiling valleys, fertile fields, and pine- and oak-clad hills, still offer both " Health and Wealth."

(*p*) Kent having so long led the van in Britain's wars, the name, " Strong men of Kent," passed into a proverb.

There are, besides these, a number of proverbs referring to different parishes and their peculiarities. Of these we give the following :—

(*a*)   " Long, lazy Lewisham, little Lee,
                Dirty Deptford, and Greenwich free."

(*b*)   " Sutton for mutton, Kirby for beef,
                South Darent for gingerbread, and Dartford for a thief."

This refers to the fertile meadows along the Darenth, and the fair at Dartford.

The next three refer to the unwholesome marsh-parishes of the Medway and the Swale.

(*c*)   " If you'd live a little while
                  Go to Bapchild."

(*d*)   " He that would not live long,
                  Let him live in Murston, Teynham, or Tong."

(*e*)   " He that rides in the hundred of Hoo,
                  Besides pilfering seamen, will have dirt enoo'."

(*f*)   " He that would go to a church miswent,
                  Let him go to Cuxton in Kent."

This appears to allude to the altar of the church having been round a corner.

(*g*) " Starv'em, Rob'em, and Cheet'em,"—Strood, Rochester, and Chatham.

(*h*) A Rochester portion : " Two torn smocks and what Nature gave you." The reason for this one cannot find, as indeed for a great number of these sayings, such as the following :—

(*i*) " You've got no calves to your legs like the Pluckley girls, and are obliged to wear straight stockings."

(*k*) " Huckinge glass breeches where rats run on tiptoe."

(*l*)
> "Go to Monk's Horton,
> Where pigs play on the organ."

(*m*) The reason for

> "Proud Town Malling, poor people ;
> They built a church to their steeple,"

we have explained in a former part of this book.

(*n*) "Poor Lenham."
This refers, no doubt, to the barrenness of the soil.

(*o*)
> "Smoky Charing lies in a hole,
> It had but one bell and that was stole."

(*p*)
> "Surly Ashford, proud Wye,
> And lousy Kennington lieth hard bye."

(*q*) We have the following two specially referring to Canterbury:—
"Canterbury is the higher rack, but Winchester the better manger." This is a reference, no doubt, to the bishopric of Winchester being better endowed than the archbishopric of Canterbury.

> "Canterbury is in decay,
> God help who may."

(*r*) Together with the neighbouring towns the city is classed as follows:—

> "Deal savages, Canterbury parrots,
> Dover sharks, and Sandwich carrots."

(*s*) Deal and Dover enjoy also the unenviable reputation of the next lines:—

> "Deal, Dover, and Harwich,
> The devil gave his daughter in marriage,
> And, by a codicil of his will,
> He added Helveot and Brill."

(*t*) Dover by itself is the subject of three proverbs:—

> "As sure as there is a dog in Dover."

> "From Berwick to Dover,
> Is three hundred miles over."

> "When it's dark in Dover
> 'Tis dark all the world over."

(*u*) "Conscience is drowned in Sandwich haven," does not speak well for that port.

Another proverb of Sandwich is, "Proud Wingham, wicked Ash, and lazy Sandwich."

(*v*)
> "The vale of Holmesdale,
> Was never won nor ever shall !"

(*w*)
> "Ramsgate capons, Peter's lings,
> Broadstairs scrubs, and Margate kings,"

speaks of the various products of Thanet.

(*x*)
> " When England wrings,
> The island sings,"

is a reference to the dry nature of Thanet.

(*y*) " Let him set up a shop on Goodwin Sands," is a reference to anybody doing anything very foolish.

(*z*) " Tenterden church steeple is the cause of Goodwin Sands " refers to the story that the men of Kent, being forced to build this church by the archbishop, forgot to repair the sea wall of the isle of Loamey, and the sea burst in ; and there has been this dangerous shoal there ever since.

(*a*) " Earl Godwin and his court are hungry." An expression used when a storm is blowing up, by the Deal fishermen, referring to the swallowing up of the vessels by the Goodwins, which were supposed to be his property, and under which they used to believe he still kept court.

(*β*) " Cowden play " means any silly way of playing.

(*γ*) " Get on anyhow, as they do at Rainham." This village, between Chatham and Sittingbourne, is not only thus a byword, but the author has also heard, " Why, you've only two sticks and a piece of paper, like a Rainham fire."

(*δ*)
> " A north-east wind in May
> Makes the Shotover man a prey."

Shotover are mackerel. This proverb refers to the danger of this wind in the Channel at this season.

(*ε*) Folkestone enjoys an unenviable notoriety for silliness. Folkestone washerwomen are rain-clouds.

(*ζ*) Greenwich geese are the old pensioners that used to be at the hospital.

(*η*) " Frindsbury Clubs " refers to the lads of Frindsbury coming to Rochester as a penance with staves. The penance was enjoined because, there being no rain, the monks of Rochester were directed to make a pilgrimage to Frindsbury, when the men assaulted them and beat them.

(*θ*) The Weald of Kent we are told, is—

> " Bad for the rider,
> Good for the abider."

This refers to the heavy nature of the soil.

(*ι*) " To be married in Finglesham church." As there is no church at this hamlet, this refers to people living together who are not married.

(*κ*) " For company, as Kit went to Canterbury." This proverb speaks for itself.

(*λ*) " Born down Ryarsh Sandpits " signifies that the person is illegitimate.

Other sayings besides these are of the cherries :—

> "If they blow in April
> You'll have your fill ;
> But if in May
> They'll all go away."

Of the weather :—

> "A drip in June
> Keeps things in tune."

In addition, the following proverbs appear to be purely Kentish :—

(*a*) "Pour water into the Thames."

(*b*) "Ducks fare well in the Thames."

(*c*) "You are as yellow as a peigle," *i.e.* a cowslip.

(*d*) "You prick up your ears like an old sow in beans."

(*e*) "That would make a donkey run away from his beans."

(*f*) "When the sage blooms, there will be mischief."

(*g*) "He wants that as much as a toad wants a side pocket."

(*h*) "Don't go groping about like a blind hen looking for a worm in a hedge."

The following two are personal : "As a thorn produces a rose, so Godwin begat Editha."

> "At Betshanger a gentleman, at Fredville a squire,
> At Bonnington a noble knight, and at London a liar."

Certain things have become quite proverbial, as "Folkestone dried beef" for dogfish, "Great Church" for Canterbury Cathedral, and "Rumbold whiting," alluding to their being best in season about St. Rumbold's day. Faversham and Milton oysters, Fordwich trout, and Medway shrimps, and smelts, have also almost become proverbial for their superiority.

CHAPTER XIV.

NO work could be at all complete on any part of Kent without some attention was called to the large extent to which every part of this county is represented in the fauna and flora of the British Isles. Let us first examine the mammals.

Great Bat or Noctule (Vespertilio noctula).—This bat has been frequently seen.

Common Bat or Pipistrelle (Vespertilio pipistrellus).—This also is a common Kentish bat.

Reddish-grey Bat (Vespertilio nattereri).—Has been seen round Chislehurst.

The Serotine Bat (Vespertilio serotinus).—Has been seen near London.

Whiskered Bat (Vespertilio mystacinus).—Has been discovered at Chislehurst.

Long-eared Bat (Plecotus auritus). —Frequently seen.

The Barbastelle (Barbastellus daubentonii).—Has been found near Dartford.

Greater Horseshoe Bat (Rhinolophus ferrum equinum).—Frequently seen.

The Mole (Talpa vulgaris).

The Hedgehog (Erinaccus Europacus).

The Common Shrew (Sorex araneus).

The Water Shrew (Sorex fodiens).

The Badger (Meles Taxus).—By no means an uncommon animal in the county, and especially in these parts. By his old name of "grey" he is frequently, as we have already shown, spoken of in the parish records.

Common Otter (Lutra vulgaris).— An animal still occasionally seen in the Medway.

Common Weasel (Mustela vulgaris). —Known in Kent as the "keyn." An animal always common, and more useful in extinguishing other pests than he is given credit for. He and the stoat or puttiec (Mustela erminea) may destroy a few young pheasants, or partridges, or chickens, or hares; but their assistance to man in clearing off rats and mice should afford them protection. These little creatures are not uncommon in the county; but if they were more plentiful, we believe they would prove a great boon in extirpating rats.

Polecat (Mustela putorius). — An animal once plentiful, but now only occasionally seen in this county.

Common Marten (Martes foina).— Lord Clifton, in a very valuable paper which he contributed to an early number of the *Rochester Naturalist*, showed that this little animal had long deserted our woods. Though we can trace his existence here into this century, no reliance can be placed upon many of the reports of his occurrence in our southern woods for a later period.

Fox (Vulpes vulgaris).

Squirrel (Sciurus vulgaris).

Harvest Mouse (Mus messorius).

Long-tailed Field-mouse (Mus sylvaticus).

Common Mouse (Mus musculus).

Black Rat (Mus rattus).—The author has seen several Kentish specimens.

Brown Rat (Mus decumanus).

Water Vole (Arvicola amphibius).

Field Vole (Arvicola agrestis).

Common Hare (Lepus timidus).

Rabbit (Lepus cuniculus).

Red Deer (Cervus elaphus) and Fallow Deer (Cervus dama).—Are found in our Kentish parks, and perhaps their ancestors once ran wild in our woods : at any rate, the outlying deer seem none the worse for the time they spend out of captivity.

The Common Seal (Phoca Vitulina), the Dolphin (Delphinus Delphis), and the Porpoise (Phocaena Communis).—Have been taken in the twin estuary of the Medway and the Thames.

A specimen of the Northern or Rudolph's Rorqual (Balaenoptera borealis).—Was stranded in the Thames off Tilbury, on October 19th, 1887. It measured 35ft. 4in., and a second was caught at Gillingham on August 30th, 1888, which measured 32ft. 2in. Another whale caught at Gravesend in September, 1883, appears not to have been identified. Of course these latter have probably never come up the Medway so high as our valley, but we think that our natural history notes in general had better be extended to the county than so small a district; especially, as in this case, whales must have come here years ago, since Henry I. gave the church at Rochester all the whales caught in the Medway.

CHAPTER XV.

NATURAL HISTORY—BIRDS, REPTILES, FISHES.

A S to whether all the birds that have been traced in Kent have positively occurred in this valley it would be hard to tell, and therefore I have contented myself with naming our commoner Kentish birds, and then speaking of the rarer sorts. We have fifty-six British residents that live with us all the year :—

Merlin	Creeper	Common Thrush
Kestrel	Nuthatch	Blackbird
Sparrow Hawk	Green Woodpecker	Hedge Sparrow
Marsh Harrier	Wagtail	Redbreast
Hen Harrier	Titlark	Wren
Montague's Harrier	Rock Pipit	Gold Crest
Tawny Owl	Skylark	Wood Pigeon
Barn Owl	Corn Bunting	Stock Dove
Great Tit	Reed Bunting	Pheasant
Cole Tit	Yellow Ammer	Partridge
Blue Tit	Chaffinch	Red-legged Partridge
Marsh Tit	Sparrow	Lapwing or Peewit
Long-tailed Tit	Greenfinch	Moorhen
Kingfisher	Goldfinch	Coot
Carrion Crow	Linnet	Mute or tame Swan
Rook	Redpole	Wild Duck
Jackdaw	Bullfinch	Teal
Magpie	Starling	Black-headed Gull
Jay	Missel Thrush	

Besides these we have twenty-one summer visitors :—

Spotted Flycatcher	Sand Martin	Whitethroat
Wryneck	Grey Wagtail	Wood Warbler
Cuckoo	Yellow Wagtail	Willow Wren
Nightjar	Tree Pipit	Chiffchaff
Swift	Sedge Warbler	Turtle Dove
Swallow	Nightingale	Landrail or Corncrake
Martin	Blackcap	Common Tern

Our winter visitants are eleven in number :—

Buzzard	Brambling	Jack Snipe	Hooper
Fieldfare	Hooded Crow	Bean Goose	Woodcock
Redwing	Common Snipe	Brent Goose	

Other birds that have been detected in Kent we mention, with a note upon where they were seen :—

Erne or Sea Eagle.—High Halstow.

Golden Eagle.—Sheppey, between Canterbury and Eastwell.

Rough-legged Buzzard.—Three at Cobham in 1876, two at St. Peter's, Isle of Thanet, and one at Waltham, near Canterbury.

Honey Buzzard. — Cobham (1881), Rainham.

Osprey.—Cobham.

Kite.—Cliffe (1881), Ramsgate (1887), Rodmersham.

Gyrfalcon.—Mr. Pemberton Bartlett told Mr. Morris this bird was rare in Kent. I never met any one who had seen it, nor saw it myself.

Peregrine Falcon.—Cobham, Rainham.

Hobby.—Chattenden Roughs, Rainham.

Red-footed Falcon.—Hythe (1862).

Goshawk.—Taken at Lydd; in Mr. Jell's collection.

Orange-legged Hobby.—This bird is claimed as a native.

Short-eared Owl.—(1881), Chalk, Ramsgate.

Long-eared Owl.—This bird has been probably seen at Cobham. I saw it this last winter at Addington.

The Eagle Owl.—Is chronicled in Kent, by Morris and Yarrell.

Snowy Owl.—Frinstead (1884).

Little Owl.—Shorne, Sevenoaks.

Hawk Owl.—Labelled Kent in the Maidstone museum.

Tengmalm's Owl.—Killed in Kent (1885).

Mottled Owl.—Lord Clifton supposes that the American mottled owl was found in the larch plantations at Cuxton, but he does not appear quite certain.

Great Shrike.—Reported in Kent by Morris.

Red-backed Shrike.—Reported from Cobham, Shorne, Cuxton, Mereworth, and Great Comp.

Woodchat.—Reported twice in Kent by Morris, also seen in 1890 at Sittingbourne.

Pied Flycatcher.—At Deal (1850), Cobham (1868), Rainham, Ramsgate.

Bearded Tit.—Occasionally seen in Kent.

Roller.—Rainham.

Bee-eater.—Kingsgate (1827), according to Morris.

Hoopoe.—Ryarsh, where it built for some years.

Cornish Chough.—Used to inhabit the cliffs near Dover for some years.

Raven.—Thames marshes : destroyed at Cobham for attacking the Heronry.

Nutcracker.—Reported by Morris, as having been once killed in Kent during the last century.

Waxwing.—Seen by the author at Higham, and others round the neighbourhood of Rochester, in the winters of 1879—80, and 1880—81 ; an occasional, and very pretty winter visitor.

Greater Spotted Woodpecker.—Cobham, Erith.

Lesser Spotted Woodpecker.—Cobham.

Alpine Swift.—One was seen at Kingsgate, in the Isle of Thanet, according to Morris (1820).

Grey-headed Wagtail.—Dover (1851).

Red-throated Pipit.—Rainham.

Shore Lark.—Two killed at Down, according to Mr. Yarrell.

Snow Bunting.—Rainham.

Cirl Bunting.—Perhaps at Cobham, Ramsgate.

Ortolan Bunting.—Lord Clifton thinks he saw this at Cobham, but appears not quite certain.

Tree Sparrow.—Not uncommon.

Hawfinch.—Not uncommon.

Siskin.—A visitor at Cobham.

Crossbill.—Doddington, Dartford, Cobham, Maidstone.

Parrot Crossbill.—Once observed in a flock at Doddington in 1851.

The Rose-coloured Pastor.—One specimen labelled Kent in the Maidstone museum.

The Dipper.—Two specimens labelled Kent in the Maidstone museum.

White's Thrush.—Cobham.

Ring Ousel.—Cobham, Ramsgate.

Oriole.—According to Morris, found between Dover and Walmer in 1841, and at Elmstone, near Sandwich, in 1849.

Bluebreast.—Found at Margate. Blue-throated warbler, probably the same species, was shot at Dartford in 1881.

17

Blackstart, or Black Redstart.
The Stonechat, the Lesser Whitethroat
and the Garden Warbler fre-
quently occur.
The Whinchat.—Cobham, Cuxton,
Ifield.
The Wheatear.—Several localities.
Reed Warbler.—The marsh-districts.
Aquatic Warbler.—Cliffe (1877), in Dr.
Plomley's collection of Kentish
birds.
Moustached Sedgechat. — Ramsgate
1888.
Thrush Nightingale.—Morris gives
Dartford (1852).
Great Sedge Warbler.—Morris gives
Sittingbourne (1853).
Bonelli's Willow Wren. — Cobham
(1877).
Icterine Warbler.—Recorded by Dr.
Plomley, at Eyhorne, near
Dover.
Grasshopper Warbler.—Several occa-
sions.
Dartford Warbler.—Several occasions.
Firecrest.—Higham, Cobham.
The Quail.—Cobham.
Pallas' Sand Grouse. — Rochester
(1888).
Great Bustard.—Ashford, Romney
(1859).
Little Bustard.—Chatham, according
to Morris (1834) ; Higham.
Pratincole.—There is a stuffed speci-
men in Lord's Clifton's col-
lection, which was the property
formerly of Sivey, a bird stuffer
in Gravesend, who declared it to
have been killed in the county.
The Courser.—Morris gives Wingham
(1793).
Great Plover.—Frequently seen.
Golden Plover.—Several occasions.
Kentish Dottrell.—The name itself be-
speaks its occurrence in the
county.
Ringed Dottrell. — Frequents East
Kent.
Common Dottrell.—Rainham.
Grey Plover.—Lower Medway (1888).
Turnstone.—Rainham (1888).
Sanderling.—Rainham.
Heron.—There are heronries still in
Kent, at Chilham, Cobham and
Penshurst.
Night Heron.—In the Maidstone mu-
seum; it has occurred in the
county.

Buff-backed Heron.—Labelled Kent
in the Maidstone museum.
White Stork.—Sandwich, Romney,
Chalk.
Black Stork.—The Weald.
Bittern. — Cobham (1848) Swans-
combe (1853), Queenborough and
Stoke (1884), and Snodland and
Cooling (1890).
Spoonbill.—Sandwich, Wingham, and
Pegwell Bay (1850).
Ibis. — Swanscombe, according to
Morris.
Curlew.—Hundred of Hoo, often.
Whimbrel. — Medway, and Swale
marshes.
Knot.—Medway, and Swale marshes,
Romney marsh.
Redshank.—Frequently occurs.
Green Sandpiper.—Frequently occurs.
Dunlin or Oxbird.—Frequently occurs.
Common Sandpiper. — Occasionally
seen.
Greenshank.—Swanscombe (1848), seen
by Mr. Green, of Rainham.
Great Snipe.—Dover (1848), Dodding-
ton (1851).
Sabine's Snipe.—Rochester (1824).
Avocet.—Found in Thames marshes.
Bar-tailed Godwit.—Romney marshes
and Erith, Pegwell Bay.
Little Stint.—Erith, Strood.
Temminck's Stint.—Rainham marshes.
Wood Sandpiper. — On the Medway,
Ramsgate.
Curlew Sandpiper.—Sandwich (1836),
Deal (1850), often found at Peg-
well Bay.
Purple Sandpiper.—Several, in differ-
ent parts of Kent.
Spotted Crake. — Cliffe, Rainham.
According to Morris, abundant in
Kent marshes.
Baillon's Crake.—Two taken at Deal
(1850 and 1851).
Water Rail.—One observed between
Nursted and Cobham.
Grey Phalarope.—Once on Thames,
near Swanscombe.
White-fronted Goose.—Frequents the
marshes.
Egyptian Goose.—In 1846, five Egyp-
tian geese were seen in Romney
marsh.
Greylag Goose.—Rainham marshes.
Bewick's Swan.—Lidsing woods.
Polish Swan.—Thirty were seen, and
four shot at Snodland in 1838.

Sheldrake.—Rainham.

Shoveller.—Cobham (1881), Boughton Monchelsea (1885).

Pintail.—Cobham (1881).

The Gadwell.—Yarrell declares to be a native of Kent.

The Gargney.—Morris says is rare in Kent.

The Widgeon.—Several times discovered.

Common Scoter.—On the Medway (1888). Also reported by Mr. Green.

Velvet Scoter.—Two labelled Kent in Maidstone museum. A stuffed specimen, seen by the author, which belonged to a countryman, who captured it when it entered a farmyard at Frindsbury to feed. Mr. Green also reports it.

Tufted Duck.—Reported by Mr. Green of Rainham.

Red-crested Whistling Duck.—Eighteen were seen at Erith in 1853.

The Ferruginous or Red Duck.—Killed at Dover (1849-50).

Pochard.—Four seen at Cobham.

Scaup.—Perhaps seen at Cobham, but it appears to be very uncertain.

Golden-eye. Cobham.

Long-tailed Duck.—This bird, according to Morris, favours Kent.

Red-breasted Merganser.—Frequents the Thames.

Goosander.—Often seen on the Thames.

Hooded Merganser.—At Gravesend (1870) two were shot.

Red-breasted Grebe.—Morris reports as killed in Kent (April 1786), afterwards at Sandwich, and at Stangate Creek on the Medway, (1849).

Great Crested Grebe.—Shot on the Medway (1876), Otterham quay; two in Maidstone museum labelled Kent.

Little Grebe or Dabchick.—Breeds at Chalk, Cobham (1881), Higham (1884), Town Malling (1891), reported by Messrs. Lamb and Green, and Lord Clifton.

Great Northern Diver.—Sheerness (1842).

Black-throated Diver.—Sittingbourne (1840), Sandwich (1842), near Erith (1850), Upchurch (1888).

Red-throated Diver.—Often seen in the estuary of the Thames and Medway.

Guillemot.—Often reported on the coast.

Rotche, or Little Auk.—Favours the coast, Boxley (1879).

Sandwich Tern.—Its name after our Kentish town proclaims it to have been recorded in the county.

Lesser Tern.—Often seen in the marshes of the Hundred of Hoo and Grayne.

Black Tern.—Once plentiful in Romney marsh.

Gull-billed Tern. — Shot in Kent (1839).

Cormorant.—According to Morris, on the Thames.

Gannet.—According to Morris shot at Greenhithe (1847).

The Puffin.—Recorded by Mr. Green.

Razor Bill.—Recorded by Mr. Green in Park Wood, October (1886).

Black-headed Gull.

Common Gull.

Lesser Black-backed Gull.

Great Black-backed Gull.

Herring Gull.

} These birds are our commonest seen gulls. The Black-backed Gull goes far inland.

Little Gull.—Sheerness (1840).

Masked Gull.—Ashford (1853).

Laughing Gull.—Morris records five seen at Winchelsea in Sussex, in 1774. As this is close to the Kentish border the birds may perhaps be recorded as belonging to Kent.

Kittiwake.—Reported by Mr. Green.

Glaucous Gull.—Ramsgate (1846.)

Skua Gull.—Not unfrequently met with.

Richardson's Skua.—Not unfrequently met with.

Pomerine Skua.—Dover (1844).

Cinereous Shearwater.—An occasional visitor.

Manx Shearwater.—Erith.

Stormy Petrel.—Sometimes seen.

I have to record my obligations to Lord Clifton, to Mr. Lamb of Maidstone, and to Mr. Green of Rainham, for much valuable information that enables me to give what I hope is a fairly complete list of the birds that have occurred in this county from time to time

---

REPTILES.

Most of the reptiles that have been found in England have been reported from Kent. We subjoin a list of those that have been seen :—

Sand Lizard.
Common or Viviparous Lizard.
Blindworm.
Ringed Snake.
Common Viper.—I am inclined to think that more than one variety of this poisonous snake is to be found in this county, from what the common people say ; but I, myself, have only seen one variety.
Common Frog.

Common Toad.
Natterjack Toad.—This toad, Bell says, is common on Blackheath and in ditches not far from Deptford. The author believes that a reptile of this species was found by him at Higham. It was reported to him as occurring some years ago at Yalding.
Common Warty Newt.
Common Smooth Newt.

FISH.

As the Medway has been always a great fish river, and as it were difficult to say how far the stray fish ascend or descend a stream, we shall mention the chief fish that frequent our Kentish waters, specially observing those that we happen to have known as the fish of the Medway and its tributary-connected waters.

Perch (Perca fluviatilis).—A very frequent fish in the Medway and Thames.
Bass (Perca labrax).—A fish very frequent round the Kentish coast.
Ruffe (Perca Cernua).—I have caught this fish in several ponds in Kent, and it is found in the Thames.
Great Weaver (Trachinus draco).
Lesser Weaver (Trachinus vipera).
Striped Red Mullet (Mullus surmuletus).
Red Gurnard (Trigla cuculus).
Streaked Gurnard (Trigla lineata).
Grey Gurnard (Trigla Gurnardus).
Miller's Thumb or Bullhead (Cottus gobio).—In the upper waters of the Medway.
Armed Bullhead (Cottus cataphractus).
Father Lasher (Cottus bubalis).

Rough-tailed Three-spined Stickleback (Gasterosteus Trachurus).—Frequent in the various dykes near the Medway and Thames.
Half-armed Stickleback (Gasterosteus semiarmatus).—In the dykes as the last.
Ten-spined Stickleback (Gasterosteus pungitius).
Fifteen-spined Stickleback (Gasterosteus spinachia).
Sea Bream (Pagellus centrodontus).
Black Bream (Sparus lineatus).
Mackerel (Scomber Scombrus).
Tunny (Thynnus vulgaris).—Has been taken at the entrance of the Thames, according to Bell.
Dory (Zeus Doree).
Atherine or Sandsmelt (Atherina presbyter).
Grey Mullet (Mugil cephalus).—The

Swale estuary is frequented by this fish.

Thick-Lipped Grey Mullet (Mugil chelo).—Also found in the estuary.

Spotted Gunnel (Blennius gunnellus). —Mouth of the Thames according to Bell.

Sordid Dragonet (Callionymus Dracunculus).—The fox of the Kentish coast.

Ballan Wrasse (Labrus maculatus).

Common Carp (Cyprinus carpio).—In the Medway I have seen carp caught, when I was young, weighing from six to eight pounds.

Prussian Carp (Cyprinus gibelio).—Thames.

Crucian Carp (Cyprinus carassius).—Thames.

Barbel (Cyprinus barbus).—Thames.

Gudgeon (Cyprinus gobio).—A common species in the Medway.

Tench (Cyprinus tinca).—Many Kentish ponds.

Bream (Cyprinus bramus).—Bream weighing four and five pounds have been taken in the Medway.

Pomeranian Bream (Cyprinus buggenhagii).—As Bell speaks of finding this fish in Dagenham Breach, it may possibly find its way in the twin waters of the Thames and Medway to our Kentish shores.

Dobule Roach (Cyprinus dobula).—Caught by Bell in the Thames below Woolwich.

Roach (Cyprinus Rutilus).—Perhaps our commonest fish.

Dace (Cyprinus leuciscus).—Not a common Medway fish by any means.

Chub (Cyprinus cephalus).—The chub often reaches a large size in the Medway.

Rudd (Cyprinus erythrophthalmus). —Have caught this fish in the Swale marshes.

Bleak (Cyprinus alburnus).—A common Medway fish.

Minnow (Cyprinus phoxinus).—Not uncommon.

Loach (Cobitis barbatula). — In streams.

Pike (Esox lucius).—Heavy fish have been taken in the different rivers and ponds of the county.

Garfish (Esox belone).

Salmon (Salmo salar).—Though both the Medway and Thames are not likely to be salmon rivers, still an occasional wanderer has been found in either stream.

Common Trout (Salmo fario).—The trout of Kentish rivers are generally good; those of the Stour at Fordwich have been rendered famous by the renowned Isaac Walton. A trout was caught in 1884 near Rochester bridge.

Smelt (Salmo eperlanus).—The smelts of the Medway, which are caught from Snodland to Rochester, have long been considered to surpass others in flavour.

Herring (Clupea harengus).

Pilchard (Clupea pilcardus).

Sprat (Clupea sprattus).—Thames and Medway.

Whitebait (Clupea alba).—This, as every one knows, is the epicure's dish from the Thames.

Twaite Shad (Clupea alosa).—Thames.

Alice Shad (Alosa communis).—River Thames.

Anchovy (Clupea eucrasicolus).—Has been found in the Thames.

Haddock (Gadus æglefinus).—Has been caught at Rochester bridge.

Whiting Pout (Gadus luscus).—Thames and Medway mouth.

Whiting (Gadus merlangus).

Hake (Gadus merlucius).

Five-bearded Rockling (Gadus mustela).

Plaice (Pleuronectes platessa).—Rivers Medway and Thames.

Flounder (Pleuronectes plesus).—Thames and Medway.

Dab (Pleuronectes limanda).—Thames and Medway.

Smooth Dab (Pleuronectes levis).

Turbot (Pleuronectes maximus).

Brill (Pleuronectes rhombus).

Muller's Top-knot (Pleuronectes punctatus).—This fish has been taken in the Medway.

Sole (Pleuronectes solea).

Lemon Sole (Solea pegusa).

Bimaculated Sucker (Cyclopterus bimaculatus).—Has been taken on the coast of Kent.

Sharp-nosed Eel (Muræna anguilla), Common Eel.—Common in the Medway.

Broad-nosed Eel (Anguilla latirostris).
—About as common as the other.
Conger Eel (Conger vulgaris).—Medway, Swale, Thames.
Great Pipe Fish (Syngnathus acus).
Short-nosed Hippocampus or Sea-Horse, (Hippocampus brevirostris).
The Sturgeon (Acipenser sturio).—The sturgeon caught in the Thames and the Medway, are considered royal fish, and as such are supposed to belong to the sovereign of right, and are expected to be forwarded by the catcher to the king or queen.
Small Spotted Dog-fish (Squalus canicula).
Common Tope (Squalus galeus).

Smooth Hound (Squalus Mustela).
Picked Dog fish (Squalus acanthias).
Skate (Raia batis).
Thornback (Raia clavata).
Lamprey (Petromyzon mannus).—Has been caught in the Thames and Medway.
Lampern (Petromyzon fluviatilis).—This little fish the author used frequently to catch in the Upper Medway, together with his schoolfellows, simply by watching where there were a number of them, and then taking off boots and stockings, and dipping their feet in the water, when the fish used to adhere to the naked feet and allow themselves thus to be drawn out of the water.

CHAPTER XVI.

PLANTS.

THE author will now conclude this short sketch of the Natural History of Kent by giving the flora. The greater number of plants that are mentioned are found in the district he is describing, and have been identified by himself. Where otherwise, the authority will be given.

Clematis vitalba.—Wild clematis.

Thalictrum flavum.—Meadow rue.

Anemone pulsatilla.—Pasque flower; reported to the author as occurring at Wrotham.

Anemone ranunculoides. Hooker tells us is to be found at Wrotham in Kent.

Adonis autumnalis. Mr. Hepworth of Rochester has kindly added this plant, with several others I shall mention with his name, to my list.

Ranunculus circinatus.—Rigid-leaved water crowfoot.

Ranununculus fluitans.—River crowfoot.

Ranunculus Drouetii.—Drouet's water crowfoot.

Ranunculus peltatus.—Water crowfoot.

Ranunculus hederaceus.—Ivy-leaved crowfoot.

Ranunculus sceleratus.—Celery-leaved crowfoot.

Ranunculus flammula.—Spearwort.

Ranunculus lingua. — Great spearwort.

Ranunculus auricomus.—Goldilocks.

Ranunculus acris.—Upright meadow buttercup.

Ranunculus bulbosus.—Bulbous buttercup.

Ranunculus repens.—Creeping buttercup.

Ranunculus parviflorus.—Small-flowered crowfoot.

Ranunculus arvensis.—Corn buttercup.

Ranunculus ficaria.—Lesser celandine.

Caltha palustris.—Marsh marigold.

Trollius Europaeus.—Globe flower; a plant common in Kentish gardens, but I cannot find that it has been discovered wild in this county.

Helleborus viridis.—Green hellebore. Two localities.

Helleborus foetidus.—Stinking hellebore. Frequent on the chalk.

Eranthis hyemalis.—Winter aconite. Perhaps naturalised in one or two districts, but certainly not wild.

Aquilegia vulgaris.—Wild columbine. This graceful plant is the ornament of many a Kentish wood.

Delphinium consolida.—Field larkspur. The London catalogue gives Ajacis only. I have had the Larkspur forwarded from East Kent. Mr. Hepworth of Rochester has detected it. The Faversham Floral, published many years ago, mentions it, and Hooker also claims it for Kent.

Aconitum napellus.—Monk's hood, common wolf bane. I have seen this plant growing where I had reason to think it a native, but, as it is a very common garden plant, it may have been an escape.

Berberis vulgaris.—Common barberry. Occasional.

Nymphaea alba.—White water lily.

Nuphar lutea.—Yellow water lily.

Papaver somniferum.—Opium poppy.

263

Very common in parts of the county.

Papaver rhoeas.—Common red poppy.

Papaver dubium.—Long smooth-headed poppy.

Papaver argemone. — Long rough-headed poppy.

Papaver hybridum.—Round rough-headed poppy.

Glaucium flavum.—The horned poppy.

Chelidonium majus.—Greater celandine.

Corydalis lutea.—Yellow corydalis: communicated by Mr. Oliver of West Malling.

Corydalis claviculata.—The climbing corydalis.

Fumaria pallidiflora.—Pale fumitory.

Fumaria officinals.—Common fumitory.

Fumaria parviflora.—Least-flowered fumitory. Hooker in his "British Flora," and Babington in his "Manual of British Botany," claim this plant for Wouldham.

Matthiolia incana.—Hoary shrubby stock; communicated by Mr. Hepworth.

Cheiranthus cheiri.—Wallflower.

Nasturtium officinale. — Common watercress.

Nasturtium sylvestre.—Creeping yellow cress.

Barbarea vulgaris.—Bitter winter yellow rocket.

Cardamine amara.—Large-flowered bitter cress.

Cardamine pratensis.—Lady's smock; cuckoo flower.

Cardamine hirsuta.—Hairy bitter cress; communicated by Professor Holmes.

Cardamine flexuosa.—Bending bitter cress; communicated by Professor Holmes.

Cardamine bulbifera.—Bulbous bitter cress.

Alyssum incanum.—White alyssum; communicated by Mr. Oliver.

Alyssum maritimum.—Sea-side alyssum; communicated by Mr. Hepworth.

Erophila vulgaris. This appears to have taken the place of Draba Verna or common whitlow grass in the last edition of the London catalogue; which is, of course, a very common plant.

Cochlearia officinalis. — Common scurvy grass.

Cochlearia Danica.—Danish scurvy grass.

Cochlearia Anglica.—English scurvy grass; communicated by Mr. Hepworth.

Cochlearia armoracia.—Horse radish. In several districts, where, if an escape originally from gardens, it was difficult to trace how it found its way.

Sisymbrium thaliana.—Common thale cress.

Sisymbrium officinale. — Common hedge mustard.

Sisymbrium Sophia.—Flixweed.

Sisymbrium Alliaria.—Common garlick mustard; Jack-by-the-hedge.

Erysimum cheiranthoides. — Worm-seed treacle mustard; in Mr. Lamb's (Maidstone) little pamphlet we find this plant.

Erysimum Orientale. — Hare's-ear treacle mustard; communicated by Professor Holmes.

Camelina sativa.—Common gold of pleasure; communicated by Professor Holmes.

Brassica Oleracea.—Sea cabbage.

Brassica napus.—Rape.

Brassica rutabaga.—Swede.

Brassica rapa.—Turnip.

Brassica nigra.—Common mustard.

Brassica sinapis.—Charlock.

Brassica alba.—White mustard: communicated by Mr. Hepworth.

Diplotaxis tenuifolia.—Wall rocket.

Diplotaxis muralis.—Sand rocket; according to Mr. Lamb of Maidstone.

Capsella bursa pastoris.—Shepherd's purse.

Senebiera coronopus.—Swinecress.

Lepidium ruderale.—Narrow-leaved pepper-wort; contributed by Messrs. Holmes and Hepworth.

Lepidium campestre.—Common Mithridates pepper-wort.

Lepidium Smithii.—Smooth field pepper-wort; according to Mr. Lamb.

Lepidium draba.—Whitlow pepper-wort. According to Mr. Lamb.

Thlaspi arvense.—Penny cress.

Iberis Amaris.—Bitter candytuft; communicated by Mr. Hepworth, who adds "probably an escape."

Cakile maritima.—Purple sea rocket ; Professor Holmes.
Raphanus raphanistrum.—Radish.
Reseda lutea.—Wild mignonette.
Reseda luteola.—Dyers' rocket ; yellow weed.
Helianthemum chamæcistus.—Rock rose.
Viola palustris.—Marsh violet.
Viola odorata.—Sweet violet.
Viola alba.—White violet.
Viola hirta.—Hairy violet.
Viola sylvatica.—Wood violet.
Viola canina.—Dog violet.
Viola tricolor.—Heart's-ease or pansy.
Viola arvensis.—Field viola ; communicated by Professor Holmes.
Polygala vulgaris.—Common milkwort.
Polygala amara.—Bitter polygala ; late Mr. Hanbury.
Frankænia lævis. — Smooth seaheath; communicated by Professor Holmes.
Dianthus armeria.—Deptford pink.
Dianthus deltoides.—Maiden pink.
Dianthus caryophyllus.—Clove pink, clove gilliflower, carnation. On the ruins of Rochester castle, and other castles in the county. From this flower came the name Sweet William. The tradition is that a Scotch baker named William set out on a pilgrimage, but was murdered at Rochester. His shrine in the cathedral was attended by hundreds of pilgrims after he had been canonised. The flower growing near his shrine got the name of St. William, or St. William's flower; hence Sweet William.
Dianthus prolifer.—Proliferous pink ; found at Dover according to Hooker.
Saponaria vaccaria.—Cow-wort ; an importation discovered some years ago near East Malling ; and, according to Hooker, in the Isle of Dogs.
Saponaria officinalis.—Common soapwort.
Silene maritima.—Sea campion; communicated by Professor Holmes.
Silene cucubalus.—Bladder campion.
Silene conica.—Striated corn campion ; Deal, according to the late Mr. Hanbury.

Silene nutans.—Nottingham catchfly ; communicated by Professor Holmes.
Silene Italica. — Italian catchfly ; communicated by Professor Holmes.
Silene noctiflora. — Night-flowering catchfly.
Cucubalus baccifer.—Berried cucubalus, Isle of Dogs ; Holmes, Babington.
Silene armeria.—Common catchfly ; Yalding, according to Hooker.
Lychnis alba.—White lychnis.
Lychnis diurna.—Red lychnis, or red campion.
Lychnis flos-cuculi.—Ragged-robin.
Lychnis githago.—Corn cockle.
Holosteum umbellatum.—Umbelliferous jagged chickweed; communicated by Mr. Hepworth of Rochester.
Cerastium tetrandrum. — Four-cleft mouse-ear chickweed; communicated by Mr. Hepworth.
Cerastium semidecandrum. — Little mouse-ear chickweed.
Cerastium glomeratum.—Broad-leaved mouse-ear chickweed.
Cerastium triviale. — Narrow-leaved mouse-ear chickweed.
Stellaria aquatica.—Water stitchwort.
Stellaria nemorum.—Wood stitchwort ; communicated by Mr. Hepworth.
Stellaria media.—Common chickweed.
Stellaria holostea.—Greater stitchwort.
Stellaria graminea.—Lesser stitchwort.
Stellaria uliginosa.—Bog stitchwort.
Arenaria trinervis. — Three-nerved sandwort.
Arenaria serpyllifolia.—Thyme-leaved sandwort.
Sagina maritima. — Sea pearlwort ; communicated by Mr. Hepworth.
Sagina apetala.—Small-flowered pearlwort ; communicated by Mr. Hepworth.
Sagina procumbens. — Procumbent Pearlwort.
Spergula arvensis.—Corn spurrey.
Lepigonum rubrum.—Field sandwort spurrey.
Lepigonum salinum.—Sea sandwort spurrey.
Tamarix gallica.—The tamarisk.
Hypericum androsæmum.—Tutsan.

Hypericum calycinum. — Large-flowered St. John's-wort ; communicated by Mr. Oliver.
Hypericum perforatum.—Perforate St. John's-wort.
Hypericum quadrangulum.—Imperforate St. John's-wort,
Hypericum quadratum. — Square-stalked St. John's-wort.
Hypericum humifusum.—Trailing St. John's-wort.
Hypericum pulchrum.—Small St. John's-wort.
Hypericum hirsutum.—Hairy St. John's-wort.
Hypericum montanum.—Mountain St. John's-wort ; communicated by Mr. Hepworth.
Althæa officinalis.—Common marsh mallow.
Althæa hirsuta.—Rough althæa.
Malva moschata.—Musk mallow.
Malva sylvestris.—Common mallow.
Malva rotundifolia.—Dwarf mallow.
Malva borealis.—Northern mallow ; a garden weed near Sevenoaks.
Tilia intermedia.—Intermediate lime or linden tree.
Tilia parviflora.—Small flowered lime or linden.
Linum catharticum.—Purging flax.
Linum angustifolium.—Narrow-leaved flax.
Linum usitatissimum.—Common flax ; communicated by Mr. Hepworth.
Geranium phæum.—Dusky crane's-bill; communicated by Mr. Hepworth.
Geranium sylvaticum.—Wood crane's-bill.
Geranium pratense.—Blue meadow crane's-bill.
Geranium pyrenaicum. — Mountain crane's-bill.
Geranium molle.—Dove's-foot crane's-bill.
Geranium pusillum.—Small-flowered crane's-bill.
Geranium dissectum.—Jagged-leaved crane's-bill.
Geranium columbinum.—Long-stalked crane's-bill.
Geranium lucidum.—Shining crane's-bill.
Geranium Robertianum. — Stinking crane's-bill or herb Robert.
Erodium cicutarium.—Hemlock stork's bill.

Erodium moschatum.—Musky stork's bill ; according to Mr. Lamb.
Oxalis acetosella.—Wood sorrel.
Ilex aquifolium.—Holly.
Euonymus europæus.—The spindle tree.
Rhamnus catharticus. — Common buckthorn.
Rhamnus frangula.—Alder buckthorn.
Acer pseudo-platanus.—Sycamore.
Acer campestre.—Maple.
Genista anglica.—Needle green weed.
Genista tinctoria.—Dyers' green weed; communicated by Professor Holmes.
Ulex Europæus.—Common gorse.
Ulex galii.—Dwarf furze ; according to Professor Holmes.
Ulex nanus.—Dwarf furze.
Cytisus scoparius.—Broom.
Ononis repens.—Common rest harrow.
Ononis spinosa.—Spinous rest harrow ; communicated by Mr. Oliver.
Medicago sativa.—Lucerne.
Medicago lupulina.—Black medick.
Medicago maculata.—Spotted medick.
Medicago minima.—Little bur medick ; according to Mr. Hanbury found at New Romney.
Mellilotus altissima.—The melilot.
Mellilotus alba.—White melilot.
Trifolium subterraneum.—Subterranean trefoil.
Trifolium pratense.—Common purple clover.
Trifolium medium.—Zigzag trefoil.
Trifolium maritimum.—Teasel-headed trefoil.
Trifolium incarnatum.—Crimson clover; undoubtedly introduced.
Trifolium arvense.—Hare's-foot trefoil.
Trifolium striatum.—Soft-knotted trefoil.
Trifolium scabrum.—Rough trefoil ; according to Mr. Lamb.
Trifolium glomeratum.—Round-headed trefoil.
Trifolium hybridum.—Hybrid trefoil.
Trifolium repens.—White clover.
Trifolium fragiferum. — Strawberry-headed trefoil.
Trifolium resupinatum.—Reserved trefoil ; according to Mr. Hepworth.
Trifolium procumbens.—Hop trefoil.
Trifolium dubium.—Lesser yellow trefoil.
Trifolium filiforme.—Slender yellow trefoil.

Anthyllis vulneraria.—Lady's fingers or kidney vetch.

Lotus corniculatus.—Common bird's-foot trefoil.

Lotus Tenuis.—Slender bird's-foot trefoil.

Lotus villosus.—Hairy bird's-foot trefoil ; Higham, Kent, according to Mr. Hooker.

Lotus pilosus.—Narrow-leaved bird's-foot trefoil.

Ornithopus perpusillus. — Common bird's-foot.

Astragalus glycyphyllos.—Sweet milkwort ; communicated by Mr. Hepworth.

Hippocrepis comosa. — Horse-shoe vetch.

Onobrychis sativa. — Common sainfoin.

Vicia hirsuta.—Hairy tare.

Vicia tetrasperma.— Slender tare.

Vicia cracca.—Tufted vetch.

Vicia orobus.— Wood bitter vetch ; communicated by Mr. Hepworth.

Vicia sylvatica.—Wood vetch.

Vicia sepium.—Bush vetch.

Vicia sativa.—Common vetch ; communicated by Mr. Hepworth.

Vicia angustifolia. — Narrow-leaved vetch.

Vicia segetalis.—Corn vetch.

Vicia lathyroides.—Spring vetch.

Vicia bithynica.—Rough-podded purple vetch ; communicated by Mr. Hepworth.

Vicia bobartii.—Bobart's vetch.

Lathyrus nissolia.—Crimson or grass vetchling.

Lathyrus hirsutus. — Rough-podded vetchling ; communicated by Professor Holmes.

Lathyrus pratensis.—Meadow vetchling.

Lathyrus tuberosus.—Tuberous bitter vetchling ; communicated by Mr. Hepworth.

Lathyrus sylvestris.—Narrow-leaved vetchling.

Lathyrus macrorhizus.—Tuberous bitter vetchling ; communicated by Professor Holmes.

Prunus communis.—Common sloe.

Prunus insititia.—Common bullace.

Prunus domestica. — Common wild plum.

Prunus avium.—Wild cherry.

Prunus cerasus.—Morello cherry.

Prunus padus.—Bird cherry.

Spiræa ulmaria.—Meadow-sweet.

Spiræa filipendula.—Common dropwort.

Rubus Idæus.—Common raspberry.

Rubus corylifolius. — Hazel-leaved bramble ; communicated by Mr. Hepworth.

Rubus cæsius.—Dewberry.

Rubus carpinifolius. — Hornbeam-leaved bramble.

Geum urbanum.—Common avens.

Geum rivale.—Water avens ; according to Mr. Lamb.

Fragaria vesca.—Wood strawberry.

Fragaria elatior.—Hautboy strawberry.

Potentilla fragriastrum. — Strawberry-leaved cinquefoil.

Potentilla tormentilla.—Tormentil.

Potentilla reptans.—Creeping cinquefoil.

Potentilla anserina. — Silver-weed, goose grass.

Potentilla argentea.—Hoary cinquefoil.

Potentilla comarum. —Marsh cinquefoil.

Alchemilla arvensis.—Field lady's mantle or parsley piert.

Alchemilla vulgaris.—Common lady's mantle.

Agrimonin Eupatoria.—Common agrimony.

Poterium sanguisorba.—Common salad burnet.

Poterium officinale.—Great burnet.

Rosa spinosissima. — Burnet-leaved rose.

Rosa tomentosa.—Downy-leaved rose.

Rosa rubiginosa.—Sweet briar ; common on the chalk hills.

Rosa canina.—Dog rose.

Rosa stylosa systyla. — Close-styled rose.

Rosa arvensis.—Trailing dog rose.

Pyrus torminalis.—Wild service tree.

Pyrus Aria.—White beam tree ; an ornament to our chalk hills.

Pyrus aucuparia.—Mountain ash, or rowan tree.

Pyrus communis.—Wild pear; communicated by Mr. Hepworth.

Pyrus malus.—Wild apple.

Pyrus malus accrba.—Wild codlin.

Pyrus malus mitis.—Wild sweet apple. This latter is, I think, probably an escape.

Pyrus Germanica.—The medlar. Hooker gives Bidborough, Kent, as one of its habitats.

Cratægus oxyacantha. All four species of hawthorn have been detected in Kent by my friend Mr. Hepworth.

Saxifraga tridactylites.—Rue-leaved saxifrage.

Saxifraga granulata.—White meadow saxifrage.

Chrysosplenium oppositifolium.— Common golden saxifrage ; discovered by Mr. Oliver.

Chrysosplenium alternifolium.—Alternate-leaved saxifrage.

Ribes grossularia. Gooseberry ; communicated by Professor Holmes.

Ribes rubrum.—Red currant.

Cotyledon umbilicus.—Wall pennywort.

Sedum rhodiola.—Rose-root stonecrop ; communicated by Messrs. Oliver and Hepworth.

Sedum telephium.—Livelong or orpine.

Sedum album.—White stonecrop ; communicated by Professor Holmes.

Sedum acre.—Biting stonecrop, or wall-pepper.

Sedum sexangulare.—Tasteless yellow stonecrop. Sheppey and Greenwich Park, according to Hooker.

Sedum reflexum.—Crooked yellow stonecrop ; communicated by Professor Holmes.

Sempervivum tectorum.—House Leek.

Drosera rotundifolia.—Round-leaved sundew ; communicated by Mr. Oliver.

Hippuris vulgaris.—Common mare's tail.

Myriophyllum verticillatum.—Whorled water milfoil.

Myriophyllum spicatum. — Spiked water milfoil.

Callitriche vernalis.—Vernal water starwort.

Callitriche stagnalis.—Stagnant water starwort ; communicated by Mr. Hepworth.

Callitriche truncata.—Truncated water starwort ; communicated by Professor Holmes.

Lythrum salicaria.—Spiked purple loose strife.

Peplis portula.— Water purslane.

Epilobium angustifolium.—Rose bay willow herb ; common in gardens but occasionally wild.

Epilobium hirsutum.—Great hairy willow herb.

Epilobium parviflorum.—Small flowered willow herb.

Epilobium montanum.—Broad smooth-leaved willow herb.

Epilobium roseum.— Pale smooth-leaved willow herb.

Epilobium tetragonum.—Square-stalked willow herb.

Epilobium palustre.—Narrow-leaved marsh willow herb.

Œnothera biennis.—Common evening primrose.

Circæa lutetiana.—Enchanter's nightshade.

Bryonia dioica.—Red-berried bryony.

Hydrocotyle vulgaris.—Marsh pennywort.

Eryngium maritimum.—Sea-holly.

Eryngium campestre.—Field eryngo ; communicated by Prof. Holmes.

Sanicula Europæa.—Wood sanicle.

Conium maculatum.—Hemlock.

Smyrnium olusatrum.—Alexanders.

Apium graveolens.—Wild celery.

Apium nodiflorum. — Procumbent marshwort.

Cicuta virosa.—Cowbane or water-hemlock.

Carum segetum.—Corn parsley.

Carum carui.—Corn caraway.

Sison amomum.—Bastard stone parsley.

Sium latifolium.—Broad-leaved water parsnep.

Sium angustifolium.—Narrow-leaved water parsnep.

Œgopodium podagraria.—Gout weed.

Pimpinella saxifraga.—Common burnet saxifrage.

Pimpinella major.—Great burnet saxifrage.

Pimpinella dissecta.—Divided burnet saxifrage ; communicated by Messrs. Hepworth and Holmes.

Conopodium denudatum (Bunium Flexuosum).—Pignut.

Chaerophyllum temulentum. — The chervil.

Scandix Pecten-Veneris.—Shepherd's needle.

Anthriscus vulgaris.—Common beaked parsley.

Anthriscus sylvestris.—Wild beaked parsley.

Fœniculum officinale. — Common fennel.
Crithmum maritimum.—Samphire.
Œnanthe fistulosa.—Common water dropwort.
Œnanthe pimpinelloides. — Callous-fruited water dropwort.
Œnanthe lachenalii.—Parsley water dropwort.
Œnanthe crocata. — Hemlock water dropwort.
Œnanthe phellandrium.—Fine-leaved water dropwort.
Œnanthe fluviatilis.— River water dropwort.
Œthusa cynapium.—Common fool's parsley.
Silaus pratensis.—Pepper saxifrage.
Angelica sylvestris.—Wood angelica.
Peucedanum officinale. — Sea hog's fennel; communicated by Professor Holmes.
Peucedanum sativum.—Common Parsnep.
Heracleum sphondylium.—Hogweed.
Daucus carota.—Wild carrot.
Daucus gummifer. Communicated by Professor Holmes.
Caucalis arvensis.—Spreading hedge parsley.
Caucalis anthriscus.—Upright hedge parsley.
Caucalis nodosa. — Knotted hedge parsley.
Hedera helix.—Ivy.
Cornus sanguinea.—Wild Cornel.
Adoxa moschatellina.—Moschatel.
Sambucus nigra.—Common elder.
Sambucus ebulus.—Dwarf elder.
Viburnum opulus.—Wild guelder-rose.
Viburnum lantana.—Traveller's tree.
Lonicera caprifolium.—Pale perfoliate honeysuckle; communicated by Mr. Hepworth.
Lonicera periclymenum. — Honey-suckle.
Galium cruciatum.—Cross-leaved bed straw.
Galium verum.—Yellow bed straw.
Galium mollugo.—Great hedge bed straw.
Galium saxatile.—Smooth heath bed straw.
Galium palustre.—White water bed straw.
Galium Witheringii.—Withering's bed straw; communicated by Mr. Hepworth.

Galium uliginosum.—Rough marsh bed straw.
Galium aparine.—Goose grass or cleavers.
Asperula odorata.—Woodruff.
Asperula cynanchica. — Squinancy wort.
Sherardia arvensis.—Field madder.
Valeriana dioica. — Small marsh valerian.
Valeriana officinalis. — Great wild valerian.
Centranthus ruber.—Red spur valerian.
Valerianella olitoria.—Lamb's lettuce.
Valerianella auricula.—Sharp-fruited corn salad; according to Mr. Lamb.
Valerianella dentata.—Narrow-fruited corn salad; according to Mr. Lamb.
Dipsacus sylvestris.—Wild teasel.
Dipsacus pilosus.—Small teasel.
Scabiosa succisa.—Devil's-bit scabious.
Scabiosa columbaria.—Small scabious.
Scabiosa arvensis.—Field scabious.
Eupatorium cannabinum. — Hemp agrimony.
Solidago virga aurea.—Golden rod.
Bellis perennis.—Daisy.
Aster tripolium.—Sea aster, or sea star-wort.
Aster linosyris. Communicated by Mr. Hepworth.
Erigeron Canadense.—Canada flea-bane.
Erigeron acre.—Blue fleabane.
Filago Germanica.—Common filago.
Filago minima.—Least filago.
Gnaphalium uliginosum.—Marsh cud-weed.
Gnaphalium sylvaticum.—Wood cud-weed.
Inula conyza.—Ploughman's spike-nard, fleabane.
Inula crithmoides.—Golden samphire; communicated by Mr. Hepworth.
Pulicaria dysenterica.—Common flea-bane.
Pulicaria vulgaris.—Small fleabane; communicated by Mr. Hepworth.
Bidens cernua.—Nodding bur mari-gold.
Bidens tripartita.—Trifid bur mari-gold.
Achillæa millefolium. — Common yarrow.
Achillæa ptarmica.—Sneeze-wort.
Anthemis cotula.—Stinking chamo-mile.

Anthemis arvensis.—Corn chamomile.
Anthemis nobilis.—Common chamomile.
Chrysanthemum segetum.—Corn marigold.
Chrysanthemum leucanthemum.—Oxeye daisy, wild marguerite.
Chrysanthemum parthenium.—Common feverfew.
Matricaria inodora.—Scentless mayweed.
Matricaria chamomilla.—Wild chamomile.
Tanacetum vulgare.—The tansy.
Artemisia absinthium. — Common wormwood.
Artemisia vulgaris.—Common mugwort.
Artemisia maritima.—Sea wormwood.
Tussilago farfara.—Colt's-foot.
Petasites vulgaris.—Butterbur.
Senecio vulgaris.—Common groundsel.
Senecio sylvaticus.—Mountain groundsel.
Senecio viscosus.—Stinking groundsel.
Senecio erucifolius.—Hoary rag-wort.
Senecio Jacobæa. — Common rag-wort.
Senecio aquaticus.—Marsh rag-wort.
Senecio saracenius. — Broad-leaved groundsel; communicated by Mr. Hepworth.
Carlina vulgaris. — Common cotton thistle.
Arctium majus.—Greater burdock.
Arctium minus.—Lesser burdock.
Carduus pycnocephalus.—Many-headed thistle; communicated by Mr. Hepworth.
Carduus nutans.—Musk thistle.
Carduus crispus.—Crisp thistle.
Carduus acanthoides.—Welted thistle.
Cnicus lanceolatus. — Spear plume thistle.
Cnicus eriophorus. — Woolly-headed plume thistle.
Cnicus palustris. — Marsh plume thistle.
Cnicus pratensis.—Meadow plume thistle; communicated by Mr. Hepworth.
Cnicus acaulis.—Dwarf plume thistle.
Cnicus arvensis. — Creeping plume thistle.
Onopordum acanthium. — Common cotton thistle; communicated by Messrs. Hepworth and Holmes.

Silybum marianum.—Milk thistle; communicated by Mr. Hepworth.
Serratula tinctoria.—Common sawwort; according to Mr. Lamb.
Centaurea nigra.—Black knapweed.
Centaurea scabiosa.—Greater knapweed.
Centaurea cyanus.—Corn bluebottle.
Centaurea calcitrapa.—Common star thistle.
Centaurea solstitialis.—Yellow star thistle.
Cichorium intibus.—Wild chicory.
Lapsana communis.—Nipple-wort.
Picris hieracioides.—Hawkweed picris.
Picris echioides.—Bristly ox-tongue.
Crepis taraxacifolia. — Dandelion hawk's beard.
Crepis setosa.—Bristly hawk's beard; communicated by Professor Holmes.
Crepis virens.—Smooth hawk's beard.
Crepis biennis.—Rough hawk's beard.
Crepis hieracioides.—Hawkweed-like hawk's beard; communicated by Professor Holmes.
Crepis paludosa. — Marsh hawk's beard.
Hieracium pilosella.—Common mouseear hawkweed.
Hieracium murorum.—Wall hawkweed; communicated by Mr. Oliver.
Hieracium vulgatum.—Wood hawkweed.
Hieracium umbellatum.—Narrowleaved hawkweed.
Hieracium boreale.—Shrubby broadleaved hawkweed; communicated by Professor Holmes.
Hypochæris radicata.—Long-rooted cat's ear.
Leontodon hirtus.—Hairy thrincia.
Leontodon hispidus.—Rough hawkbit.
Leontodon autumnalis.—Autumnal hawkbit.
Taraxacum officinale.—Common dandelion.
Taraxacum erythrospermum.—Redseeded dandelion; communicated by Professor Holmes.
Lactuca virosa.—Strong-scented lettuce.
Lactuca muralis.—Ivy-leaved lettuce.
Sonchus oleraceus.—Common annual sow thistle.
Sonchus asper.—Sharp-fringed annual sow thistle.

Sonchus arvenis.—Corn sow thistle.
Sonchus palustris.—Marsh sow thistle.
Tragopogon pratensis minus.—Lesser
yellow goat's-beard.
Jasione montana.—Annual sheep's-bit;
according to Mr. Lamb.
Campanula glomerata.—Clustered bell
flower.
Campanula trachelium.—Nettle-leaved
bell flower.
Campanula latifolia.—Giant bell
flower; communicated by Mr.
Hepworth.
Campanula rapunculoides.—Creeping
bell flower; communicated by
Professor Holmes.
Campanula rotundifolia.—Harebell.
Specularia Hybrida.—Corn bell flower;
according to Mr. Lamb.
Vaccinium myrtilus.—Bilberry.
Arbutus unedo.—Strawberry tree or
Arbutus. Though common in some
parts of Ireland as a wild tree,
this shrub has never been acknowledged as an English native.
In Kent, though only found in
gardens and shrubberies, it, nevertheless, with two or three other
trees (the evergreen or holm oak,
the deodara, the Chilian or Arau-
canian pine, the cedar of Lebanon,
and others), flourishes as if this
were its native home. Perhaps it
is merely reintroduced into what
was once its original habitat.
Calluna vulgaris.—Common ling.
Erica tetralix.—Cross-leaved heath.
Erica cinerea.—Common heath.
Pyrola rotundifolia.—Round-leaved
winter green; communicated by
the late Rev. A. J. Woodhouse,
vicar of Ide Hill, near Sevenoaks.
Hypopithys multiflora.—The bird's
nest.
Statice limonium.—Sea lavender.
Statice auriculæfolia occidentalis.—
Auricula-leaved thrift; commu-
nicated by Professor Holmes.
Statice rariflora.—Remote-flowered
sea-lavender; according to Hooker.
Armeria maritima.—Common thrift or
sea gilliflower.
Hottonia palustris.—Water violet.
Primula vulgaris.—Common primrose.
Primula acaulis.—Stalkless primrose.
Primula caulescens.—Stalked primrose.
Primula veris.—Cowslip.
Primula elatior.—Oxlip.

Primula hybrida.—Hybrid primrose;
the hybrid primroses between
veris and vulgaris, veris and
elatior, vulgaris and elatior, have
all been found by the author in
the woods around Cobham.
Cyclamen hederæfolium.—Sow bread;
both Hooker and Babington report
this plant at Sandhurst and Goudhurst, and Lamb mentions it as
a plant growing round Maidstone
in 1839.
Lysimachia vulgaris.—Great yellow
loose strife.
Lysimachia nummularia.—Moneywort
or creeping Jenny.
Lysimachia nemorum.—Wood loose
strife.
Glaux maritima.—Sea milkwort.
Anagallis arvenis.—Scarlet pim-
pernel.
Anagallis cærulea.—Blue pimpernel.
Anagallis tenella.—Bog pimpernel;
according to Mr. Lamb.
Samolus valerandi.—Brookweed.
Fraxinus excelsior.—Ash.
Ligustrum vulgare.—Privet.
Vinca major.—Great periwinkle.
Vinca minor.—Lesser periwinkle.
Blackstonia perfoliata.—Perfoliate
yellow-wort.
Erythræa centaurium.—Common cen-
taury.
Gentiana amarella.—Small-flowered
gentian.
Gentiana campestris.—Field gentian.
Menyanthes trifoliata.—Buckbean.
Polemonium cæruleum.—Jacob's lad-
der; this plant has been found
once or twice in Kent, but it seems
doubtful whether it has a claim
to be considered a native.
Cynoglossum officinale.—Common
hound's tongue.
Cynoglossum montanum.—Green-leav-
ed hound's tongue.
Symphytum officinale.—Comfrey.
Borago officinalis.—Borage; probably
an escape.
Anchusa officinalis.—Common al-
kanet; probably an escape.
Lycopsis arvensis.—Small bugloss.
Myosotis cæspitosa.—Tufted water
scorpion grass.
Myosotis palustris.—Water scorpion
grass or forget-me-not.
Myosotis repens.—Creeping water-
scorpion grass.

Myosotis sylvatica.—Upright wood scorpion grass.
Myosotis arvensis.—Field scorpion grass.
Myosotis collina.—Early field scorpion grass.
Myosotis versicolor.—Yellow and blue scorpion grass.
Lithospermum purpuro-cæruleum.—Purple gromwell ; according to Hooker found at Darenthwood and Greenhithe, where the author has failed to detect it. He was, however, assured by Mr. Carrington-Ley that he had found it near Halling.
Lithospermum officinale.—Common gromwell ; according to Mr. Lamb.
Lithospermum arvense.—Corn gromwell.
Echium vulgare.—Common viper's bugloss.
Calystegia sepium.—Greater bindweed.
Convolvulus arvensis.—Small bindweed.
Cuscuta epithymum.—Lesser dodder.
Cuscuta trifolii.—Clover dodder.
Solanum dulcamara.—Woody nightshade or bitter sweet.
Solanum marinum.—Sea nightshade.
Solanum nigrum.—Common nightshade.
Atropa belladonna.—Deadly nightshade or dwale.
Datura strymonium.—Common thorn apple.
Hyoscyamus niger.—Common henbane.
Verbascum thapsus.—Great mullein.
Verbascum pulverulentum.—Yellow hoary mullein.
Verbascum lychnitis.—White mullein.
Verbascum nigrum.—Dark mullein.
Verbascum blattaria.—Moth mullein ; according to Hooker.
Verbascum virgatum.—Large-flowered primrose mullein ; according to Hooker.
Verbascum hybridum.—Hybrid mullein. Of the hybrid mulleins, that between thapsus and lychnitis and that between nigrum and pulverulentum have been found by the author.
Linaria cymbalaria.—Ivy-leaved toad flax.

Linaria elatine.—Sharp-pointed toad flax.
Linaria spuria.—Round-leaved toad flax ; according to Mr. Lamb.
Linaria repens.—Creeping toad-flax.
Linaria vulgaris.—Common toad-flax.
Linaria viscida.—Lesser toad flax.
Linaria origanifolia. — Marjoram-leaved toad flax. This plant, a native of Spain, was found growing luxuriantly on the walls of the old Abbey at Malling. It was identified in London by Professor Holmes of the Pharmaceutical Society: it is a native of Spain, and is found growing nowhere else, we believe, in England, except on Wells cathedral.
Antirrhinum majus.—Greater snapdragon.
Antirrhinum orontium.—Lesser snapdragon.
Scrophularia aquatica.—Water figwort.
Scrophularia nodosa.—Knotted figwort.
Mimulus luteus.—Yellow mimulus ; found by the author in one place, and reported to him by some ladies as growing in another ; probably an escape.
Digitalis purpurea.—Foxglove.
Veronica hederæfolia.—Ivy-leaved speedwell.
Veronica polita.—Gray field speedwell.
Veronica agrestis.—Green procumbent speedwell.
Veronica Buxbaumii.—Buxbaum's speedwell. I conclude this is "Persica" of the London catalogue.
Veronica arvensis.—Wall speedwell.
Veronica serpyllifolia.—Thyme-leaved speedwell.
Veronica officinalis.—Common speedwell.
Veronica chamædrys.—Germander speedwell.
Veronica montana.—Mountain speedwell ; according to Mr. Lamb.
Veronica scutellata.—Marsh Speedwell.
Veronica anagallis.—Water speedwell.
Veronica beccabunga.—Brooklime.
Euphrasia officinalis.—Eyebright.
Bartsia odontites.—Red bartsia.
Pedicularis palustris.—March louse wort.

Pedicularis sylvatica.—Pasture louse wort.
Melampyrum pratense.—Cow-wheat.
Rhinanthus cristagalli.—Yellow rattle.
Orobanche major.—Greater broom rape.
Orobanche caryophyllacea.—Clove-scented broom rape; communicated by Professor Holmes.
Orobanche elatior.—Tall broom rape.
Orobanche picridis.—Picris broom rape; communicated by Professor Holmes.
Orobanche minor.—Lesser broom rape.
Orobanche flavescens.—Yellow broom rape.
Orobanche amethystea.—Bluish broom rape; found by Mr. Hepworth.
Utricularia vulgaris.—Greater bladder-wort.
Verbena officinalis.—Common vervain.
Mentha sylvestris.—Horse mint.
Mentha hirsuta.—Water capitate mint.
Mentha sativa.—Marsh whorled mint; communicated by Mr. Oliver.
Mentha arvensis.—Corn mint.
Mentha pulegium.—Penny royal; communicated by Professor Holmes.
Lycopus Europæus.—Gypsy wort.
Origanum vulgare.—Sweet marjoram.
Thymus serpyllum.—Wild thyme.
Calamintha clinopodium.—Wild basil.
Calamintha arvensis.—Common basil thyme.
Calamintha officinalis.—Wood cala-minth.
Melissa officinalis.—Balm; found by Dr. Morton and Mr. Hepworth.
Salvia verbenaca.—Wild sage or clary.
Salvia pratensis.—Meadow sage.
Nepeta cataria.—Cat mint.
Nepeta glechoma.—Ground ivy.
Scutellaria galericulata.—Common skull cap.
Scutellaria minor.—Lesser skull cap.
Prunella vulgaris.—Self-heal.
Marrubium vulgare.—Common white horehound.
Stachys betonica.—Wood betony.
Stachys palustris.—Marsh woundwort.
Stachys sylvatica.—Hedge wound-wort.
Stachys arvensis.—Corn woundwort.
Stachys annua.—Annual woundwort; communicated by Professor Holmes.
Galeopsis ladanum.—Red hemp nettle.

Galeopsis speciosa.—Showy hemp nettle.
Galeopsis tetrahit.—Common hemp nettle.
Lamium amplexicaule.—Henbit dead nettle.
Lamium intermedium.—Intermediate dead nettle; communicated by Professor Holmes.
Lamium purpureum.—Red dead nettle.
Lamium maculatum. — Variegated dead nettle.
Lamium album.—White dead nettle.
Lamium galeobdolon.—Yellow weasel snout.
Ballota negra.—Black horehound.
Teucrium scorodonia.—Wood german-der or wood sage.
Ajuga reptans.—The bugle.
Ajuga chamæpitys.—Ground pine.
Plantago major.—Greater plantain.
Plantago media.—Hairy plantain.
Plantago lanceolata.—Ribwort plan-tain.
Plantago maritima.—Sea-side plan-tain.
Plantago coronopus.—Buck's horn plantain.
Scleranthus annuus.—Annual knawel.
Chenopodium vulvaria. — Stinking goosefoot; communicated by Mr. Hepworth.
Chenopodium album.—White goose-foot.
Chenopodium album paganum. Com-municated by Mr. Holmes.
Chenopodium urbicum. — Upright goosefoot.
Chenopodium rubrum.—Red goose-foot.
Chenopodium bonus Henricus.—Good king Henry.
Beta maritima.—Sea beet.
Atriplex littoralis.—Grass-leaved sea orache; communicated by Mr. Hepworth.
Atriplex patula angustifolia.—Halbert-leaved spreading orache; com-municated by Professor Holmes.
Atriplex deltoidea. — Triangular-leaved orache; communicated by Professor Holmes.
Atriplex portulacoides.—Sea purslane.
Atriplex pedunculata.—Stalked sea orache; communicated by Profes-sor Holmes.
Salicornia herbacea.—Jointed glass-wort.

18

Suæda maritima.—Annual sea blite ; communicated by Mr. Hepworth.
Salsola kali.—Prickly saltwort.
Polygonum convolvulus. — Climbing bistort.
Polygonum dumetorum.—Copse bistort ; communicated by Professor Holmes.
Polygonum aviculare. — Common knapweed.
Polygonum hydropiper.—Biting persicaria.
Polygonum persicaria.—Spotted persicaria.
Polygonum lapathifolium. — Pale-flowered persicaria.
Polygonum amphibium.—Amphibious persicaria.
Polygonum bistorta.—Snakeweed.
Rumex conglomeratus.—Sharp dock.
Rumex obtusifolius. — Broad-leaved dock.
Rumex crispus.—Curled dock.
Rumex hydrolapathum.—Great water dock.
Rumex acetosa.—Common sorrel.
Rumex acetosella.—Sheep's sorrel.
Aristolochia clematitis.—Birthwort. My father, the late Dr. Fielding, found this on the walls of Allington castle ; but I have not been able to identify it, either by friends or personally, in Kent.
Daphne mezereum.—Common mezereon ; according to Mr. Lamb.
Daphne laureola.—Spurge laurel.
Hippophæ rhamnoides.—Sea buckthorn ; communicated by Professor Holmes and Mr. Hepworth.
Viscum album.—Mistletoe.
Euphorbia peplis.— Purple spurge.
Euphorbia helioscopia.—Sun spurge.
Euphorbia amygdaloides. — Wood spurge.
Euphorbia cyparissias. — Cypress spurge ; discovered by Mr. Oliver.
Euphorbia peplus.—Petty spurge.
Euphorbia exigua.—Dwarf spurge.
Buxus sempervirens.—Box.
Mercurialis perennis.—Dog's mercury.
Mercurialis annua.—Annual mercury.
Ulmus montana.—Common elm.
Ulmus campestris.—Wych elm.
Humulus lupulus.—The hop.
Urtica dioica.—Great nettle.
Urtica urens.—Small nettle.
Parietaria officinalis.—Common pellitory of the wall.

Betula alba.—Common birch.
Alnus glutinosa.—Alder.
Carpinus betulus.—Hornbeam.
Corylus avellana.—Hazel.
Quercus rober.—Common British oak.
Quercus rober pedunculata.
Quercus rober sessiliflora.
Castanea sativa.—Chestnut.
Fagus sylvatica.—Beech.
Salix alba.—White willow.
Salix viminalis.—Common osier.
Salix cinerea.—Grey sallow.
Salix caprea.—Great round-leaved sallow.
Populus alba.—Great white poplar.
Populus canescens.—Grey poplar.
Populus tremula.—The aspen.
Populus nigra.—Black poplar.
Ceratophyllum demersum.—Common hornwort.
Ceratophyllum submersum. — Unarmed hornwort.
Juniperus communis.—Common juniper.
Juniperus nana.—Dwarf juniper.
Taxus baccata.—The yew.
Pinus sylvestris.—Scotch fir or pine.
Hydrocharis morsus-ranæ.—Frog-bit.
Neottia nidus-avis.—Bird's-nest orchis.
Listera ovata.—Twayblade.
Spiranthes autumnalis.—Ladies' tresses.
Cephalanthera rubra.—Purple helleborine.
Cephalanthera ensifolia. — Narrow-leaved helleborine.
Cephalanthera pallens.—Pale helleborine.
Epipactis latifolia. — Broad-leaved helleborine.
Epipactis palustris.—Marsh helleborine.
Orchis hircina.—Lizard orchis.
Orchis pyramidalis.—Pyramidal orchis.
Orchis ustulata.—Dwarf orchis.
Orchis purpurea.—Dark orchis.
Orchis militaris.—Military orchis.
Orchis simia.—Monkey orchis ; my father, the late G. H. Fielding, Esq., M.D., found this near Dartford in 1848.
Orchis morio.—Fool's orchis.
Orchis mascula.—Early purple orchis.
Orchis incarnata.—Crimson orchis.
Orchis latifolia.—Marsh orchis ; reported to me, but, though I have found it elsewhere, I never discovered it in Kent.

Orchis maculata.—Spotted orchis.
Aceras anthropophera.—Man orchis.
Ophrys apifera.—Bee orchis.
Ophrys aranifera.—Early spider orchis.
Ophrys arachnites.—Late spider orchis; my father had specimens of this forwarded from near Walmer.
Ophrys fucifera.—Drone orchis; my father had specimens of this.
Ophrys muscifera.—Fly orchis.
Herminium monorchis.—Musk orchis; I have had this shown me by several friends.
Habenaria conopsea.—Gadfly orchis.
Habenaria albida.—Whitish orchis.
Habenaria viridis.—Frog orchis.
Habenaria bifolia.—Butterfly orchis.
Habenaria chloroleuca.—Great butterfly orchis.
Iris fœtidissima.—Stinking iris.
Iris pseudacorus.—Common iris.
Narcissus pseudo narcissus.—Daffodil.
Narcissus poeticus.—The poet's narcissus; according to Hooker.
Narcissus biflorus.—Pale narcissus; according to Hooker. A friend declared he had found this on a common, near Rochester, but though I know the place I could never discover it.
Galanthus nivalis.—Snowdrop.
Tamus communis.—Black bryony.
Ruscus aculeatus.—Butcher's broom.
Convallaria majalis.—Lily of the valley.
Allium oleraceum.—Streaked field garlick.
Allium ursinum.—Broad-leaved garlick.
Muscari racemosum.—Starch grape hyacinth; communicated by Professor Holmes.
Scilla nutans.—Wild hyacinth.
Lilium Martagon.—Martagon lily; found by the late Dr. Morton and Mr. Hepworth.
Fritillaria meleagris.—Fritillary.
Tulipa sylvestris.—Wild tulip.
Colchicum autumnale.—Meadow saffron; communicated by Mr. Hepworth.
Narthecium ossifragum.—Bog asphodel; communicated by Mr. Oliver.
Paris quadrifolia.—Herb Paris.
Juncus bufonius.—Toad rush.

Juncus glaucus.—Hard rush; communicated by Mr. Hepworth.
Juncus conglomeratus. — Common rush.
Juncus maritimus.—Lesser sharp sea rush; communicated by Mr. Hepworth.
Juncus lamprocarpus.—Shining-fruited jointed rush.
Juncus acutiflorus. — Sharp-flowered rush.
Luzula forsteri.—Narrow-leaved hairy wood rush.
Luzula pilosa.—Broad-leaved hairy wood rush.
Luzula maxima.—Great hairy wood rush.
Luzula campestris.—Field wood rush.
Luzula multiflora. — Many-flowered wood rush.
Typha latifolia.—Great reed mace or bull rush.
Typha angustifolia.—Lesser reed mace or cat's tail; communicated by Messrs. Hepworth and Holmes.
Sparganium ramosum; branching bur reed.
Sparganium simplex.—Unbranched upright bur reed.
Arum maculatum.—Common arum, lords and ladies.
Lemna trisulca.—Ivy-leaved duckweed.
Lemna minor.—Lesser duckweed.
Lemna gibba.—Gibbous duckweed.
Alisma plantago.—Great water plantain.
Sagittaria sagittifolia.—Arrowhead.
Butomus umbellatus. — Flowering rush.
Triglochin palustre. — Marsh arrow grass.
Triglochin maritimum.—Seaside arrow grass.
Potamogeton natans.—Sharp-pointed broad-leaved pond weed.
Potamogeton rufescens. — Reddish pond weed; communicated by Mr. Hepworth.
Potamogeton perfoliatus.—Perfoliate pond weed; communicated by Mr. Hepworth.
Potamogeton crispus.—Curly pond weed; communicated by Mr. Hepworth.
Potamogeton densus. — Opposite-leaved pond weed.

Potamogeton pectinatus. — Fennel-leaved pond weed.
Zannichellia palustris. — Common horned pond weed.
Zostera marina.—Broad-leaved grass wrack.
Zostera nana. — Dwarf grass; this plant, according to Hooker, is found on Dover Beach.
Scirpus cæspitosus. — Scaly-stalked club rush.
Scirpus maritimus.—Salt marsh club rush.
Scirpus sylvaticus.—Wood club rush; communicated by Mr. Hepworth.
Carex paniculata.—Great panicled sedge ; communicated by Professor Holmes.
Carex vulpina.—Great sedge ; communicated by Professor Holmes.
Carex muricata.—Greater prickly sedge; communicated by Professor Holmes.
Carex divulsa.—Grey sedge ; communicated by Professor Holmes.
Carex elongata.— Elongated sedge ; communicated by Professor Holmes.
Carex ovalis.—Oval spiked sedge ; communicated by Mr. Hepworth.
Carex Goodenowii.—Common sedge.
Carex glauca.—Glaucous heath sedge ; communicated by Messrs. Holmes and Hepworth.
Carex præcox.—Vernal sedge.
Carex pendula.—Great pendulous sedge ; communicated by Messrs. Holmes and Hepworth.
Carex sylvatica.—Pendulous wood sedge ; communicated by Messrs. Holmes and Hepworth.
Carex distans.—Loose sedge.
Carex hirta.—Hairy sedge ; communicated by Professor Holmes.
Carex pseudo-cyperus.—Cyperus-like sedge ; communicated by Mr. Hepworth.
Carex paludosa.—Lesser common sedge ; communicated by Mr. Hepworth.
Carex riparia.—Greater common sedge ; communicated by Mr. Hepworth.
Carex depauperata.—Starved wood sedge ; according to Hooker.
Setaria glauca.—Glaucous bristle grass ; communicated by Professor Holmes.

Phalaris canariensis.—Canary grass.
Phalaris arundinacea.—Reed canary grass.
Anthoxanthum odoratum.—Sweet-scented vernal grass.
Alopecurus agrestis.—Slender foxtail.
Alopecurus geniculatus.—Floating fox tail.
Alopecurus bulbosus.—Tuberous fox-tail
Alopecurus pratensis.—Meadow fox-tail.
Millium effusum.—Spreading millet grass.
Phleum pratense.—Timothy grass.
Agrostis alba.—Marsh bent grass.
Agrostis vulgaris.—Fine bent grass.
Calamagrostis epigeion.—Wood small reed.
Gastridium lendigerum.—Awned nit grass.
Aira caryophyllea.—Silver hair grass.
Aira præcox.—Early hair grass.
Deschampia cæspitosa.—Tufted hair-grass.
Holcus mollis.—Creeping soft grass.
Holcus lanatus.—Meadow soft grass.
Trisetum flavescens.—Yellow oat.
Avena pubescens.—Downy oat.
Avena pratensis.—Perennial oat.
Avena fatua.—Wild oat.
Phragmites communis.—Common reed.
Kœleria cristata.—Crested kœleria : communicated by Professor Holmes.
Catabrosia aquatica.—Water whorl grass ; communicated by Messrs. Hepworth and Holmes.
Melica uniflora.—Wood melic grass.
Dactylis glomerata. — Cock's-foot grass ; communicated by Messrs. Hepworth and Holmes.
Briza media.—Common trembling or quaking grass.
Poa annua.—Annual meadow grass.
Poa nemoralis.—Wood meadow grass.
Poa pratensis.—Smooth-stalked meadow grass.
Poa trivialis.—Roughish meadow grass.
Glyceria fluitans.—Floating meadow grass.
Glyceria aquatica.—Reed meadow grass ; communicated by Professor Holmes.
Glyceria maritima. — Creeping sea meadow grass.

Festuca rigida.—Stiff fescue ; communicated by Messrs. Hepworth and Holmes.

Festuca loliacea.—Darnel fescue ; communicated by Messrs. Hepworth and Holmes.

Festuca sciuroides.—Barren fescue.

Festuca ovina.—Sheep's fescue.

Festuca ovina capillata. — Headed sheep's fescue.

Festuca rubra arenaria.—Red sand fescue.

Festuca elatior pratensis. — Tall meadow fescue ; communicated by Professor Holmes.

Bromus asper.—Hairy wood brome grass.

Bromus erectus.—Upright brome grass.

Bromus sterilis.—Barren brome grass.

Bromus commutatus.—Tumid brome grass.

Bromus mollis.—Soft brome grass.

Bromus arvensis.—Taper brome grass ; communicated by Messrs. Hepworth and Holmes.

Brachypodium sylvaticum. — Slender false brome grass.

Lolium perenne.—Beardless rye grass.

Lolium perenne Italicum. — Italian beardless rye grass.

Lolium temulentum.—Darnel.

Agropyron caninum.—Fibrous-rooted wheat grass.

Agropyron repens.—Couch grass.

Agropyron junceum.—Rush sea wheat grass.

Lepturus filiformis.—Sea hard grass.

Hordeum pratense.—Meadow barley.

Hordeum murinum.—Wall barley.

Hordeum maritimum.—Sea-side barley.

Pteris aquilina.—Common bracken.

Lomaria spicant.—Hard fern.

Asplenium fontanum.—Smooth rock spleenwort.

Asplenium adiantum nigrum.—Black-stalked spleenwort.

Asplenium trichomanes. — Common wall spleenwort.

Asplenium ruta muraria.—Wall rue.

Athyrium-filix fœmina.—Lady fern.

Scolopendrium vulgare. — Hart's tongue fern.

Polystichum aculeatum.—Soft prickly shield fern.

Polystichum angulare. — Angular-leaved shield fern.

Lastræa thelypteris. — Marsh shield fern.

Lastræa oreopteris.—Mountain shield fern.

Lastræa filix mas.—Male shield fern.

Lastræa dilatata.—Toothed fern.

Polypodium vulgare.—Common polypody.

Ophioglossum vulgatum. — Adder's tongue.

Equisetum maximum.—Greatest horse tail.

Equisetum arvense.—Field horse tail.

Equisetum sylvaticum.—Wood horse tail.

Equisetum palustre.—Marsh horse tail.

Equisetum limosum.—Smooth horse tail.

Lycopodium clavatum.—Common club moss ; communicated by Professor Holmes.

Chara vulgaris.—Common chara ; communicated by Professor Holmes.

Nitella opaca.—Dark nitella ; communicated by Professor Holmes.

I cannot but here express my gratitude to those through whom I am enabled to give this information on Natural History. As regards birds, Lord Clifton, Mr. Green of Rainham, Mr. Lamb of Maidstone, and the late Francis Plomley, Esq., M.D., of Maidstone, have greatly assisted me. In the matter of the flora I have to thank my two first teachers, my own father, the late George Hunsley Fielding, Esq., M.D., and the late Reverend Alfred Joseph Woodhouse, M.A., Vicar of Ide Hill, near Sevenoaks, besides the late Doctor Henry Morton, of Brompton, Chatham, Professor Holmes of the Pharmaceutical Society,* Mr. Hepworth of Rochester, and Mr. Oliver of Town Malling, and several of the members of the

* Of Great Britain.

Rochester Naturalist's Club. I shall now proceed to give a few Kentish expressions for animals, birds, flowers, etc. :—

Anthony pig.—The youngest of the litter.
Aps.—An aspen tree.
August bug, May bug, June bug, July bug. — Used indiscriminately of annoying insects according to their time of appearance.
Ball squab.—A just hatched or featherless young bird.
Bargoose.—The sheldrake.
Bearbine or bearbind.—Wild convolvulus.
Blue shutters.—Jelly fish.
Boar cat.—Tomcat.
Bread and cheese.—The young shoots of the hawthorn.
Bull huss.—Dog fish.
Bull rout.—The goby.
Bunting.—A shrimp.
Cadlock, challock.—Charlock.
Capons.—Red herrings.
Cecksies.—A piece of elder hollowed out to catch earwigs.
Cheese bug, pea bug, slater.—The wood louse.
Cove keys.—Cowslips.
Crack nuts.—Hazel or Spanish nuts.
Cuckoo bread.—Wood-sorrel.
Cuckoo corn.—Spring corn.
Cuckoo's eyes.—Speedwell.
Cuckoo flower.—Lady's smock, or wild arum.
Cuckoo's mate.—Wryneck.
Cuckoo.—A fool.
Culver key.—A cowslip.
Devil's thread, and devil's root, and hell root.—The dodder and the broom rape.
Dicky hedge poker, and Jimmy hedge moper.—A hedge sparrow.
Didapper.—The dabchick.
Dishwasher, generally with Peggy.—The wagtail.
Dog daisy.—The oxeye.
Droke.—Duckweed.
Dumbledore.—The dor beetle.
Ess.—A large worm.
Eyelebourne or nailbourn.—An intermittent spring.
Fat Hen.—Good King Henry.
Flitty mouse or flinty mouse.—The bat.—The "y" is frequently added by Kentish people to all words.
Folkestone beef.—Dried dog fish.
French May.—Lilac.

Gads.—Rushes.
Galls.—Jelly fish.
Gaskin.—A wild plum.
Gatteridge tree.—Spindle tree.
Gazels.—Black currants.
Gol.—A gosling.
Golding or fly golding.—A lady-bird.
Granada.—A golden pippin.
Grandmother's nightcap.—The aconite.
Greybird.—Thrush.
Gut weed.—Sow thistle.
Heaver.—A crab.
Hen and chickens.—Ivy-leaved toad flax.
Hoplog.—A caterpillar.
Horsebuckle.—A cowslip.
Horsenails.—Tadpoles.
Keyn.—Weasel.
Kite legs.—Orchis mascula.
Kitty come down the lane and kiss me.—Wild arum.
Kitty run the streets.—Wild pansy.
Kitty Herne, or Kate Hern.—Heron.
Ladies' fingers.—Orchis mascula.
Lady keys.—Cowslips.
Longtails.—Kentish wild oats.
Man sucker.—A cuttle fish.
Mazzard.—The wild plum.
Measuring bug.—Caterpillar.
Milk jugs.—Greater stitchwort.
Miller's eyes.—Jelly fish.
Mollie.—The hedge sparrow.
Moses.—A young frog.
Mother of thousands.—Linaria Cymbalaria ; ivy-leafed toad flax.
Nimble Dick.—A horsefly.
None so pretty.—London pride.
Old man.— Southernwood.
Old woman's orchis.—Orchis fusca.
Our Saviour's flannel.—The great mullein.
Oxbird.—The dunlin.
Poot bird.—Spotted flycatcher.
Pretty Betsy.—Spur valerian.
Punjer.—A large crab.
Puttice.—A stoat.
Quicken.—The mountain ash.
Rabbits' mouths, or bunnies' mouths.—Snapdragon.
Ragged Jack.—The ragged robin.
Red butchers.—Orchis mascula.
Red petticoat.—The wild poppy.
Ruddock.—The redbreast.
Runnet.—The yellow bed straw.
Scags or skegs.—Wild plums.

Screech owl.—The common swift.
Sea cob.—The sea gull.
Sea grapes.—Eggs of cuttle fish.
Sea Kitty.—Sea gull.
Sea nettles.—Jelly fish.
Sea snail.—The periwinkle.
Sea starch.—Jelly fish.
Shiny bug.—Glow-worm.
Shorn bug.—Stag beetle.
Shotover.—Mackerel.
Simson.—Common groundsel.
Skegs.—Wild plums.

Snags.—Hard insects.
Snodgog.—Gooseberry.
Targrass.—Wild vetch.
Timnail.—Vegetable marrow.
Tuke.—The redshank.
Uncle owl.—The skate.
Wax dolls.—The wild fumitory.
Winder.—Widgeon.
Wireweed or grass.—Knapweed.
Yellow bottle, blue bottle, or red
   bottle.—For certain flowers of this
   colour.

CHAPTER XVII.

CHARITIES.

BESIDES the usual doles of money, bread, coals, etc., which are found in most English parishes, certain parishes of the district we are considering are specially benefited.

Ditton was, as we have already stated, one of the manors of Sir Thomas Pope, who, in 1554, founded Trinity College, Oxford. He ordered that all persons, who were natives of the parish over which his manors claimed, should be considered as having preference before other students for the emoluments of his foundation.

Tonbridge School, founded by Sir Andrew Judde in 1553, was largely endowed by Sir Thomas Smythe and other benefactors. Amongst the exhibitions of the school five are offered annually in July, of £90, £80, £70, £60, and £21, each tenable for four years at the universities, medical schools, agricultural colleges, School of Mines, South Kensington, Woolwich, Sandhurst, Cooper's Hill, or other approved place of higher education. Foundation scholarships give free tuition in the school. All boys whose parents or guardians are living in Kent within ten miles of Tonbridge are on the Foundation. The town of West Malling, and perhaps parts of the parishes of East Malling, Ditton, Allington, Addington, Aylesford, Leybourne, Offham and Ryarsh, are within the above-mentioned distance.

Dr. Plume, by his will, made in the year 1704, left a certain sum of money to augment the incomes of incumbents in the ancient diocese of Rochester which are under £300 a year. At the present time the parishes of

Addington	East Malling	Snodland
Allington	Halling	Trottescliffe
Aylesford	Leybourne	West Malling
Birling	Offham	
Ditton	Ryarsh	

are all thus able to be increased.

Cobham College was founded, in Edward III.'s reign, by a certain Lord Cobham, for five priests or chaplains. In Henry VIII.'s reign the master and brethren, foreseeing that what had happened to many religious houses might happen to theirs, sold their college and lands and possessions to George Lord Cobham. His son, Sir William Broke, founded what was henceforth known as the New College of Cobham, to which twelve parishes have the right of nomination. It is to keep a married couple as a rule. The wife, however, though admitted into the college, has no right to remain after her husband's death unless she is re-elected for the vacancy caused by her husband's death. The rules enjoin that, notice being given on the Sunday or Sunday-week after notice of the vacancy has been received, the incumbent, churchwardens, sidesmen, and overseers, elect, directly after evening service on the same Sunday, a person who has lived in that parish at least three years. There are twenty inmates of the college, one of whom is warden and another sub-warden. Amongst the different parishes that nominate, Halling alone of our parishes has the right to nominate one person perpetually.

Sir Robert Brett, by his will, not only left certain rentals to be given in doles to the poor, but also directed that half a sovereign should be paid to a lecturer, on market-days, at Malling. Market-days there having become a thing of the past, permission has been granted to transfer this lecture or sermon to Friday evening.

The Rev. Everard Home left a charity to educate the poor of Leybourne, East Malling, and Southborough, which is administered by a board of governors. Exhibitions are offered for persons from these parishes and Ditton.

The almshouses at Aylesford have on them this inscription :—

"This house was founded by the Right Honourable Sir William Sedley, heir and sole executor unto his brother, John Sedley, Esq., for the behalf of the poore persons, with like allowance for ever; six at the charge of the said John Sedley and the residue of the said Sir William Sedley; finished primo Aprilis, Anno Dni. 1607. Annoque Regni Regis nostri Jacobi Quinto. Gloria Soli Deo, Deo Patriæ Tibi," with the Sedley arms. And in addition round a triangle runs, "Sacrum Deo Uni"; within, "et Trino," as the

diagram shows. The revenues are raised from the Sedley, Savage, and Faunce charities, the resources of which are derived from lands in Aylesford, East Malling and Frittenden, all in the county of Kent. It is for twelve poor people.

Dr. Milner appears to have first endowed the schools at Aylesford, and Mr. Betts built the Infant school, and the late Mr. Brassey largely aided in building the National school.

The West Malling schools were endowed by Mr. Tresse, Dr. Kennard, Rev. G. F. Bates, Mr. Bell, and Mr. Peter Sutton.

Snodland schools have an endowment given them by Mr. May.

Rev. Paul Baristow left, in 1711, money for establishing a charity school at Trottescliffe.

At East Malling Miss Smith left provision to keep five poor widows in almshouses, with an annuity of £12 each; and for this same parish Lady Jane Twisden, in 1702, left a rental of £4 4s. to apprentice children.

Edward Godden left land, in 1661, rented at £10 a year, to apprentice children belonging to Snodland; and Edward Godden, in the 14th of Charles II., left land bringing in a rental of £10 a year to apprentice the children of Birling.

The Rochester Bridge Wardens give preference in their exhibitions in the Rochester Grammar School for Girls to girls who are residents in the following parishes amongst others: Halling, Snodland including Paddlesworth, Trottescliffe.

The Maidstone Grammar School for Girls, which is also endowed by the same trust, gives preference to girls that are resident in Aylesford, Birling, Ditton, Leybourne, Malling East, Malling West, and Offham.

It is a curious thing that though the parishes of Addington, Allington, and Ryarsh are among the parishes bound to contribute to repairing the piers of Rochester bridge, for some reason they have been omitted both from the Maidstone and Rochester scheme in the offer of exhibitions for the Girls' Grammar Schools.

These parishes were probably omitted because they are not mentioned by name as bound to repair Rochester bridge. The ninth pier, however, has to be repaired by Snodland, Paddlesworth, and " *the men in that valley* "—a phrase no doubt intended to signify the people of these parishes.

MALLING
AND IT'S
VALLEY.

SCALE
One Inch to One Mile

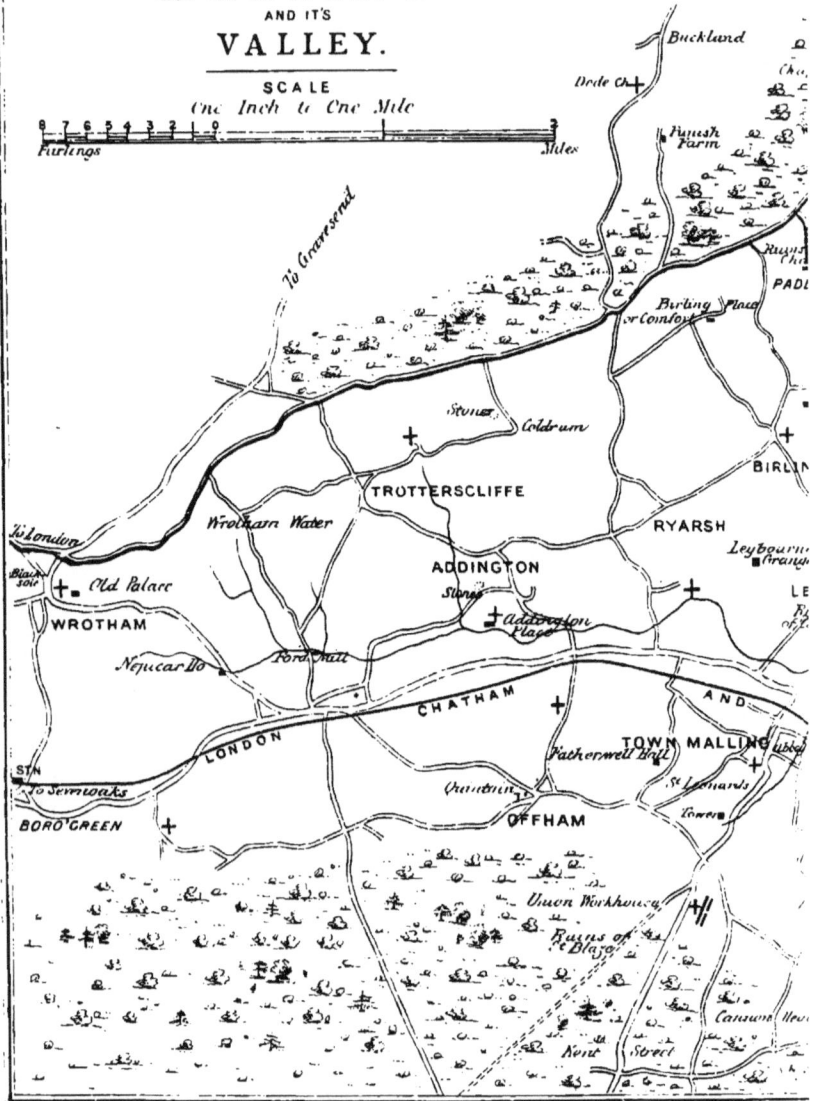

8 7 6 5 4 3 2 1 0   3

Furlongs   *Miles*

Buckland

Dode Ch.

Trnish Farm

To Gravesend

Ruins Ch.

PADL

Birling Place or Comfort

Stones   *Coldrum*

BIRLI

TROTTERSCLIFFE

Wrotham Water

RYARSH

Leybourne Grange

ADDINGTON

To London

Stones

Black soic   *Old Palace*

Addington Place

LE

WROTHAM

Nepicar Ho

Ford Hall

CHATHAM

A N D

TOWN MALLING

Bbey

Fatherwell Hall

LONDON

St Leonards

STN

To Sevenoaks

Quintain

Towers

BORO'GREEN

OFFHAM

Union Workhouse

Ruins of St Blaze

Canwon How

Kent Street

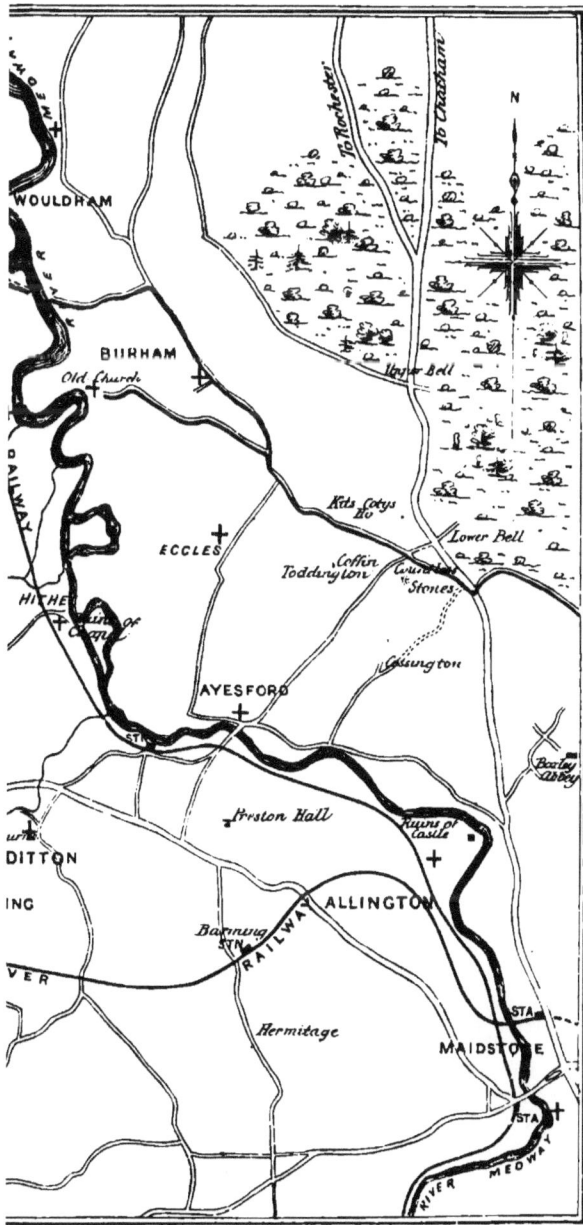

WOULDHAM

BURHAM
Old Church

ECCLES

Kits Cotys Ho

Coffin
Toddington

HITHE
Ruins of
Chapel

AYESFORD

ST

Preston Hall

DITTON

ING

Barming
STN

ALLINGTON

Hermitage

RAILWAY

RIVER

N

To Rochester

To Chatham

Upper Bell

Lower Bell

Countless
Stones

Cossington

Boxley
Abbey

Ruins of
Castle

Ruins of
Castle

MAIDSTONE

STA

STA

RIVER MEDWAY

A.F. BOWKER C.E. F.R.G.S. DEL.

INDEX.

283

19